# CULTURE OF HUMAN TUMOR CELLS

# Culture of Specialized Cells

Series Editor

## R. Ian Freshney

**CULTURE OF EPITHELIAL CELLS, SECOND EDITION**
R. Ian Freshney and Mary G. Freshney, Editors

**CULTURE OF HEMATOPOIETIC CELLS**
R. Ian Freshney, Ian B. Pragnell and Mary G. Freshney, Editors

**CULTURE OF IMMORTALIZED CELLS**
R. Ian Freshney and Mary G. Freshney, Editors

**DNA TRANSFER TO CULTURED CELLS**
Katya Ravid and R. Ian Freshney, Editors

# CULTURE OF HUMAN TUMOR CELLS

**Editors**

**Roswitha Pfragner**
Karl Franzens Universität Graz, Austria

**R. Ian Freshney**
University of Glasgow, Scotland

**WILEY-LISS**

A JOHN WILEY & SONS, INC., PUBLICATION

Published by John Wiley & Sons, Inc., Hoboken, New Jersey.
Published simultaneously in Canada.

For general information on our other products and services please contact our Customer Care Department within the U.S. at 877-762-2974, outside the U.S. at 317-572-3993 or fax 317-572-4002.

Wiley also publishes its books in a variety of electronic formats. Some content that appears in print, however, may not be available in electronic format.

*Library of Congress Cataloging-in-Publication Data:*

Culture of human tumor cells / edited by Roswitha Pfragner,
   R. Ian Freshney.
    p.   cm.
   ISBN 0-471-43853-7
   1. Cancer cells. 2. Human cell culture. I. Pfragner, Roswitha
II. Freshney, R. Ian.
   RC267.C776   2004
   616.99′4027—dc21             2004011003

Printed in the United States of America

10 9 8 7 6 5 4 3 2 1

# Contents

# Preface

This is the sixth book in the Culture of Specialized Cells Series, and, like the others, seeks to provide practical help in the culture of particular cell types. As before the objective has been to provide established methods that have been tried and tested in the author's laboratory, to compare these to other methods, and to supplement them with information on applications, some of which are derived from the author's own work.

This book concentrates on tumor cells and provides detailed stepwise protocols accompanied by lists of the necessary reagents, and sections on the preparation of reagents and media. To avoid unnecessary duplication, suppliers' names are not given following items in the text but are provided in a Sources of Materials table at the end of each chapter, with the details of suppliers provided at the end of the book. Abbreviations are explained the first time that they used, and a complete list is provided at the beginning of the book. We have followed the conventions used in earlier books in the series in that tissue culture grade water is referred to as UPW (ultra-pure water) regardless of the method of preparation, and the abbreviation PBSA is used for Dulbecco's phosphate buffered saline lacking $Ca^{2+}$ and $Mg^{2+}$, on the assumption that the abbreviation PBS should be used for the complete formulation.

We realize that there are many types of tumor not covered in this book; it would have taken several volumes to do so. We have also focused on human tumors as the most relevant models for the study of aberrations in genetic expression, growth regulation, and potential modes of therapy. It was not our intention to produce a comprehensive survey, but to provide protocols for the commonest types of tumor, in the hope that this will satisfy the needs of most workers. Furthermore, these protocols may provide the basis for methods applicable to other tumors, with minor modifications. Readers are also referred to the previous book in the series, Culture of Epithelial Cells, 2$^{nd}$ Ed, by R.I. Freshney and M.G. Freshney [2003], which contains protocols for several different types of normal epithelial cells, and to Human Cell Culture, Vols I & II, John R.W. Masters and Bernard Palsson (Eds) (1999), Dordrecht, Kluwer Academic Publishers, for a more comprehensive review of cultured cell lines from human tumors.

We are grateful to all our contributing authors for providing their specialized knowledge and the time to prepare these detailed protocols and review other work in their respective fields.

**Roswitha Pfragner**
**R. Ian Freshney**

# Contributors

**Mary L. Alpaugh**, Department of Pathology and Revlon/UCLA Breast Center, UCLA School of Medicine, Los Angeles, CA

**Sanford H. Barsky**, Department of Pathology and Revlon/UCLA Breast Center, UCLA School of Medicine, Los Angeles, CA

**Annemarie Behmel**, Department of Medical Biology and Human Genetics, Medical University Graz, Graz, Austria

**Isabella J. Berry**, Beatson Institute for Cancer Research, Bearsden, Glasgow, Scotland

**Robert K. Bright**, Department of Microbiology and Immunology, Southwest Cancer Center, Texas Tech University Health Sciences Center, Lubcock, TX

**Julie E. Burns**, YCR Cancer Research Unit, Department of Biology, University of York, York, UK

**Deborah Burt**, CRC Department of Immunology, Paterson Institute for Cancer Research, Christie Hospital NHS Trust, Manchester, UK

**Louise J. Clark**, Southern General Hospital, Glasgow, Scotland, UK

**Lesley W. Coggins**, Q-One Biotech Ltd., Todd Campus, West of Scotland Science Park, Glasgow, G20 OXA, Scotland, UK

**John L. Darling**, Division of Biomedical Science, School of Applied Sciences, University of Wolverhampton, Wolverhampton, UK

**Hans G. Drexler**, DSMZ-German Collection of Microorganisms and Cell Cultures, Braunschweig, Germany

**Kirsten G. Edington**, Beatson Institute for Cancer Research, Bearsden, Glasgow, Scotland

**Vivian X. Fu**, Department of Surgery, Section of Urology, University of Wisconsin-Madison Medical School, Madison, WI

**Ruth Halaban**, Department of Dermatology, Yale University School of Medicine, New Haven, CT

**Haruo Iguchi**, Division of Biochemistry, National Kyusyu Cancer Research Institute, Fukuoka, Japan

**Elisabeth Ingolic**, Research Institute for Electron Microscopy, Technical University Graz, Graz, Austria

**David F. Jarrard**, Department of Surgery, Section of Urology, University of Wisconsin-Madison Medical School, Madison, WI

**Hee-Sung Kim**, National Cancer Center, Goyang, Gyeonggi, Korea

**Akira Kono**, Transgenic Inc., Chvougai, Kumamoto, Japan

**Ja-Lok Ku**, Laboratory of Cell Biology, Korean Cell Line Bank, Cancer Research Center and Cancer Research Institute, Seoul National University College of Medicine, Seoul, Korea

**Jennifer D. Lewis**, Laboratory of Prostate Cancer Biology, Robert W. Franz Cancer Research Center, Earle A. Chiles Research Institute, Portland, OR

**Roy Mitchell**, Beatson Institute for Cancer Research, Bearsden, Glasgow, Scotland

**Kazuhiro Mizumoto**, Department of Surgery and Oncology, Graduate School of Medical Sciences, Kyushu University, Maedashi, Fukuoka, Japan

**Margaret O'Prey**, Beatson Institute for Cancer Research, Bearsden, Glasgow, Scotland

**Jae-Gahb Park**, Laboratory of Cell Biology, Korean Cell Line Bank, Cancer Research Center and Cancer Research Institute, Seoul National University College of Medicine, Seoul, Korea; and National Cancer Center, Goyang, Gyeonggi, Korea

**So-Yeon Park**, National Cancer Center, Goyang, Gyeonggi, Korea

**E. Kenneth Parkinson**, Beatson Institute for Cancer Research, Bearsden, Glasgow, Scotland

**Roswitha Pfragner**, Department of Pathophysiology, Medical University Graz, Graz, Austria

**Catherine A. Reznikoff**, Department of Surgery, Section of Urology, University of Wisconsin-Madison Medical School, Madison, WI

**Gerry Robertson**, Department of Plastic Surgery, Canniesburn Hospital, Bearsden, Glasgow, Scotland

**Michael J. Rutten**, Department of Surgery, Oregon Health Sciences University, Portland, OR

**Steven R. Schwarze**, Department of Surgery, Section of Urology, University of Wisconsin-Madison Medical School, Madison, WI

**Masaki Shono**, Department of Surgery and Oncology, Graduate School of Medical Sciences, Kyushu University, Maedashi, Fukuoka, Japan

**David Soutar**, Department of Plastic Surgery, Canniesburn Hospital, Bearsden, Glasgow, Scotland

**Valerie Speirs**, Molecular Medicine Unit, University of Leeds, Clinical Sciences Building, St. James's University Hospital, Leeds, UK

**Peter Stern**, CRC Department of Immunology, Paterson Institute for Cancer Research, Christie Hospital NHS Trust, Manchester, UK

**Soichi Takiguchi**, Division of Biochemistry, National Kyusyu Cancer Center Research Institute, Notame, Fukuoka, Japan

**Catherine West**, Experimental Radiation Oncology, Paterson Institute for Cancer Research, Christie Hospital NHS Trust, Manchester, UK

**Robert H. Whitehead**, Novel Cell Line Development Facility, Vanderbilt University, Nashville, TN

**Anne P. Wilson**, Woodbine Terrace, Stanton, Near Ashbourne, Derbyshire, UK

**Gerhard H. Wirnsberger**, Department of Internal Medicine, Medical University Graz, Graz, Austria

**Reen Wu**, Center for Comparative Respiratory Biology and Medicine, Schools of Medicine and Veterinary Medicine, University of California at Davis, Davis, CA

# List of Abbreviations

| | |
|---|---|
| ADH | atypical ductal hyperplasia |
| aFGF | acidic fibroblast growth factor (FGF-1) |
| ALL | acute lymphoblastic leukemia |
| $\alpha$1-AT | $\alpha$1-antitrypsin |
| 5-AzaC | 5-azacytidine |
| BCP | B-cell precursor |
| BEGM | bronchial epithelial growth medium |
| bFGF | basic fibroblast growth factor (FGF-2) |
| BHE | bovine hypothalamus extract |
| B-LCL | B-lymphoblastoid cell line |
| BPE | bovine pituitary extract |
| BPH | benign prostatic hyperplasia |
| BSA | bovine serum albumin |
| cAMP | cyclic adenosine monophosphate |
| CAT | chloramphenicol acetyl transferase |
| CEA | carcinoembryonic antigen |
| CHAPS | (3-[(3-cholamidopropyl) dimethylammonio]-1-propane-sulfonate |
| CHX | cycloheximide |
| CIS | carcinoma in situ |
| CGH | comparative genomic hybridization |
| CM | conditioned medium |
| CPTX | immortalized prostate tumor derived epithelial cells |
| CT | cholera toxin |
| CTL | cytotoxic T cells |
| dB-cAMP or dbcAMP | $N^6,2'$-O-dibutyryladenosine $3':5'$-cyclic monophosphate |
| DCIS | ductal carcinoma in situ |

| | |
|---|---|
| DCS | donor calf serum |
| Dct | DOPAchrome tautomerase |
| Dex | dexamethasone |
| DMEM | Dulbecco's modification of Eagle's medium |
| DMSO | dimethyl sulfoxide |
| DSMZ | Deutsche Sammlung von Mikroorganismen und Zellkulturen |
| EBV | Epstein Barr Virus |
| EDNRB | endothelin-β receptor |
| EDTA | ethylenediaminetetraacetic acid |
| EGF | epidermal growth factor |
| EGTA | ethylene glycol-O, O′-bis(α-amino-ethyl)-N,N,N′,N′-tetraacetic acid |
| EPO | erythropoietin |
| ER | estrogen receptor |
| ET | endothelin |
| FACS | fluorescence-activated cell sorter |
| FBS | fetal bovine serum |
| FCS | fetal calf serum |
| FISH | fluorescence in situ hybridation |
| G-CSF | granulocyte colony-stimulating factor |
| GM-CSF | granulocyte-macrophage colony-stimulating factor |
| HB-EGF | heparin-binding epidermal growth factor |
| HBSS | Hanks' buffered salt solution |
| HGF | hepatocyte growth factor |
| HHV | human herpes virus |
| HIF-1 | hypoxia inducible factor |
| HITES | RPMI 1640 supplemented with hydrocortisone, insulin, transferrin, estradiol, and selenium |
| HIV | human immunodeficiency virus |
| HLA | human leukocyte antigen |
| HMEC | human mammary epithelial cells |
| HPV16 | human papilloma virus serotype 16 |
| HRE | hypoxia response element |
| HRT | hormone replacement therapy |
| HTLV | human T-cell leukemia virus |
| HUVEC | human umbilical vein endothelial cells |
| IBMX | isobutyl-1-methylxanthine |
| IFα, | interferon alpha |
| IFN | interferon |
| IGF | insulin-like growth factor |
| IL | interleukin |
| IMDM | Iscove's modified Dulbecco's medium |
| iNOS | nitric oxide synthetase |

| | |
|---|---|
| KGM | keratinocyte growth medium |
| K-SFM | keratinocyte serum-free medium |
| LOH | loss of heterozygosity |
| LXSN16E6E7 | recombinant retrovirus encoding the E6 and E7 genes of HPV-16 |
| M/SCF | mast/stem-cell factor |
| MEM | minimal essential medium |
| MHC | major histocompatibility antigens |
| MITF | microphthalmia transcription factor |
| MMP | matrix metalloproteinase |
| MRA | mycoplasma removal agent |
| MSH | melanotropin (melanocyte stimulating hormone) |
| MUC-1 | mucin-1 |
| Na-But | sodium butyrate |
| NK | natural killer |
| NPTX | immortalized normal prostate epithelial cells |
| NSCLC | non-small cell lung cancer |
| PA317 | retrovirus producer cell line |
| PAI | plasminogen activator inhibitor |
| PBL | peripheral blood lymphocyte |
| PBSA | Dulbecco's phosphate buffered saline solution A (without $Ca^{2+}$ and $Mg^{2+}$) |
| PCR | polymerase chain reaction |
| PD-ECGF | platelet-derived endothelial growth factor |
| PDGF | platelet-derived growth factor |
| Pen/Strep | penicillin-streptomycin |
| PHA | phytohemagglutinin |
| PIN | prostate-intraepithelial neoplasia, a precursor to prostate carcinoma |
| PMA | phorbol 12-myristate 13-acetate |
| PNII/APP | protease nexin II/β-amyloid precursor protein |
| PSA | prostate specific antigen |
| RA | retinoic acid |
| RFLP | restriction fragment length polymorphisms |
| RPMI | Rosewell Park Memorial Institute |
| SCF | stem cell factor |
| SCLC | small cell lung cancer |
| SCID | severely compromised immune-deficient |
| SEM | scanning electron microscopy (Chapter 15) |
| SEM | surrogate end-point marker (Chapter 10) |
| SF | scatter factor |
| SFM | serum-free medium |
| SKY | spectral karyotyping |
| SPRRI | small proline-rich protein 1 |

| | |
|---|---|
| TCC | transitional cell carcinoma |
| TEM | transmission electron microscopy |
| TGF-$\alpha$ | tumor-derived growth factor alpha |
| TGF-$\beta$ | tumor-derived growth factor beta |
| TIL | tumor infiltrating lymphocytes |
| TIMP | tissue inhibitor of metalloproteinase |
| TNF-$\alpha$ | tumor necrosis factor alpha |
| TPA | 12-O-tetradecanoyl phorbol-13-acetate (PMA) |
| TRP1, TRP2 | tyrosinase-related protein 1 or 2 |
| uPA | urokinase-type plasminogen activator |
| UPW | ultrapure water, e.g. reverse osmosis or glass distillation combined with carbon filtration and high grade deionization |
| UVB | ultraviolet B irradiation |
| VEGF | vascular endothelial growth factor |
| vWf | von Willebrand factor |
| X3T3 | lethally-irradiated Swiss 3T3 |

# 1

# Growth of Human Lung Tumor Cells in Culture

Reen Wu

*Center for Comparative Respiratory Biology and Medicine, Schools of Medicine and Veterinary Medicine, University of California at Davis, Davis, California, rwu@ucdavis.edu*

*Culture of Human Tumor Cells*, Edited by Roswitha Pfragner and R. Ian Freshney.
ISBN 0-471-43853-7 Copyright © 2004 Wiley-Liss, Inc.

## 1.  INTRODUCTORY REVIEW

### 1.1.  Cell Culture and Lung Cancer Development

Lung cancer is the most frequent cause of cancer deaths in both men and women [Landis et al., 1998]. Environmental air pollutants, such as tobacco smoke and asbestos, are accepted as the major causes of the disease [Parkin et al., 1994]. Most lung cancers are of epithelial origin, with a few exceptions, such as mesothelioma, which is derived from a non-epithelial cell type (mesothelial cells) attributable to asbestos exposure [Mossman et al., 1990; Rom et al., 1991; Jaurand, 1991; Pass and Mew, 1996]. Epithelial lung cancers are generally classified into two major types: small cell lung cancer (SCLC, about 25% of lung cancers) and non-small cell lung cancer (NSCLC, about 75% of lung cancers). Tumors of the former type disseminate widely and early and are seldom cured by surgical resection, whereas the latter may be cured by surgery if diagnosed early. In addition, SCLC tumors initially have much better response to cytotoxic therapies than do NSCLC. NSCLC consists of at least three major cell types—squamous carcinoma, adenocarcinoma, and large cell carcinoma—in addition to several subtypes. The cell type origin of these cancer cells is still unknown.

Like many other tumors, the development of lung cancer is a multistep process the nature of which has not been fully elucidated. Although the total annual number of cases has declined recently, probably due to a decreased trend in cigarette consumption, the incidence and mortality rates of lung cancer have increased alarmingly, especially among the female population and populations in developing countries. The nature of this increase is poorly understood. In addition, the overall cure rate for lung cancer remains low, at only about 13%. Reasons for such a low cure rate are many. One of the major reasons is related to the difficulty in the early detection of the disease with current technologies, which are not sufficiently sensitive. It is assumed that the earlier the detection of these tumors, the better the chances for successful treatment. The development of routine techniques for the isolation and in vitro maintenance of these transformed cells, especially at their early stages of transformation, will provide a significant advance in this research area.

## 1.2. Requirements for Growth Factors and Hormones for Lung Cancer Cells to Grow

Despite deficiency in the development of an in vitro system for studying the multiple stages of lung cancer cell development, cultivation of primary cells from normal and lung cancer specimens has brought some success. These advances may be useful in the future for the development of an in vitro cell culture model, allowing lung cancer cells at different stages of transformation to be studied. Such a system would help to generate biomarkers for the early detection of lung cancer. We, as well as other researchers, initially used a serum-supplemented medium to establish primary cultures of normal specimens and lung tumor specimens. These attempts have proven to be unsuccessful [Wu, 1986; Lechner et al., 1986]. Cell proliferation is either inhibited or unresponsive to growth factors. Furthermore, fibroblast overgrowth is a constant concern whenever a mixed population of cells is grown in serum-containing medium. Deficiency in growth factors or presence of cell growth inhibitors in serum-supplemented medium could explain the poor in vitro growth of these lung cancer and normal epithelial cells. To eliminate the problems created by adding serum to medium, serum-free and hormone- and growth factor-supplemented defined media were developed. Many years ago, Sato and colleagues proposed that most cells in vivo are regulated by hormones and growth factors in interstitial fluid, and not by serum [Barnes and Sato, 1980]. In pursuance of this concept, their laboratories developed various defined media for culturing many different types of cells [Hayashi and Sato, 1976; Bottenstein et al., 1979]. Based on their findings and by subsequent use of trial and error, our laboratory [Wu, 1986] and Lechner's [Lechner et al., 1986] were the first to develop a serum-free, hormone-supplemented medium for primary human airway epithelial cell growth and long-term maintenance in culture. The principal supplements in this medium are insulin, transferrin, hydrocortisone, epidermal growth factor (EGF), retinoid, and a supplement essential for cyclic adenosine monophosphate (cAMP) generation. Later, this medium was modified further, and, with the incorporation of collagen gel substratum and culture at the air–liquid interface [Wu, 1997], the primary cells can now differentiate properly, similar to the epithelium seen in vivo. In this regard, retinoid is a very important supplement for cell differentiation as well as for cells to grow on a collagen gel.

Using a similar trial-and-error approach, Simms et al. [1980] were the first to develop a serum-free, hormone-supplemented defined medium that supported the continuous replication of established SCLC cell lines. This defined medium initially contained five supplemental growth factors: hydrocortisone, insulin, transferrin, estradiol, and selenium (HITES). Carney et al. [1981] used HITES-supplemented medium to establish cell lines directly from SCLC tumor samples. While HITES-supplemented medium was superior to the routine 10% serum-supplemented medium for the establishment of SCLC cultures, addition of small amounts of serum to HITES-supplemented medium resulted in an increased growth advantage [Gazdar et al., 1990]. With these media, HITES and HITES plus 2% serum, more than 70% of tumor cells derived from SCLC cancers could be cultivated in vitro [Oie et al., 1996].

The culturing of NSCLC presented a different set of problems from the culturing of SCLC. Most surgically removed NSCLC specimens usually contained large amounts of tumor cells with a heterogeneous phenotypes. Before the incorporation of hormones and growth factors into the medium, very few NSCLC cell lines were established in medium supplemented with serum. Most of these cell lines were relatively or completely undifferentiated and were not representative of the tumors from which they were derived. In addition, phenotypic differences among different tumor cells required different culture media and growth supplements. It was realized relatively early that it was easier to establish cultures from adenocarcinomas and large cell undifferentiated carcinomas than from squamous cell carcinomas. By trial and error, Brower et al. [1986] were the first to develop ACL-3 medium for the growth of lung adenocarcinoma cells. This medium was modified further to produce ACL-4, which, with or without serum, is now quite suitable for culturing many types of adenocarcinomas with a success rate of more than 55% [Gazdar and Minna, 1996]. We have initially used the serum-free defined medium [Lechner et al., 1986; Wu, 1986] (without retinoid), originally designed for culturing primary human airway epithelial cells, to grow cells derived from various lung adenocarcinoma specimens with limited fibroblast overgrowth. After this initial manipulation, serum was added to selectively promote the growth of the cancer cells because normal epithelial cells are more sensitive to the inhibition by serum. With this approach, several adenocarcinoma cell lines were established from these limited specimens [Yang et al., 1992; Chu et al., 1997].

The major problem with culturing squamous carcinoma cells of

NSCLC is that fully differentiated epidermal cells do not replicate. The difficulty is in finding the right medium to balance cell differentiation with cell growth. Retinoid is an important regulator that can inhibit terminal differentiation of squamous cells but promote mucous cell differentiation. However, most squamous cancer cells are inhibited by treatment with retinoids. Another approach is to reduce the calcium level in the medium because high calcium is required for the expression of terminal differentiation [Levitt et al., 1990].

3T3 fibroblasts, whose proliferation has been arrested by irradiation or mitomycin C, have been used as a feeder layer to support the continuous cultures of various epidermal keratinocytes [Rheinwald and Beckett, 1981; Allen-Hoffmann and Rheinwald, 1984]. A similar supportive role was seen with feeder layers in culturing squamous cancer cells from lung. By trial and error, it was determined that NSCLC cells derived from squamous cell carcinoma required serum, cholera toxin, and EGF, in addition to the common requirements for insulin, transferrin, and hydrocortisone [Gazdar and Oie, 1986]. Despite this progress, the successful rate in culturing lung squamous carcinoma cells in vitro remains low, at the 25% level [Gazdar and Minna, 1996].

Table 1.1 summarizes a comparison of the growth requirements for normal lung and various lung cancer cells. Several of the supplements are commonly required for these cells, regardless of their origin. These supplements are insulin, transferrin, hydrocortisone, selenium, and various nutrients, as well as ethanolamine and phosphoethanolamine, which are involved in lipid metabolism. However, some supplements are required specifically for individual cell types. For instance, the critical difference between cancer and normal cells is the requirement for retinoid. Retinoids are cytotoxic for most cancer cells, whereas normal cells are dependent on them for cell differentiation and growth on a collagen substratum. Between SCLC and NSCLC cells, the most critical distinction is with EGF and bombesin supplementation. EGF is required for all NSCLC cultures, but is not required by SCLC cells. In contrast, bombesin is needed for SCLC cells, but not required for culturing NSCLC cells. SCLC cells produce bombesin; however, these cells still require the bombesin supplement when they are plated at low cell density. Thus, there is an autocrine/paracrine mechanism involved in the growth of SCLC cells. A major difference between squamous and other NSCLC cells is in the requirement for low calcium to sustain cell proliferation. A high calcium level enhances terminal cell differentiation of squamous cell type.

From this comparison, we can draw a simple conclusion that there are two basic approaches for culturing various lung cancer cells: one for SCLC and the other for NSCLC. In the former approach, the HITES-based supplemented medium is sufficient to initiate primary

**Table 1.1.** A comparison of growth responses of human normal airway epithelial cells and SCLC and NSCLC cells to nutrients and growth factor/hormone supplements.

| Medium components | Normal airway epithelial cells[1,3] | SCLC[3] | NSCLC[3] Adeno/Large | Squamous |
|---|---|---|---|---|
| *Nutrients* | | | | |
| Calcium: low or high[4] | low | low/high | low/high | low |
| Ethanolamine | + | + | + | + |
| Phosphorylethanolamine | + | + | + | + |
| Selenium | + | + | + | + |
| *Hormones/Growth factors* | | | | |
| Insulin | + | + | + | + |
| Transferrin | + | + | + | + |
| EGF | + | − | + | + |
| Bombesin | − | + | − | − |
| Cholera toxin | + | + | + | + |
| Fetal bovine serum *(5%)* | − | − | + | + |
| Hydrocortisone | + | + | + | + |
| Triiodothyronine | + | − | + | − |
| Vitamin A | + | − | − | − |
| 17-β-Estradiol | − | + | − | − |

[1,2] Based on Wu, 1986 and Lechner et al., 1986.
[3] Based on Oie et al., (1996)
[4] Low calcium: <0.1 mM.
[5] Cholera toxin equivalent chemicals, like arginine vasopressin and epinephrine, were used for SCLC and primary human airway epithelial cultures, respectively.

cultures from SCLC tumors. Additional factors, such as bombesin and cholera toxin, or serum, can be added to stimulate cell growth. This approach has been used before with great success [Oie et al., 1996]. The second approach is the use of the serum-free defined medium that was developed for culturing of primary normal airway epithelial cells. This type of medium contains combined supplements ("C" supplements), which are similar to the supplements in the ACL-4 medium developed by Oie et al. [1996], except for the additionl of bovine hypothalamus extract (BHE). Using this serum-free and vitamin A-depleted medium, both normal and cancer epithelial cells from NSCLC tissue specimens will survive and proliferate, but not stromal cells. After an initial period of serum-free incubation, serum can be added to induce normal epithelial cells to express terminal differentiation, because the differentiation of most of NSCLC cells is less responsive to serum. The addition of serum may also have growth stimulatory effects on NSCLC cells. This approach has been successful for culturing cells derived from adenocarcinoma [Yang et al., 1992, Chu et al., 1997]. For squamous carcinoma cells, low calcium medium, such as the bronchial epithelial growth medium (BEGM)

**Table 1.2.** Two types of culture medium for culturing human lung tumor cells.

|  | HITES-Based Medium | C-Based Medium |
| --- | --- | --- |
| *Basal medium*[1] | RPMI 1640 | DMEM/F 12 or BEGM |
| Sodium selenite | 30 nM | 30 nM |
| Ethanolamine | 10 μM | 10 μM |
| Phosphorylethanolamine | 10 μM | 10 μM |
| Sodium pyruvate | — | 0.5 μM |
| Glutamine | 2 mM | 2 mM |
| Adenine | — | 0.18 mM |
| HEPES[2] | 15 mM | 15 mM |
| *Supplement* |  |  |
| Insulin | 5 μg/ml | 5 μg/ml |
| Transferrin | 5 μg/mi | 5 μg/ml |
| EGF | — | 10 ng/ml |
| Bombesin | 0.1 μM | — |
| Cholera toxin | 10 ng/ml | 10 ng/ml |
| Hydrocortisone | 10 nM | 100 nM |
| Triiodothyronine | — | 0.1 nM |
| Bovine hypothalamus extract | — | 15–30 μg/mi |
| Bovine serum albumin | 0.2% | 0.5% |
| (Fetal bovine serum)[3] | — | (5%) |

[1] Antibiotics added in these media are: penicillin (100 U/ml), streptomycin (100 μg/ml), gentamicin (5 μg/ml), and fungizone (5 μg/ml).
[2] HEPES stock buffer is at 1.5 M, pH 7.2–7.3.
[3] Fetal bovine serum can be used after primary culture to enhance NSCLC cancer cell proliferation in culture.

developed by the commercial company Clonetics, can be used as the basal nutrient medium.

## 2. PREPARATION OF MEDIA AND REAGENTS

Table 1.2 summarizes the compositions of HITES- and C-based media. Most of these chemicals, except bovine hypothalamus extract (BHE), are available commercially (see Section 6). The commercially equivalent products to BHE are endothelial cell growth supplement (ECGS) and pituitary extract (PE).

### 2.1. Preparation of HITES-Based Medium

#### 2.1.1. Nutrient Medium Preparation

(i) Supplement RPMI 1640 medium with penicillin G (100 U/ml), streptomycin (100 μg/ml), gentamicin (5 μg/ml), fungizone (5 μg/ml), [N-(2-hydroxyethyl)piperazine-N′-ethanesulfonic acid] buffer (HEPES, 15 mM, pH. 7.2), ethanolamine (10 μM), phosphoryl-ethanolamine (10 μM), and bovine serum albumin (BSA, 2 mg/ml).

(ii) Sterilize through a 0.2 μm membrane filter and store at 4°C.

(iii) Before use, add glutamine at 2 mM final concentration.

### 2.1.2. Preparation of Hormone and Growth Factor Supplements

See Sources of Materials table at end of chapter for suppliers of hormones and growth factors.

(i) Dissolve hydrocortisone at 0.05 mM in 95% alcohol or in phosphate-buffered saline (PBSA) if a water-soluble analogue is used.

(ii) Dissolve insulin in 1 mM acetic acid at 2.5 mg/ml.

(iii) Dissolve transferrin in PBSA at 2.5 mg/ml.

(iv) Dissolve 17-$\beta$-Estradiol in 95% alcohol at 0.1 mM.

(v) Prepare sodium selenite at 10 mM in PBSA.

(vi) Dissolve cholera toxin in PBSA at 10 μg/ml.

### 2.1.3. Additional Growth Supplements for HITES-Based Medium

Prepare bombesin at 1 mM in PBSA containing BSA, 1 mg/ml.

### 2.1.4. Complete HITES-Based Medium

(i) Before use, add aseptically complete RPMI-1640 nutrient medium from above preparation (2.1.1) with various hormones and growth factors for the final concentrations as shown in Table 1.2.

(ii) Add bombesin (diluted to 0.1 mM in PBSA with BSA, 1 mg/ml) to a final concentration of 0.1 μM, if needed.

(iii) Store this complete HITES-based medium at 4°C for a maximum of one week.

## 2.2. Preparation of C and C-Modified Media

The major difference between C and C-modified nutrient media is the calcium level. C medium is a 1:1 mixture of Dulbecco's Modified Eagle's medium (DMEM) and Ham's F12, whereas C-modified medium is a low-calcium-based bronchial epithelium growth medium (BEGM) medium developed by Clonetics. These media are prepared and supplemented with antibiotics, HEPES, ethanolamine, phosphoethanolamine, glutamine, and BSA as described (see section 2.1.1).

### 2.2.1. Preparations of Hormone and Growth Factor Supplements

Prepare insulin, transferrin, hydrocortisone/dexamethasone, and sodium selenite (see section 2.1.2.). In addition, dissolve 100 μg EGF in 10 ml PBSA to give 10 μg/ml, and triiodothyronine (T3) in PBSA at 10 μM.

### 2.2.2. Preparation of Bovine Hypothalamus Extract (BHE)

Prepare BHE as described previously [Robinson and Wu, 1991], summarized as follows:

(i)  Briefly homogenize 100 g of bovine hypothalamus in 200 ml ice-cold PBSA at 4°C in a blender for approximately 3–6 min.

(ii)  Pass the homogenates through 2 to 3 layers of cheesecloth, and stir for 1–2 h at 4°C.

(iii)  Centrifuge the homogenates at 13,800 $g$ for 40 min at 4°C, and collect the supernatants and add streptomycin sulfate at 0.5 g/ml.

(iv)  Stir the mixture at 4°C for an additional 1 to 2 h.

(v)  Centrifuge the mixture again at 13,800 $g$ for 40 min. This process can remove the lipid fraction in the extract.

(vi)  Filter-sterilize the supernatant through a 0.2 μm filter membrane.

(vii)  Store aliquots of 5 ml per vial at −20°C for 2 to 3 months.

### 2.2.3.  Complete C and C-Modified Media

Before use for cell culture, C or C-modified nutrient media containing various antibiotics, HEPES, ethanolamine, phosphorylethanolamine, glutamine, and BSA as described above in section 2.1.1(i), is supplemented with various hormones, growth factors, and BHE as described in Table 1.2. The complete C and C-modified media can be stored at 4°C for no more than one week.

## 3.  SAFETY PRECAUTIONS

Normally, lung cancer tissues are obtained from patients, with consent, by surgery or by bronchoscopic biopsies for medical diagnosis. The pathological stage of the tissues and status of the patient is not always clear. To avoid the transmission of diseases or infectious agents associated with the tissue, extra care in handling these materials, the wastes, and the culture, is needed. Great care should be taken to handle these biohazardous materials using good laboratory practice. Consult national and local safety guidelines before initiating this work [See also Caputo, 1996].

## 4.  STEP-BY-STEP PROTOCOLS

### 4.1.  Tissue Handling and Shipping

Immediately after surgical removal of tissues from patients, tissues should be immersed in a nutrient medium containing antibiotics and be kept cold while being transported to the lab. Normally, we use minimal essential medium (MEM). Others used RPMI-1640 or L-15 [Leibovitz, 1986]. We found no difference in the choice of medium for this step. Immersion in cold medium seems to be the most critical requirement for keeping cells viable.

## Protocol 1.1.  Processing of Lung Tumor Samples

### Reagents and Materials
*Sterile*
- ❑  MEM: Eagle's MEM without bicarbonate
- ❑  Petri dishes, 10 cm
- ❑  Forceps, fine
- ❑  Scissors, fine

### Protocol
(a)  Clean tumor tissue further, once it has arrived at the lab, by gently swirling around with MEM.
(b)  Place the tumor specimen in a sterile 10 cm Petri dish, and add a small amount of medium to keep the tissue wet.
(c)  Remove necrotic areas, fatty tissue, blood clots, and connective tissue with forceps and scissors or crossed scalpels.
(d)  Rinse the remaining tumor tissue further with fresh MEM.

### 4.2.  Cell Dissociation

For cell isolation, MEM is supplemented with BSA, 10 mg/ml, to prevent cell damage caused by mechanical and enzymatic treatments. The presence of BSA can also serve as a detoxification reagent to neutralize toxins potentially released by dead or dying cells. Two methods for cell dissociation will be described and, if the tissue size is large enough, both these cell-disaggregating methods should be tried. The first method is the simplest, fastest, and least traumatic method of obtaining cells from tumor tissue for culture by the mechanical spillout method [Leibovitz, 1986]. The second method is an enzymatic method.

## Protocol 1.2.  Harvesting Cells from Lung Tumor by Mechanical Spillout

### Reagents and Materials
*Sterile*
- ❑  Fine scissors or scalpels, #11
- ❑  MEM-BSA medium: Eagle's MEM without bicarbonate, supplemented with BSA, 10 mg/ml
- ❑  Ficoll/metrizoate solution, e.g., Lymphosep, density, 1.077 g/cc
- ❑  Universal containers or 20 ml centrifuge tube

*Protocol*

(e)   Mince the tumor into very small pieces with sterile scissors or by cross-cutting with two scalpels. This mechanical teasing approach will release tumor cell aggregates, which usually adhere loosely to stromal tissue.

(f)   After repeated resuspension with a wide-bore pipette and further mincing, tumor cells are released into the medium.

(g)   Collect tumor cells by centrifugation at 200 g for 5 min.

(h)   Resuspend pellet in 5 ml medium.

(i)   Perform Ficoll/metrizoate centrifugation to eliminate erythrocytes, dead cells, and other tissue debris.

(j)   Add 5 ml Ficoll/metrizoate to universal container or centrifuge tube

(k)   Layer cell suspension carefully, without mixing, over 5 ml Ficoll/metrizoate.

(l)   Centrifuge for 10 min at 100 g.

(m)   Collect cells from interface.

(n)   Dilute cells to 10 ml in medium and centrifuge at 100 g for 5 min.

(o)   Suspend the final cell pellet at $2 \times 10^5 - 1 \times 10^6$ cells/ml in the complete HITES-based, C-based, or C-modified medium.

(p)   Seed into 25-cm$^2$ flask(s) according to the pathological status of the tumor specimen (see section 4.4).

**The Enzymatic Method**

Because tumor cells are loosely adherent to the stromal layer, these cells can be dislodged by incubating tissue fragments with 0.1% collagenase (in MEM).

## Protocol 1.3.   Disaggregation of Lung Tumor Tissue in Collagenase

*Reagents and Materials*

*Sterile*

❑   Collagenase (type IV, Sigma cat #C5138 or equivalent) solution, 0.1% in MEM

❑   MEM-BSA medium: MEM without bicarbonate, supplemented with BSA, 10 mg/ml

❑   Growth medium, e.g., complete HITES-based, C-based, or C-modified medium

❑   Ficoll/metrizoate

❑   Scalpels

❑   Universal containers or 20 ml centrifuge tubes

### Protocol

(a) Cut tissues into small pieces (about 2–5 mm diameter), and immerse in the collagenase solution.

(b) After 30 min incubation at 37°C, mince these tissues as described in the mechanical spillout method and rinse with MEM-BSA medium.

(c) Collect these media and centrifuge (200 g, 5 min) and resuspend in the medium.

(d) Carry out a Ficoll/metrizoate gradient centrifugation (see Step (e) in Protocol 1.2) to remove cell debris, erythrocytes, and tissue fragments, etc.

(e) Collect the cell pellet by centrifugation and suspend into various culture media.

There are other enzymes, such as neutral protease, elastase, trypsin, and glycoconjugate-degrading enzymes, and chelating chemicals, such as ethylene diamine tetraacetate (EDTA) or ethyleneglycol-bis($\beta$-aminoethyl ether)-N,N,N′,N′-tetraacetic acid (EGTA), which are used in cell dissociation from tissue. Some of these enzymes and chemicals cause cytotoxic effects, but some of them do not. Thus, the selection of these dissociating enzymes and chemicals is based on trial-and-error.

### 4.3. Cell Culture Conditions

To initiate primary lung cancer cell culture, it is advisable to culture serum-free to avoid the overgrowth of stromal cell types. Normally, if the pathological status of the tumor tissue is known, one can use HITES-based medium for SCLC tumor cells, C medium for adenocarcinoma, and large cell carcinoma cells or C-modified medium for squamous cell carcinoma. However, if the pathological status is unclear, tumor cell suspensions should be divided and suspended into these three types of medium for culture to maximize the chance of growing tumor cells.

### Protocol 1.4.   Culture of SCLC Cells

#### Reagents and Materials

*Sterile*
- ❏ Serum-free HITES-based medium (see Section 2.1)
- ❏ Fetal bovine serum (FBS)
- ❏ Dimethyl sulfoxide (DMSO)
- ❏ Cryo-tubes

*Protocol*

(a) Following disaggregation seed the tumor cell suspension at $1 \times 10^6$ cells/ml in serum-free, HITES-based medium.

(b) After 1–2 weeks incubation, most SCLC cells grow as "floating aggregates," whereas normal cells will attach to the dish to proliferate. Thus, it is very easy to recognize if the primary culture has SCLC cells. Recover these floating aggregates from the medium by centrifugation at 200 $g$ for 5 min).

(c) If too much cell debris contaminates the culture, recover the viable cells by Ficoll/metrizoate centrifugation (see Step (e) in Protocol 1.2).

(d) Occasionally, SCLC cells adhere to the dish surface. Detach these cells by gently rapping the flasks and recovering the floating cells.

(e) During passage, it is necessary to avoid a low seeding density. Normally, maintain the seeding density at $5–10 \times 10^5$ cells/ml for each passage.

(f) To boost cell proliferation in subsequent culture, supplement the HITES-based medium with 2–5% FBS.

(g) Preserve SCLC cells (at $1–5 \times 10^6$ cells/ml) cryogenically in the HITES-based medium supplemented with 10% DMSO and 10–50% FBS, and store in a liquid $N_2$ tank.

## Protocol 1.5.   Culture of Tumor Cells from Adenocarcinoma and Large Cell Carcinoma

*Reagents and Materials*

*Sterile*

❑ C-based medium
❑ Trypsin, 0.1% solution in PBSA with 1 mM EDTA
❑ Trypsin inhibitor, 1% solution in C medium

*Protocol*

(a) Plate primary tumor cells in a 25 cm²-flask in C-based medium. The seeding density should be $5 \times 10^6$ cells/ml. After 2–4 days incubation, most of the tumor cells, as well as normal epithelial cells and stromal cells, adhere to the surface of the flask. Normally, there are few stromal cells seen in this initial culture. However, if present, these cells either will not grow or will have a very low proliferative activity. In contrast, normal epithelial cells can grow rapidly.

(b) After a week or two in culture, trypsinize the adherent cells with 0.1% trypsin, 1 mM EDTA solution in PBSA.

(c) After the cells have been detached, add an equal volume of 1%

trypsin inhibitor solution prepared in C medium and recover cells by centrifugation (200 g, 5 min).

(d) Suspend cell pellet in a fresh C medium at $1 \times 10^6$ cells/ml, and seed into tissue culture dishes or flasks.

(e) Change medium the next day, and then change medium continuously once every other day until confluence is reached.

(f) Passage confluent cultures again as before, and seed in serum-free C medium.

(g) To eliminate normal epithelial cells that can contaminate the cultures, treat cultures with C medium supplemented with 10% FBS.

(h) After a one-week incubation, passage cultures again and maintain in the serum-free C medium.

Depending on the outcome of the culture, additional serum-treatment may be needed. This approach should preserve most of adenocarcinoma and large cell carcinoma cell types in culture. If there is contamination with stromal cells in culture, the following two procedures are recommended. One is to trypsinize the culture partially, by using a low-concentration trypsin solution (<0.05%), because fibroblasts are easier to detach than epithelial cells. The second approach is to perform differential attachment, because fibroblasts attach to the dish surface faster than do epithelial cells. Plate cells briefly in a tissue culture dish (2–6 h), after which the unattached cells are recovered and plated into a new dish. If the fibroblast contamination persists, these two procedures should be used repeatedly. Otherwise, a cell cloning approach (see Section 4.4.3), such as using limited dilution and clonal plating, should be used. After several passages, these tumor cells can be cryogenically preserved in C medium supplemented with 10% DMSO and 10–50% FBS and stored in a liquid $N_2$ tank.

### 4.4.1. Culturing Tumor Cells from Squamous Carcinoma

The procedure to culturing tumor cells from squamous carcinoma tumor is the same as for tumor cells from adenocarcinoma and large cell carcinoma, except that a low-calcium (0.1 mM), serum-free, C-modified medium is used. After several passages, serum (10% FBS) can be added to the squamous cancer cell cultures to boost cell proliferation.

### 4.4.2. Culture of Tumor Cells from Mixed Types of Lung Tumors

Occasionally, the lung tumor obtained contains mixed types of carcinoma. For growing these cells, these three types of media should be used: HITES-based and C and C-modified media. The culture conditions are the same as described above.

### 4.4.3. Cloning

After several passages, lung cancer cultures can be purified further by clonal selection. Clonal selection is also used for cultures with mixed morphologies.

### Protocol 1.6.  Cloning Cultures from Lung Carcinoma

**Reagents and Materials**
- ❏ Trypsin
- ❏ Trypsin inhibitor
- ❏ Conditioned medium (CM): fresh medium 1:1
- ❏ 96-well plate
- ❏ 24-well plate
- ❏ 6-well plate

**Protocol**
(a) To achieve high efficiency in clonal selection, prepare CM from confluent cultures of the same cells. Briefly, media are conditioned in confluent cultures for 1–2 days. Filter these CM to remove cell debris, and then dilute with fresh culture medium at a 1:1 ratio. Use this 50% CM medium for cloning.
(b) Dilute trypsinized cells serially from 1000 cells/ml to 1 cell/ml and plate into 96-well plates (200 μl/well).
(c) After 1–2 weeks' incubation, harvest colonies from those wells in each 96-well plate carrying one colony, and replate into separate wells of a 24-well plate.
(d) After confluence, transfer the 24-well plate cultures to 6-well plates, and then to 10 cm dishes, and so on. This gradual dilution will enhance the survival of tumor cells during each transfer.

In some cases, when the serial dilution approach cannot be used; selection has to be performed on a confluent culture.

(a) Place a drop of trypsin solution on top of the colony of interest.
(b) After a brief incubation, pipette a drop of medium on top of the colony from a Pasteur pipette carrying culture medium with trypsin inhibitor.
(c) Using the same pipette to scrape the colony, recover the detached cells and plate in a 24- or 96-well plate.

### 4.4.  Characterization of Neoplastic and Differentiated Functions

Cultured cells should be characterized for their epithelial nature, neoplastic phenotype, and differentiated functions. The epithelial nature can be characterized by anti-keratin antibody or by transmission electron microscope (TEM). Briefly, for floating cell aggregates, such as SCLC cultures, cells are cytocentrifuged onto glass slides and fixed with ice-cold methanol or other fixatives. For adherent cells, a direct fix with methanol is sufficient. These fixed cells are stained with anti-keratin antibody (from Sigma) according to standard immunohisto-chemical procedure. Fluorescence staining is preferred because a cytoskeletal filament structure is more apparent. Epithelial cell types are positive for cytokeratin staining. The staining should be negative for all stromal cells. For TEM, cultured cells are fixed in gluta-raldehyde and processed for TEM according to the standard protocol. The epithelial cell type should have keratin fibers, tonofilaments, and desmosomes.

Neoplastic properties should be determined in two ways. One is an in vitro, soft agar colony-forming assay; the other is an in vivo tumorigenicity assay in either severe combined immunodeficiency (SCID) mice or immune-deficient nude mice. For the colony-forming assay on soft agar, exponentially growing cells are suspended in complete growth medium containing 0.3% agar and overlaid on 1% agar-ose in 10 cm tissue culture dishes ($1 \times 10^3 - 1 \times 10^4$ cells/dish). These dishes are maintained at 37°C in a $CO_2$ incubator for 2 weeks. The number of visible colonies is counted to determine mean values for colony-forming efficiency. Normal cells cannot grow on soft agarose, and they should be used as a negative control in this assay. Known tumor cells and cell lines that can grow on soft agarose should be included in the assay as positive controls. For in vivo tumorigenicity assay, harvest cells and suspend in PBSA at $2.5 \times 10^7$ cells/ml. Inject 0.2 ml from each tumor cell line dorsally into each of five 6- to 8-week old SCID or nude mice. These mice are then observed for 8 weeks for the development of tumors. The tumor can be characterized further for morphology and immunohistochemical analyses. Tumors are fixed briefly in formaldehyde and are embedded in paraffin. The histological sections of tumors are examined for cell morphology, nuclear atypia, and the presence of duct-like structures. In addition, various anti-bodies can be used to characterize the nature of the tumor. These antibodies, such as anti-small proline-rich protein (SPRR1B) antibody [Lau et al., 2000] and anti-involucrin antibody (Pierce), can be used to identify squamous cells. This can be supported further by the finding of the formation of a cornified envelope, which is insoluble in the presence of sodium dodecyl sulfate (SDS) and reducing agent. Anti-mucin antibody [Lin et al., 1989] and Alcian blue-PAS stain will con-firm identification as adenocarcinoma.

Antibodies specific to L-dopa decarboxylase production [Baylin et al., 1980] and to bombesin-like immunoreactivity [McMahon et al., 1984] are used to confirm an SCLC type of cancer. Where these are not expressed, i.e. in variant SCLC cell lines, creatine kinase BB isozymes (CKBB) and neuron specific enolase (NSE) may be used [Carney et al., 1985; Gazdar et al., 1985].

## 4.5. Variations

Other enzymatic procedures have been used successfully for lung epithelial cell isolation. These are trypsin, pronase (neutral protease), and elastase. Trypsin is used routinely for passaging cell cultures, and it can dissociate cancer cells from stromal tissues. However, reports indicate that the viability of dissociated cells is poor. Pronase and the neutral protease (e.g., Sigma's Type XIV protease) have been used for the isolation of epithelial cells from lung tissues. However, we have found that the treatment is best performed at 4°C overnight, because higher temperature treatment decreases cell viability. Elastase is used routinely for alveolar epithelial cell isolation [Dobbs and Gonzales, 2002], and it can be potentially useful if tumor cells adhere to stroma containing elastin.

The rationale for using serum-free medium instead of serum-supplemented medium is that the growth of stromal cells is limited in the former growth condition. However, this approach may also limit the growth of cancer cells, so, after several passages in serum-free medium, serum can be added to boost the proliferation of the cancer cells. However, this treatment must be performed with some caution, as the treatment may also stimulate terminal differentiation, especially of squamous cells. The other possibility is that cancer cells may produce growth factors, such as platelet-derived growth factor, to stimulate fibroblast growth. In that case, fibroblast contamination is inevitable. It is then necessary to perform a cloning or differential adherence approach to select cancer cells.

It was found from the SCLC cultures that cells can change their differentiated functions. For example, SCLC cells can be changed from a bombesin-secreting type to a non-secreting type [Carney et al., 1985; Gazdar et al., 1985]. This change is associated with the virulence of the cancer cells, which also become refractory to chemotherapy. This change also occurs in vivo. The same is true for NSCLC cells, which can be changed between a mucin-secreting cell type and a squamous cell type or can become undifferentiated. The nature of these changes is not clear and could be spontaneous or due to further mutations. For instance, c-*myc* amplification is associated with the change from classical SCLC cells to variant cells [Johnson et al.,

1986]. Because of this phenomenon, it is necessary to characterize periodically the differentiated functions of the cultured cells.

The other variation applicable to this protocol is in the in vivo tumorigenesis assay. It has been shown previously that cancer cells can repopulate a denuded tracheal graft. "Denuded" means that the lining epithelium has been removed [Terzaghi and Klein Szanto, 1980]. Klein Szanto and his colleagues showed that cancer cells will maintain some differentiated features while expressing malignancy in the graft [Momiki et al., 1991]. The information generated from this approach is particularly relevant and useful because airway material was used in.

## 5. DISCUSSION

The development of an in vitro model system for lung cancer cells permits a direct examination of the life cycles, differentiated functions, and genetics of these tumor cells. The life cycle study allows the understanding of growth rate, growth behavior, growth factor requirement(s), and cell cycle control mechanisms of these tumor cells. These analyses, when they are compared with normal epithelial cells, may yield information as to why the tumor cells have an unlimited life span and unregulated growth control.

Study of the differentiated function will allow an understanding of the intrinsic and ectopic phenotypic expression, the regulation of expression, particularly of the biomarkers unique to these tumor cells. The biomarkers are useful not only for early detection of these tumor cells but also as targets for immuno-therapeutic treatments. Genetic studies will permit the understanding of the nature of the mutations associated with the development of these tumors; for example, which chromosomal loci are involved. Genetic analysis will also allow the elucidation of which oncogenes and tumor suppressor genes are up- and down-regulated, and what is the control mechanism involved in the regulation of gene expression in these tumor cells. Thus, much information can be generated from the successful culturing of tumor cells in vitro. Such information may be difficult to obtain from in vivo procedures because of inaccessibility and limited cell recovery. Furthermore, the heterogeneity of cell types in vivo also make interpretation of data difficult.

In addition to obtaining fundamental information on regulatory mechanisms, the in vitro system also allows the initiation of various studies on the development of therapeutic treatment for these cancer cells. One can test directly the effects of various drugs on defined target cells and provide a rational basis for selection of drugs for chemotherapy. In addition, results based on in vitro chemotherapy studies of SCLC cells have led to the conclusion that there are both variant and classic types in SCLC, a finding which has significance in the treatment of this type of cancer.

Lastly, cell culture information may also help in the development of culture conditions suitable for the growth and maintenance of a pre-neoplastic cell population. Theoretically, these cells should be poised between the normal and the cancer cell types, and their growth behavior and other properties, except malignancy, characteristic of this lineage. Using pre-neoplastic cell cultures may give a better understanding of the process of carcinogenesis, and its regulation. In addition, the identification of pre-neoplastic cell populations could provide the means to identify environmental risk factors associated with particular genetic aberrations.

## ACKNOWLEDGMENTS

The author thanks Dr. Cheryl Soref for editing this manuscript.

## SOURCES OF MATERIALS

| Item | Supplier |
| --- | --- |
| Agar | Difco (Becton Dickinson) |
| BEGM | Clonetics (BioWhittaker) |
| BHE | see ECGS and PE |
| Bovine hypothalamus | Pel Freeze |
| BSA | Sigma |
| Cholera toxin | List Biologics |
| ECGS | Collaborative Research (Becton Dickinson) |
| EGF | Upstate Biotechnology |
| Ethanolamine | Sigma |
| Ficoll/metrizoate | Amersham Pharmacia, ICN, Nygaard |
| Fungizone | Sigma |
| Gentamycin | Sigma |
| Growth factors | Sigma |
| HEPES | Sigma |
| Hormones | Sigma |
| Lymphosep | ICN |
| MEM | Gibco (Invitrogen) |
| Penicillin | Gibco (Invitrogen) |
| Pituitary extract (PE) | Gibco (Invitrogen) |
| Phosphoethanolamine | Sigma |
| RPMI 1640 | Gibco (Invitrogen) |
| Streptomycin | Sigma |
| Triiodothyronine (T3) | Sigma |
| Trypsin | Gibco (Invitrogen) |
| Trypsin inhibitor | Sigma T0800 |

# REFERENCES

Allen-Hoffmann, B.L., and Rheinwald, J.G. (1984) Polycyclic aromatic hydrocarbon mutagenesis of human epidermal keratinocytes in culture. *Proc. Natl. Acad. Sci. USA* 81:7802–7806.

Barnes, D., and Sato, G. (1980) Serum-free cell culture: a unifying approach. *Cell* 22:649–655.

Baylin, S.B., Abeloff, M.D., Goodwin, D., Carney, D.N., Gazdar, A.F. (1980) Activities of L-dopa decarboxylase and diamine oxidase (histaminase) in human lung cancers and decarboxylase as a marker for small (oat) cell cancer in cell culture. *Cancer Res.* 40:1990–1994.

Bottenstein, J., Hayashi, I., Hutchings, S., Masui, H., Mather, J., Mc Clure, D.B., Ohasa, S., Rizzino, A., Sato, G., Serrero, G., Wolfe, R, Wu, R. (1997) The growth of cells in serum-free hormone-supplemented media. In Jakoby, W.B., and Paston, I.H. (eds) *Methods in Enzymology.* New York, NY: Academic Press, Vol. 58, pp. 94–109.

Brower, M., Carney D.N., Oie H.K., Gazdar A.F., Minna J.D. (1986) Growth of cell lines and clinical specimens of human non-small cell lung cancer in a serum-free defined medium. *Cancer Res.* 46:798–806.

Caputo, J.L. (1996): Safety Procedures. In Freshney, R.I., Freshney, M.G., eds., Culture of Immortalized Cells, New York, John Wiley & Sons, pp. 25–51.

Carney, D.N., Bunn P.A., Gazdar A.F., Pagan J.A., Minna J.D. (1981) Selective growth in serum-free hormone-supplemented medium of tumor cells obtained by biopsy from patients with small cell carcinoma of the lung. *Proc. Natl. Acad. Sci. USA* 78:3185–3189.

Carney, D.N., Gazdar, A.F., Bepler, G., Guccion, J.G., Marangos, P.J., Moodt. T.W., Zweig, M.H., Minna, J.D. (1985): Establishment and identification of small cell lung cancer cell lines having classic and variant features. Cancer Res., 45:2913–2923.

Chu, Y.W., Yang, P.C., Yang, S.C., Shyh, Y.C., Hendrix, M.J.C., Wu, R., Wu, C.W. (1997) Selection of invasive and metastatic subpopulations from a human lung adenocarcinoma cell line. *Am. J. Respir. Cell Mo. Biol.* 17:353–360.

Dobbs, L.G., Gonzalez, R.F. (2002): Isolation and culture of pulmonary alveolar epithelial type II cells. In Freshney, R.I., Freshney, M.G., eds., Culture of Epithelial Cells, 2nd ed., New York, John Wiley & Sons, pp. 303–335.

Gazdar, A.F., and Minna, J.D. (1996) NCI series of cell lines: an historical perspective. *J. Cell Biochem.* Suppl 24:1–11.

Gazdar, A.F., and Oie, H.K. (1986) Cell culture methods for human lung cancer. *Cancer Genet. Cytogenet* 19:5–10.

Gazdar, A.F., Carney, D.N., Nau, M., Minna, J.D. (1985). Characterization of variant subclasses of cell lines derived from small cell lung cancer having distinctive biochemical, morphological, and growth properties. Cancer Res., 45:2924–2930.

Gazdar, A.F., Oie, H.K., Shackleton, G.H., Chen, T.R., Triche, T.J., Myers, C.E., Chrousos, G.P., Brennan, M.F., Stein, C.A., La Rocca R.V. (1990) Establishment and characterization of a human adrenocortical carcinoma cell line that expresses multiple pathways of steroid biosyntheses. *Cancer Res.* 50:5488–5496.

Hayashi, I., and Sato, G.H. (1976): Replacement of serum by hormones permits growth of cells in a defined medium. Nature, 259:132–134.

Jaurand, M.C. (1991) Observations on the carcinogenicity of asbestos fibers. *Ann. NY Acad. Sci.* 643:258–270.

Johnson, B.E., Battey, J, Linnoila, I., Becker, K.L., Makuch, R.W., Snider, R.H., Carney, D.N., Minna, J.D. (1986) Changes in phenotype of human small cell ling cancer cell lines after transfection and expression of the c-myc proto-oncogene. *J. Clin. Invest.* 78:525–532.

Landis, S.H., Murray, T., Bolden, S., Wingo, P.A. (1998) Cancer statistics, 1998. *CA Cancer J. Clin.* 48:6–29.

Lau, D., Xue, L., Hu, R., Liaw, T., Wu, R., Reddys, S. (2000) Expression and regulation of a molecular marker SPR1, in multistep bronchial carcinogenesis. *Am. J. Respir. Cell Mol. Biol.* 22:92–96.

Lechner, J.F., Stoner, G.D., Yoakum, G.H., et al. (1986) In vitro carcinogenesis studies with tracheobronchial tissues and cells. In Schiff L.J. (ed) *In Vitro Models of Respiratory Epithelium.* Boca Raton, FL: CRC Press, pp. 143–159.

Leibovitz, A. (1986) Development of tumor cell lines. *Cancer Genet. Cytogenet* 19:11–19.

Levitt, M.L., Gazdar, A.F., Oie, H.K., Schuller, H., Thatcher, S.M. (1990) Cross-linked envelope-related markers for squamous differentiation in human lung cancer cell lines. *Cancer Res.* 50:120–128.

Lin, H., Carlson, D.M., St. George, J.A., Plopper, C.G., Wu, R. (1989) An ELISA method for the quantitation of tracheal mucins from human and nonhuman primates. *Am. J. Respir. Cell Mol. Biol.* 1:41–48.

McMahon, J.B., Schuller, H.M., Gazdar, A.F., Becker, K.L. (1984) Influence of priming with 5-hydroxytryptophan on APUD characteristics in human small cell lung cancer cell lines. *Lung* 162:261–264.

Momiki, S., Baba, M., Caamano, J., Lizasa, T., Nakajima, M., Yamaguchi, Y., Klein Szanto, A.J.P. (1991) In vivo and in vitro invasiveness of human lung carcinoma cell lines. *Invas. Metastasis* 11:66–75.

Mossman, B.T., Bignon, J., Corn, M., Seaton, A., Gee, J.B.L. (1990) Asbestos: scientific developments and implications for public policy. *Science* 247:294–301.

Oie, H.K., Russell, E.K., Carney, D.N., Gazdar, A.F. (1996) Cell culture methods for the establishment of the NCI series of lung cancer cell lines. *J. Cell Biochem.* Suppl 24:24–31.

Parkin, D.M., Pisani, P., Lopez, A.D., Masuyer, E. (1994) At least one in seven cases of cancer is caused by smoking. Global estimates for 1985. *Int. J. Cancer* 5904:494–504.

Pass, H.I., and Mew, D.J.Y. (1996) In vitro and in vivo studies of mesothelioma. *J. Cell Biochem.* Suppl 24:142–151.

Rheinwald, J.G., and Beckett, M.A. (1981) Tumorigenic keratinocyte lines requiring anchorage and fibroblast support cultured from human squamous carcinomas. *Cancer Res.* 41:1657–1663.

Robinson, C.B., and Wu, R. (1991) Culture of conducting airway epithelial cells in serum-free medium. *J. Tiss. Cult. Meth.* 13:95–102.

Rom, W.N., Travis, W.D., Brody, A.R. (1991) Cellular and molecular basis of the asbestos-related diseases. *Am. Rev. Respir. Med.* 143:408–422.

Simms, E., Gazdar, A.F., Abrams, P.G., Minna, J.D. (1980) Growth of human small cell (oat cell) carcinoma of the lung in serum-free growth factor supplemented medium. *Cancer Res.* 40:4356–4363.

Terzaghi, M., and Klein Szanto, A.J.P. (1980) Differentiation of normal and cultured preneoplastic tracheal epithelial cells in rat: importance of epithelial-mesenchymal interactions. *J. Natl. Cancer Inst.* 65:1039–1045.

Wu, R. (1986) In vitro differentiation of airway epithelial cells. In Schiff, L.J. (ed) *In Vitro Models of Respiratory Epithelium.* Boca Raton, FL: CRC Press, pp. 1–26.

Wu, R., Zhao, Y.H., Chang, MM. (1997): Growth and differentiation of conducting airway epithelial cells in culture. *Europ. Respir. J.*, 10:2398–403.

Yang, P.C., Luh, K.T., Wu, R., Wu, C.W. (1992) Characterization of the mucin differentiation in human adenocarcinoma cell lines. *Am. J. Respir. Cell Mol. Biol.* 7:161–171.

# 2

# Culture of Normal and Malignant Gastric Epithelium

Jae-Gahb Park[1,2], Ja-Lok Ku[1], Hee-Sung Kim[2], So-Yeon Park[2] and Michael J. Rutten[3]

[1] Laboratory of Cell Biology, Korean Cell Line Bank, Cancer Research Center and Cancer Research Institute, Seoul National University College of Medicine, Seoul, Korea; [2] National Cancer Center, 809 Madu1-dong, Goyang, Gyeonggi, Korea; and [3] Department of Surgery, Oregon Health Sciences University, Portland, Oregon

Corresponding authors: Jae-Gahb Park (cancerous gastric cell culture) jgpark@plaza.snu.ac.kr; Michael J. Rutten (normal gastric cell culture) rutten@ohsu.edu

*Culture of Human Tumor Cells*, Edited by Roswitha Pfragner and R. Ian Freshney.
ISBN 0-471-43853-7   Copyright © 2004 Wiley-Liss, Inc.

# 1.  INTRODUCTION

## 1.1.  Gastric Carcinoma

Gastric cancer is a worldwide disease. Its incidence is high in Japan, Korea, China, Costa Rica, Chile, and Eastern Europe, but low in the United States and United Kingdom [Correa and Chen, 1994]. Although the incidence of gastric cancer has declined dramatically in Western countries, it is the second most common cause of cancer-related deaths worldwide [Fuchs and Mayer, 1995]. In the United States, gastric cancer was the most common cause of cancer deaths in 1930 [Neugut et al., 1996]. However, since that time, the annual mortality rate in the United States has dropped from about 38 to 7 per 100,000 for men and from 28 to 4 per 100,000 for women [Landis et al., 1998]. Nevertheless, gastric cancer remains one of the most common cancers in East Asia; in Japan for example, it is the second leading cause of cancer-related deaths, and its annual mortality rate was 53 per 100,000 for men and 28 per 100,000 for women in 1997.

Environmental influences, especially diet, are thought to be the most important offenders [Fuchs and Mayer, 1995]. Numerous studies have indicated that salted, smoked, pickled, and preserved foods rich in nitrite and preformed *N*-nitroso compounds are associated with an increased risk of gastric cancer [Palli, 2000]. Among the nondietary contributory factors, substantial evidence indicates increased risk in association with *Helicobacter pylori* infection. Infection by *H. pylori* leading to chronic gastritis and intestinal metaplasia is now thought to contribute, though is insufficient in itself, to cause gastric carcinogenesis. It seems that several factors, including diet, cigarette smoking, infection by *H. pylori*, and individual susceptibility, interact in a complex multifactorial process to determine the risk of gastric cancer development.

The prognosis of gastric cancer is poor, despite the extensive use of radical surgery in combination with other modes of treatment and the developments of early diagnostic tools, and is related directly to the pathologic stage based on the depth of tumor invasion and the extent of nodal and distant metastasis [Roder et al., 1993]. The five-year survival rate of surgically treated early gastric cancer is 90–95%. In contrast, the five-year survival rate for advanced gastric carcinoma remains below 15%.

Cell lines from human gastric-cancers are useful tools in the study of cell biology and in the development and testing of new therapeutic modalities. A large bank of well-characterized cell lines should reflect the diversity of tumor phenotypes and provide adequate models of tumor heterogeneity. In general, gastric carcinoma has been more difficult to establish in long-term culture than esophageal or colorectal-carcinoma cell lines [Park et al., 1990; Park et al., 1994; Park and Gazdar, 1996; Park et al., 1997].

Gastric carcinoma cell lines can be established from ascitic effusions, metastatic tissues (regional lymph nodes and distant metastatic sites, such as liver), and primary tumors. The establishment of cell lines from ascitic effusions has proven to be more efficient than establishment from primary tumors because the cancer cells are rich, free-floating, and primed for in vitro growth. However, effusions usually contain mesothelial cells and lymphocytes. Mesothelial cells can attach more easily to flasks than cancer cells and generally survive for more than one year [Sekiguchi and Suzuki, 1994]. Lymphocytes can inhibit the growth of cancer cells. Enzymatic digestion of solid primary tumor tissues may result in poor recovery of viable cells and overgrowth of the contaminating stromal cells rather than cancer cells. The faster growing stromal fibroblast-like cells easily overgrow the cancer cells. The mechanical spillout method provides a simpler, faster, and less traumatic method of obtaining cells for culture [Oie et al., 1996]. This method allows the minimization of stromal cell contamination because stromal cells are not easily detached from the tissue matrix by mechanical means.

Rutzky and Moyer [1990] summarized the features of key relevance for successful culture: 1) nonenzymatic or minimal dissociation of tumor tissue; 2) seeding cultures as explants and at high cell densities; 3) removal of contaminating fibroblasts, usually after they have aided culture initiation; and 4) delaying passage until high cell densities have been achieved, and plating cells at high density.

## 1.2. Normal Gastric Epithelium

Several methods have been described for isolating cells from the gastric mucosa of animals for cell culture. However, only in the past few years has there been a successful isolation and culture of normal human gastric mucous epithelial cells from biopsies [Smoot et al., 2000; Wagner et al., 1998], surgical resections [Rutten et al., 1996; Wagner et al., 1998], and fetal tissue [Basque and Menard, 1999]. The isolation and successful culture of human antral mucous and gastrin-releasing G-cells have also been reported [Richter-Dahlfors et al., 1998]. In addition, several reports on preparations of pepsinogen-secreting chief cells derived from animal gastric mucosae have been reported [Okayama et al., 1998; Heim et al., 1997b; Heim et al., 1997a; Giebel et al., 1995; Tanaka and Tani, 1995; Okayama et al., 1994; Tani et al., 1994; Fukamachi et al., 1994; Scheiman et al., 1991; Ota et al., 1990; Tani et al., 1989; Olson, 1988; Defize et al., 1985; Sanders et al., 1983; Ayalon et al., 1982; Beinborn et al., 1993] and there are also reports of successful human chief cell cultures [Basque and Menard, 2000; Defize et al., 1985]. In the past, the most difficult gastric epithelial cell type to culture has been the acid-secreting gastric parietal cell [Chew et al., 1989; Chew, 1994]. However, it is now possible to culture human parietal cells from the enzymatic digestion of fetal gastric mucosa [Basque et al., 1999]. In many of these cases, the successful culturing of normal (nontransformed) human gastric cells has come from the experience of working with animal cell cultures. Also, the advances in dissociation media, hormones, extracellular matrices [Freshney and Freshney, 2002], and antibiotics now make it possible to isolate and culture a variety of human gastric epithelial cell types.

In this chapter, we describe the culture of normal gastric epithelial cells, as well as of primary gastric cancer cell lines, and include the detailed protocols relating to establishment from ascitic effusions and primary tumor tissues by using mechanical spillout, the isolation of pure cancer cells, and the maintenance and propagations and preservation of cancer cells.

## 2. PREPARATION OF GASTRIC TUMOR SPECIMENS FOR PRIMARY CELL CULTURE

### 2.1. Growth Media

Generally, tumor-derived cells are cultured in RPMI 1640, Dulbecco's modified Eagle's medium (DMEM) and Ham's F-10 and are supplemented with 10% heat-inactivated fetal bovine serum (FBS). Although normal fibroblasts generally grow faster and overgrow cancer cells in serum-supplemented media, RPMI 1640 supplemented with 10% heat-inactivated FBS is suitable for the primary cell culture of gastric cancer specimens [Park et al., 1990; Park and Gazdar, 1996; Park et al. 1997].

### 2.2. Initiation Medium

RPMI 1640
Heat-inactivated fetal bovine serum (FBS), 10%
Sodium bicarbonate (NaHCO$_3$), 20 mM
$N$-(2-hydroxyethyl)piperazine-$N'$-ethanesulfonic acid (HEPES) buffer, 25 mM
Antibiotics (streptomycin penicillin, usually 100 μg/ml and 100 units/ml respectively)
Filter through 0.22 μm bottle-top filter

### 2.3. Procurement of Tumor Specimens

Solid tumors (primary tumors and tumors in lymph nodes) must be carefully and aseptically excised from pathologically proven gastric tumor samples and transferred to a cell culture laboratory in RPMI 1640 medium. Whenever possible, invasive areas from the serosal surface should be selected for primary tumor cultures to reduce the possibility of microbial contamination.

The transporting medium, RPMI 1640, should contain antibiotics, such as penicillin/streptomycin, to prevent bacterial contamination. Ascitic effusions should be collected in sterile glass bottles or syringes. Heparin (5 μg/ml) may be added to the collected fluids to prevent clotting, if the effusions contain many erythrocytes.

## 3. PRIMARY CELL CULTURE OF GASTRIC TUMORS

### 3.1. Procedure for Solid Tumors

The procedure is described in Figure 2.1.

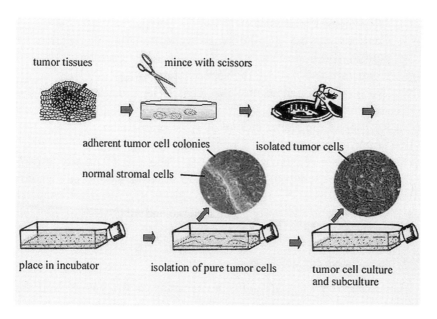

**Figure 2.1.** Procedure for primary cell culture of solid tumors.

## Protocol 2.1. Preparation of Primary Cultures from Solid Gastric Tumors

### Reagents and Materials

*Sterile*

❑ Initiation medium: RPMI 1640 supplemented with 10% heat-inactivated FBS
❑ PBSA (pH 7.0)
❑ Petri dishes, 10 cm
❑ Scissors
❑ Forceps
❑ Pipettes, 5 and 10 ml
❑ Culture flasks (25 cm$^2$) coated with attachment factors, such as collagen and laminin

### Protocol

(a) Transfer tumor tissues to fresh, sterile PBSA in a dish and rinse.
(b) Transfer tumor tissues to a new dish, and dissect off necrotic areas, fatty tissues, blood clots, and connective tissues with forceps and scissors.
(c) Finely mince the tumor tissues into very small pieces with sterile scissors.
(d) Disassociate tissue pieces into small aggregates by vigorous pipetting with 10 ml of PBSA.

(e)   Allow heavier pieces to sediment by gravity.

(f)   Carefully harvest the tumor cell aggregate-containing washes by pipette, and transfer to a 15 ml sterile centrifuge tube.

(g)   Centrifuge and remove the supernatant.

(h)   Wash by resuspending in PBSA, centrifuge, and again remove the supernatant.

(i)   Resuspend the pieces in 5 ml of initiation medium.

(j)   Transfer the pieces to a 25-cm$^2$ flask.

(k)   Harvest the remaining large pieces, place in a 15 ml sterile centrifuge tube, centrifuge, remove the supernatant and culture immediately.

(l)   Cap the flask and maintain in a humidified incubators at 37°C in an atmosphere of 5% $CO_2$ and 95% air.

### 3.2.   Procedure for Ascitic Effusions

The processing of ascitic effusions depends on the volume and the cellular content (erythrocytes, mesothelial cells, monocytes, and tumor cells). For specimens with few or no erythrocytes, the cells are centrifuged, resuspended in RPMI 1640 medium, and, without further processing, seeded into 25-cm$^2$ flasks. For specimens with large numbers of erythrocytes, the cells are centrifuged over Ficoll/metrizoate and resuspended in RPMI 1640 medium.

### Protocol 2.2.   Isolation of Gastric Cancer Cells from Ascitic Effusions by Density Sedimentation

#### Reagents and Materials
*Sterile*
❑   Initiation medium: RPMI 1640 supplemented with 10% heat-inactivated FBS
❑   PBSA (pH 7.0)
❑   Ficoll/metrizoate, e.g., Lymphosep, 1.077 g/cc
❑   Pipettes, 5 and 10 ml
❑   Culture flasks, 25cm$^2$

#### Protocol
(a)   Add 7 ml Ficoll/metrizoate to a 15-ml sterile centrifuge tube.

(b)   To the 7 ml Ficoll/metrizoate cushion (carefully layering on top of Ficoll), add 7 ml of diluted ascitic fluid by using a sterile pipette with its tip touching the side of the tube.

(c)   Centrifuge at 400 $g$ for 30–40 min in a tabletop centrifuge at room temperature.

(d)   Carefully aspirate the upper layer by using a Pasteur pipette, leaving the putative cancer cell layer undisturbed at the interface.

(e)     Using a clean Pasteur pipette, transfer the cancer cell layer to a clean centrifuge tube.

(f)     The pellets from the bottom of the Ficoll cushion usually contain a significant number of tumor aggregates. After a single wash with PBSA, resuspend the pellets in PBSA and leave to stand in a rack at room temperature, allowing heavier cell aggregates to sediment by gravity. After aspirating off the supernatant, which contains most of the erythrocytes, collect the cell aggregates.

(g)     Add at least 3 volumes (6 ml) of PBSA to the cancer cells in the centrifuge tube.

(h)     Suspend the cells by gently drawing them in and out of a Pasteur pipette.

(i)     Spin the cells at 400 $g$ for 10 min, decant the supernatant from cell pellet, add 10 ml of RPMI 1640, resuspend the cells with a Pasteur pipette, and then spin at 400 $g$ for 10 min.

(j)     Repeat this washing step.

(k)     Decant the medium, and resuspend the cell pellet in 5 ml of initiation medium.

(l)     Seed the cell pellet into 25-cm$^2$ flasks.

(m)     Cap the flask and maintain the culture in a humidified incubator at 37°C in an atmosphere of 5% $CO_2$ and 95% air.

## 4.  ENRICHMENT OF CANCER CELLS

Epithelial cells, presumed to include the cancer cells, can grow as either floating aggregates, firmly or loosely adherent colonies, or as both adherent and floating subpopulations. If the cancer cells are adherent or grow as floating aggregates, the medium is changed weekly until a substantial outgrowth of cells is observed. Isolation of the cancer cells is performed when heavy tumor-cell growth is observed. Several methods can be used to isolate pure cancer cells.

### 4.1.  Floating Aggregates

If floating aggregated cancer cells are grown in suspension, pure cancer cells can be obtained easily because most of the fibroblasts are attached to the surface of the culture flask.

### Protocol 2.3.  Enrichment of Gastric Cancer Cells by Transfer of Aggregates

#### Reagents and Materials
*Sterile*
- ❑  RPMI 1640 medium supplemented with 10% heat-inactivated FBS
- ❑  Centrifuge tubes, 15 ml
- ❑  Culture flasks, 25 cm$^2$

*Protocol*

(a) Harvest and transfer the floating aggregates to a sterile centrifuge tube.

(b) Spin the cells at 400 $g$ for 5 min, decant 2/3 of volume of supernatant from the cell pellets, and resuspend the pellets in and additional 5 ml of RPMI 1640 medium supplemented with 10% heat-inactivated FBS.

(c) Seed the cell suspension derived from the pellets into 25-$cm^2$ flasks.

(d) Cap the flasks, and maintain the culture in a humidified incubator at 37°C in an atmosphere of 5% $CO_2$ and 95% air.

### 4.2. Adherent Colonies

### 4.2.1. Rapping Method

This method is used when cancer cells are loosely attached to the surface of the culture flask and grown as adherent colonies among the stromal fibroblast cells. Of these cancer cells, some cancer cells or whole colonies are detached from the colonies and grown in suspension.

### Protocol 2.4. Enrichment of Gastric Cancer Cells by Mechanical Detachment of Aggregates

*Reagents and Materials*

*Sterile*

❑ RPMI 1640 medium supplemented with 10% heat-inactivated FBS
❑ Centrifuge tubes, 15 ml
❑ Culture flasks, 25 $cm^2$

*Protocol*

(a) Gently tap the flasks.

(b) Harvest the detached individual cells and colonies.

(c) Transfer the floating aggregates to a sterile centrifuge tube.

(d) Spin the cells at 400 $g$ for 5 min, decant 2/3 volumes of supernatant from the pellets, and add 5 ml of RPMI 1640 medium supplemented with 10% FBS.

(e) Seed the cell pellet into 25-$cm^2$ flasks.

(f) Cap the flask and maintain the culture in a humidified incubators at 37°C in an atmosphere of 5% $CO_2$ and 95% air.

### 4.2.2. Scraping Method

Usually firmly adherent colonies surrounded by stromal fibroblasts form large densely packed colonies, which continue to grow laterally,

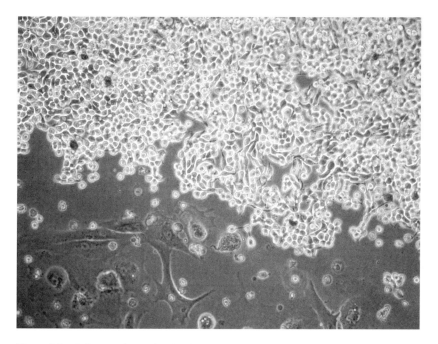

**Figure 2.2.** Primary culture of ascitic fluid from gastric cancer patients.

slowly infiltrating and displacing the adjacent fibroblasts or mesothelial cells (Figure 2.2). Eventually, the entire flask may be taken over
by these presumed tumor cells. In this case, cancer cells can be isolated
by scraping and by differential trypsinization [Oie et al., 1996]. In our
experience, when large adherent cell colonies are less impeded by surrounding fibroblasts, scraping is the best way of isolating the putative
cancer cells; the cells are less damaged and stromal contamination is
minimized [Park et al., 1990; Park and Grazdar, 1996; Park et al.,
1997].

## Protocol 2.5.   Enrichment of Gastric Cancer Cells by Scraping

### Reagents and Materials

*Sterile*

- ❏ Growth medium: RPMI 1640 medium supplemented with 10% heat-inactivated FBS
- ❏ Cell scraper
- ❏ Centrifuge tubes, 15 ml
- ❏ Culture flasks, 25 cm$^2$

*Non-sterile*

- ❏ Fine-tipped marker pen

*Protocol*
(a) Maintain the culture until the adherent colonies form well-isolated large colonies (5–10 mm in diameter). This may take several months.
(b) Select colonies and mark with marker pen on the flask base.
(c) Decant the medium and add 1 ml of growth medium.
(d) Detach the adherent colonies by scraping with a policeman or cell scraper, to allow them to float freely in the medium.
(e) Add 4 ml of growth medium and harvest the detached colonies with a pipette.
(f) Transfer the detached cells into 25-$cm^2$ flasks.
(g) Maintain cultures in humidified incubators at 37°C in an atmosphere of 5% $CO_2$ and 95% air.

### 4.2.3. Differential Trypsinization Method

Some cells grow among stromal fibroblasts and produce small colonies of cells presumed to be cancer cells. Usually these cancer cells do not overgrow the fibroblasts. In these cases, the differential trypsinization method is preferred. The method depends on differences in cell susceptibility to trypsinization [Oie et al., 1996]. Some cancer cells are more susceptible to trypsinization than fibroblasts are and vice versa.

### Protocol 2.6. Enrichment of Gastric Cancer Cells by Differential Trypsinization

*Reagents and Materials*
*Sterile*
❏ PBSA
❏ Trypsin, 0.025%, 0.05%, or 0.1% in PBSA
❏ Growth medium: RPMI 1640 medium supplemented with 10% heat-inactivated FBS
❏ Centrifuge tubes, 15 ml
❏ Culture flasks, 25 $cm^2$

*Protocol*
(a) Decant the medium, rinse with PBSA, and remove the PBSA by aspirating.
(b) Add 1.5 ml of diluted trypsin.
(c) Place into incubator at 37°C for approximately 3 min.
(d) Observe the flasks under the inverted microscope throughout incubation, and determine whether putative cancer cells or fibroblasts are detached. If both cancer cells and stromal fibroblasts remain attached, incubate the culture flasks until either the cancer cells or fibroblasts detach.

(e)   If cancer cells have detached and fibroblasts remain attached, add 5 ml of growth medium containing serum and harvest the detached cells.

(f)   If, however, cancer cells remain attached and the fibroblasts have detached, remove the medium, rinse with the growth medium, and then add 5 ml of growth medium.

(g)   Centrifuge the harvested cancer cells, remove the supernatant, and add 5 ml of growth medium.

(h)   Transfer the cells into 25-cm$^2$ flasks.

(i)   Maintain cultures in humidified incubators at 37°C in an atmosphere of 5% $CO_2$ and 95% air.

## 5.   PROPAGATION AND PRESERVATION OF CANCER-DERIVED CELLS

After cancer-derived cell lines are established, initial passages are performed when heavy epithelial cell growth and large colonies are observed presumed to be tumor-derived. Subsequent passages are performed every one or two weeks. The RPMI 1640 medium supplemented with 10% heat-inactivated FBS, which is used in primary cell culture, can be also used for the propagation of gastric cancer cell lines [Park et al., 1990; Park and Gazdar, 1996; Park et al., 1997]. This medium must be used until the isolated cells are considered as a cell line. If the generation number (>100) implies that this is a continuous cell line, this indicates that the cells are neoplastic. It is recommend that samples of early passages, are frozen in cryoprotective medium to prevent loss of the cell lines by contamination or other laboratory accidents.

### 5.1.   Floating Aggregates

Floating cell aggregates grow loosely attached, in ball-like tightly packed clumps. Loosely attached cell aggregates are subcultured after dissociation by pipetting (Figure 2.3).

### 5.2.   Tightly Packed Clumps

**Protocol 2.7.   Subculture of Gastric Cancer Cells by Dispersing Cell Aggregates**

***Reagents and Materials***
*Sterile*
❏   PBSA
❏   Trypsin/EDTA: Trypsin, usually 0.05%, EDTA, 0.5 mM, in PBSA

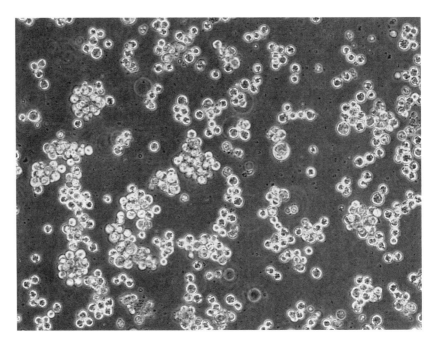

**Figure 2.3.** SNU-16 gastric carcinoma cell line. Cancer cells grow as loosely attached cell aggregates.

- Growth medium: RPMI 1640 medium supplemented with 10% heat-inactivated FBS
- Centrifuge tubes, 15 ml
- Culture flasks, 25 cm$^2$

**Protocol**

(a) Collect the clumps by pipette and transfer to a sterile centrifuge tube.

(b) Spin cells at 400 *g* for 5 min, discard the supernate from pellets, rinse with sterile PBSA, centrifuge, and again remove the supernatate.

(c) Add 2 ml of Trypsin/EDTA in PBSA, and incubate at 37°C for approximately 3 min.

(d) Add 5 ml of RPMI 1640 medium supplemented with 10% FBS and centrifuge.

(e) Decant the supernate and redisperse the cells by pipetting with the growth medium.

(f) Seed the cell pellet into 25-cm$^2$ flasks.

(g) Cap the flask and maintain the cultures in a humidified incubator at 37°C in an atmosphere of 5% $CO_2$ and 95% air.

In this case, some cells can be grown as adherent colonies or as a monolayer, and others can be grown as floating cell aggregates

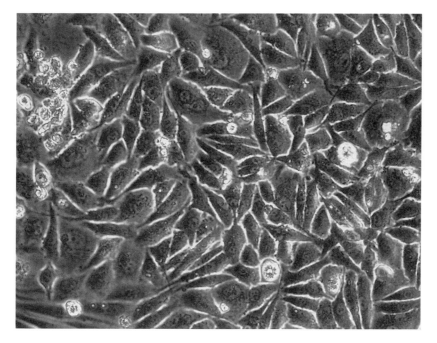

**Figure 2.4.** SNU-668 gastric carcinoma cell line. Cancer cells grow as adherent cultures; showing diffusely spreading growth of cultured tumor cells with fusiform or polygonal contours.

(Fig 2.8b). Growth as floating aggregates is an indication that the cells are neoplastic.

### 5.3. Adherent Cells

(See Figs. 2.4 and 2.6b)

## Protocol 2.7. Subculture of Adherent Gastric Cancer Cells

### Reagents and Materials

*Sterile*

- ❑ PBSA
- ❑ Trypsin EDTA: usually 0.05% trypsin and 0.5 mM EDTA, in PBSA
- ❑ Growth medium: RPMI 1640 medium supplemented with 10% heat-inactivated FBS
- ❑ Centrifuge tubes, 15 ml
- ❑ Culture flasks, 25 cm$^2$

### Protocol

(a)  Remove the supernatant and rinse with sterile PBSA.

(b)  Add 2 ml of Trypsin-EDTA in PBSA and incubate at 37°C for approximately 3 min.

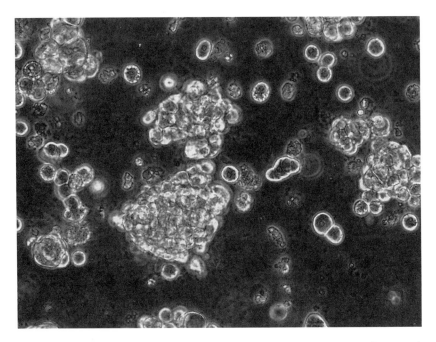

**Figure 2.5.** SNU-620 gastric carcinoma cell line. Cancer cells grow as both adherent and floating cell aggregates.

(c) Add 5 ml of RPMI 1640 medium supplemented with 10% FBS and centrifuge.

(d) Decant the supernatate and redisperse the cancer cells by pipetting with the growth medium.

(e) Seed the cell pellet into new 25-cm$^2$ flasks.

(f) Cap the flasks and maintain the cultures in humidified incubators at 37°C in an atmosphere of 5% $CO_2$ and 95% air.

If the adherent cells are not detached in the flasks, the concentrations of Trypsin/EDTA and the incubation times can be increased.

### 5.4. Cultures Containing both Adherent and Floating Subpopulations

### Protocol 2.8. Subculture of Mixed Floating and Adherent Subpopulations of Gastric Cancer Cells

***Regents and Materials***
*Sterile*
❑ PBSA
❑ Trypsin EDTA: usually 0.05% trypsin and 0.5 mM EDTA, in PBSA
❑ RPMI 1640 medium supplemented with 10% heat-inactivated FBS

❏ Centrifuge tubes, 15 ml
❏ Culture flasks, 25 cm$^2$

## Protocol
*Floating Subpopulations*
(a) Collect the floating cell aggregates, transfer to sterile centrifuge tubes, and allow to settle.
(b) Remove the supernatant, and resuspend in growth medium.
*Adherent Subpopulations*
(a) Add 2 ml of Trypsin-EDTA in PBSA and incubate at 37°C for approximately 3 min.
(b) Add 5 ml of RPMI 1640 medium supplemented with 10% FBS, and centrifuge.
(c) Decant the supernatant, resuspend in growth medium, and combine the cells with the floating subpopulations.
(d) Seed the cells into new 25-cm$^2$ flasks.
(e) Cap the flask, and maintain the culture in humidified incubators at 37°C in an atmosphere of 5% $CO_2$ and 95% air.

## 6. REPRESENTATIVE CHARACTERISTICS OF CULTURED GASTRIC CARCINOMA CELL LINES

Cell lines which became continuous are regarded as tumor cell derived, as normal gastric epithelium would not be expected to immortalize. The ability of aggregates to show cell proliferation in suspension also implies that they are neoplastic in origin.

The SNU-216 cell line derived from moderately differentiated tumor (Fig. 2.6a) grew as a diffuse monolayer in a large island pattern. Most cells grew attached to the flask in vitro (Fig. 2.6b).

The SNU-484 cell line derived from primary tumors, presented as poorly differentiated carcinoma with mild desmoplasia (Fig. 2.7a), grew as a mixture of attached and floating populations, heterogeneous in cell density (Fig. 2.7b).

SNU-520 was derived from a tumor, which showed poorly differentiated adenocarcinoma with minimal desmoplasia (Fig. 2.8a), and exhibited free-floating growth, as single cells or as an aggregate mimicking a cell ball (Fig. 2.8b). Individual cells were round to ovoid with a high nuclear-cytoplasmic ratio. The aggregated cells were arranged around intercellular lumens in a polarized pattern. Intercellular junctional complexes were noted in the lateral borders of the aggregated cells.

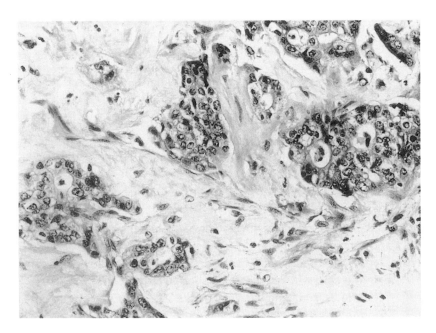

**Figure 2.6a.** Primary tumor of SNU-216 gastric carcinoma cell line.

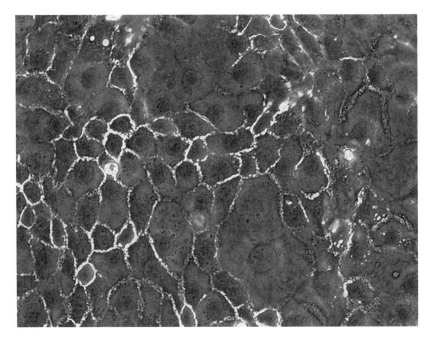

**Figure 2.6b.** SNU-216 gastric carcinoma cell line.

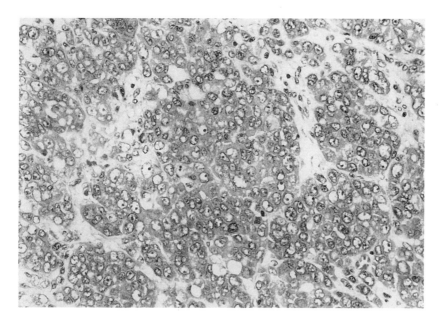

**Figure 2.7a.** Primary tumor of SNU-484 gastric carcinoma cell line.

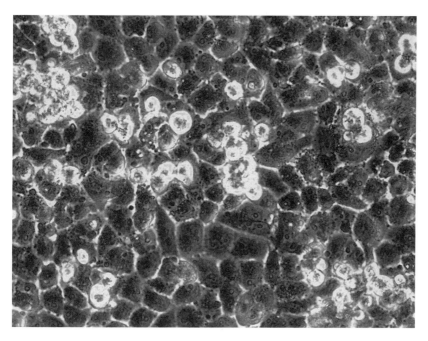

**Figure 2.7b.** SNU-484 gastric carcinoma cell line.

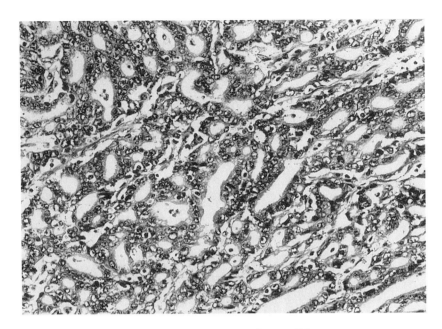

**Figure 2.8a.** Primary tumor of SNU-520 gastric carcinoma cell line.

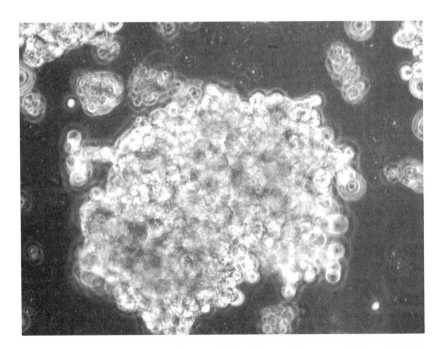

**Figure 2.8b.** SNU-520 gastric carcinoma cell line. Cancer cells grow as tightly packed cell aggregates.

# 7. GASTRIC CANCER CELL LINES IN INTERNATIONAL CELL LINE BANKS

*ATCC (American Type Culture Collection)*

(a) Web site: http://www.atcc.org

(b) Gastric cancer cell lines: AGS, KATO-III, HS746T, SNU-1, SNU-5, SNU-16

*ECACC (European Collection of Cell Cultures)*

(a) Web site: http://www.ecacc.org

(b) Gastric cancer cell lines: AGS, KATO-III, HGC-27

*JCRB (Japanese Cell Line Resource Bank)*

(a) Web site: http://cellbank.nihs.go.jp

(b) Gastric cancer cell lines: MKN-1, MKN-28, MKN-45, MKN-74, NUGC-2, NUGC-3, NUGC-4, KATO-III, AZ-521

*KCLB (Korean Cell Line Bank)*

(a) Web site: http://cellbank.snu.ac.kr

(b) Gastric cancer cell lines: SNU-1, SNU-5, SNU-16, SNU-216, SNU-484, SNU-520, SNU-601, SNU-620, SNU-638, SNU-668, SNU-719, NCI-N87

# 8. NORMAL GASTRIC EPITHELIAL CELLS

## 8.1. Equipment, Reagents, and Preparation of Media

The following equipment, reagents, and preparation of media are a compilation of items needed for all the following protocols. Everyday items, such as laminar flow hood, bench-top centrifuge, $CO_2$ incubator, and cell counter are assumed to be basic requirements and are not listed.

### 8.1.1. Equipment

Shaking water bath
DNA mini-fluorimeter, Hoefer

### 8.1.2. Plastic and Glassware

Tissue culture plates, 24- and 12-multiwell (Falcon)
Tissue culture flasks, 75 cm$^2$

Tissue culture Erlenmeyer flask, 125 ml
Centrifuge tubes, $16 \times 125$ mm
Siliconizing solution (Sigmacote)

### 8.1.3.  Cell Isolation Tools

Scalpels, disposable, with #5, #10, #11 blades
Razor blades
Forceps, dissecting
Scissors, abdominal
Sylgard plastic
Petri dishes, 6 cm
Glass slides, $(75 \times 25$ mm)
Nylon mesh
70% Ethanol or Wescodyne disinfectant

### 8.1.4.  Histological Chemicals and Materials

*Fixatives, embedding media and mountants*

Formaldehyde solution, 37%
Glutaraldehyde, 70% grade-1
Osmium tetroxide, 4% solution
Paraformaldehyde
Methanol
Bouin's
Triton-X
Electron microscopy embedding media: Epon; LR-White
Coverslips, #1
FluorSave, fluorescent mounting medium
Tissue histological mounting medium

*Histological stains*

Periodic acid Schiff (PAS)
Light-Green SF dye
Methylene Blue
Trypan Blue
Basic fuchsin
Thiocarbohydrazide
Hematoxylin-eosin
Giemsa
Janus Green
Nile Blue
Fast Red

*Antibodies*

Anti-cytokeratin
Anti-vimentin antibody
Goat-antirabbit antibody IgG
Goat-antimouse IgG

*Assay kits*

Lactate dehydrogenase assay
MTT cell viability assay
BrdU DNA cell labeling assay

### 8.1.5. Cell Culture Medium, Reagents, and Growth Matrices

*Media*

DMEM with 44 mM $HCO_3^-$, 25 mM HEPES, 25 mM glucose, and
   4 mM glutamine
Leibovitz's L-15 medium
RPMI-1640
DMEM/Ham's F-10 (1:1 mix)
Percoll cell separation media
Hanks' $10\times$ balanced salt solution (HBSS)

*Antibiotics*

Penicillin/streptomycin stock: 10,000 U/ml penicillin, 10 mg/ml strep-
   tomycin
Gentamycin stock: 10 mg/ml
Fungizone (amphotericin-B) stock: 250 µg/ml

*Enzymes*

Collagenase, type-I; (from *Clostridium histolyticum*)
Pronase (from Streptomyces griseus)
Matrisperse
Dispase
Deoxyribonuclease
Trypsin/EDTA

*Media supplements*

Bovine serum albumin powder (BSA), RIA-grade
FBS
Vitamins
Amino acids

PBSA
TCM, serum-free supplement

### Growth factors
Epidermal growth factor, (human recombinant EGF)
Transforming growth factor α, (human recombinant TGF-α)

### Matrices
Type-I rat tail collagen
Fibronectin, human
Collagen, Type-I rat tail
Matrigel

### Coated cell culture inserts
Falcon Cyclopore or Corning Transwell COL: porous filters coated
with fibronectin, Matrigel, type IV collagen, type I collagen

### 8.1.6.  Preparation of Media and Cell Isolation Solutions

Use standard sterilization techniques for all glassware. During the
cell isolation procedure, if there are any problems with cells adhering
to the glassware, then the glassware should be siliconized. Add 5 ml of
siliconizing solution to each piece of glassware, Parafilm the top of each,
swirl the solution to coat the inside of the glassware, pour out the siliconizing solution, and air dry.

### (a)  Media Preparation [Rutten et al., 1996]

(i)  *Cell isolation medium:* To 451 ml of DMEM/HEPES media
add 100 U/ml, 100 µg/ml of penicillin/streptomycin, 12.5 µg/ml of
gentamycin, 2 µg/ml of Fungizone, 10 ml of Celox TCM supplement,
10 ml of MEM amino acids, 5 ml of BME of vitamin solution, 50 ng/
ml of EGF.

(ii)  *Cell culture medium:* The same medium as indicated above but
with 10% fetal calf serum (FCS).

(iii)  *Cell isolation solutions* [Rutten et al., 1996]: Collagenase solution: To a siliconized 125 ml screw-cap Erlenmeyer flask add 20 ml of
serum-free culture media, 20 mg of type I collagenase, and 20 mg BSA
powder. Sterilize by filtration.

(iv)  *Iso-osmotic percoll solution:* to a 50 ml sterile Corning centrifuge tube, mix 36 ml of Percoll with 4 ml of 10 HBSS, and 1.6 ml
of 0.1 $N$ HCl (osmolarity is 310 mosm/l, density of 1.125 g/ml,
pH 7.4).

**(b) Media Preparation [Smoot et al., 2000]**

(i) *Cell isolation medium:* Leibovitz's L-15 medium containing 1% penicillin/streptomycin stock to give 100 U/ml and 100 µg/ml respectively.

(ii) *Cell culture medium:* Ham's F-12 medium containing 1% penicillin/streptomycin stock to give 100 U/ml and 100 µg/ml respectively, and 10% serum.

**(c) Media Preparation [Wagner et al., 1998]**

(i) *Cell isolation medium:* 70 mM NaCl, 20 mM NaHCO$_3$, 1.5 mM Na$_2$HPO$_4$, 5 mM KCl, 1.5 mM MgCl$_2$, 1 mM CaCl$_2$, 11 mM glucose, 50 mM HEPES.

(ii) *Cell culture medium:* RPMI 1640 media with 10% heat-inactivated FBS, gentamycin (100 µg/ml), penicillin (100 IU/ml), and amphotericin-B (2.5 µg/ml).

**(d) Media Preparation from Menard and Colleagues [Basque et al., 1999].**

(i) *Cell isolation medium:* Leibovitz L-15 plus garamycin and mycostatin (40 µg/ml each)

(ii) *Cell culture medium:* DMEM/Ham's F-12 (1:1) with penicillin (50 U/ml), streptomycin, (50 µg/ml), and 10% serum (Celect Gold FBS).

**(e) Media Preparation from Buchan and Colleagues [Richter-Dahlfors et al., 1998]**

(i) *Cell isolation medium:* DMEM/Ham's F-10 (1:1 mix), 1.0 mM Ca$^{2+}$, 8 µg/ml insulin, 50 µg/ml hydrocortizone, 5% FBS, 10 µg/ml penicillin, 10 µg/ml streptomycin, and 10 µg/ml gentamycin

(ii) *Cell culture medium:* DMEM/Ham's F-10 (1:1 mix), 1.0 mM Ca$^{2+}$, 8 µg/ml insulin, 50 µg/ml hydrocortisone, 10 µg/ml penicillin, 10 µg/ml streptomycin, 10 µg/ml gentamycin, and 10% FBS

## 8.2. Tissue Procurement and Safety

The following tissue procurements and safety procedures apply to all of the following protocols.

### 8.2.1. Tissue Procurement

The procurement of all human biopsy or surgical specimens for cell culture studies must be performed under the authorization of a university human studies review committee or an institutional review board. This review process is necessary to primarily safeguard the privacy of the patient. In most cases, after approval by the appropriate committee, an investigator will be able to obtain human biopsy or surgical specimens from patients undergoing routine endoscopy or elective surgery. In addition, all of the human gastric mucosa used in the following studies were free of *H. pylori* and were identified as histologically normal during pathological analysis. The success of any tissue procurement process also depends on a trained individual who can work with the surgeons and pathologists to acquire the human tissue sample within a reasonable amount of time without compromising the surgical or diagnostic process.

### 8.2.2.  Lab Safety

All laboratory personnel working with human tissue must assume that the tissue specimen may harbor latent viruses or other pathogenic organisms. All work should be performed in a Biosafety Level II laminar flow hood, which should be inspected yearly for HEPA filter replacement and proper laminar air flow. The biological safety cabinet must not be cluttered with lab items that could block exhaust grills and therefore restrict airflow.

(i)  An automatic pipettors should always be used for media changes on cells.

(ii)  Eating or drinking in the tissue culture area should be prohibited all times and all personnel involved in the tissue culture process should wash their hands before and after tissue culture work.

(iii)  All laboratory work surfaces must be disinfected before and after use with 70% alcohol or Wescodyne disinfectant.

(iv)  All personnel that handle human tissue must have training in a biohazard safety course. All rules must be strictly followed on the disposal of sharps, biohazardous material, and autoclaved waste.

(v)  A university or other institutional biosafety official must be notified concerning any questions on a specific issues tissue disposal.

### 8.3.  Human Gastric Mucous Cell Isolation and Culture

### Protocol 2.9.  Primary Culture of Human Gastric Mucous Epithelium Using Collagenase Digestion and Cell Separation on Percoll

This procedure is taken from the method of [Rutten et al. (1996].

***Reagents and Materials***
*Sterile*
❑  Serum-free medium (see Section 8.1)
❑  Culture media with 10% serum (see Section 8.1)

- ❏ Collagenase, 1 mg/ml of type I with 0.1% bovine albumin
- ❏ Percoll, isosmotic (see 8.1.6a, iii)
- ❏ Conical centrifuge tubes, 50 ml
- ❏ Erlenmeyer flask, 125 ml with screw-cap
- ❏ Petri dishes, 6 cm, with polymerized Sylgard
- ❏ Glass microscope slides
- ❏ Razor blades
- ❏ Syringe, 50 ml with an attached 15-gauge Luer-stub adapter
- ❏ Nylon mesh, 200 μm, in suitable holder
- ❏ Multiwell dishes, 12-well, or porous filters, with extracellular matrix coatings consisting of fibronectin or Type-I collagen or coated with 10 μg/ml/cm$^2$ Type-I collagen or 10 μg/ml/cm$^2$ fibronectin.
- ❏ Gas cylinder with $CO_2$, 5% in $O_2$, 95%

### Protocol

*(1)  Cell Isolation*

(a)  Procure human surgical tissues and wash twice in serum-free medium by placing the specimens in 50-ml conical tubes with 40 ml of serum-free media and spinning them at 50× *g* for 3 min.

(b)  After washing the resected stomach samples, pin them down (luminal side of tissue facing up) with 18-gauge needles in a 6-cm Petri dish containing polymerized Sylgard.

(c)  Scrape the luminal surface with a glass slide held at a 45° angle to remove the epithelium from the underlying muscle layers (use 2–3 glass slides for this process).

(d)  Transfer the scraped tissue to a new 6-cm Sylgard Petri dish, and mince with razor blades.

(e)  Wash all samples three times in serum-free cell culture media with centrifugation at 100 *g* for 3 min.

(f)  Put the pellets from the last wash step into a siliconized 125-ml screw-cap Erlenmeyer flask containing 20 ml of serum-free culture media containing 1 mg/ml of type I collagenase with 0.1% bovine albumin.

(g)  Gas the flask with 95% $O_2$:5% $CO_2$ and put into a 37°C shaking water bath, and gyrate at 120 oscillations/min. Incubate the surgical specimens for 45 min and the gastric biopsy samples for 90 min.

(h)  Remove the flask every 15 min from the water bath, triturate the digestion mixture six times with a 10-ml pipette, gas the flask with 95% $O_2$:5% $CO_2$ and put back into the water bath.

(i)  At the end of the final incubation period, pour the collagenase digestion mixture into a 50-ml syringe with an attached 15-gauge Luer-stub adapter.

(j)     Push the collagenase mixture out of the syringe and over a 200-μm nylon mesh.

(k)     Wash the nylon mesh filtered suspension, and spin twice in serum-free culture media at 100 $g$ for 3 min.

(l)     Resuspend the pellet in 15 ml of serum-free culture media, and take two 200-μl aliquots for cell counts in an electronic cell counter or hemocytometer.

(m)     Mix 5-ml aliquots of the resuspension with 5 ml of isosmotic Percoll in Falcon 16 × 125 mm tubes, and centrifuge for 15 min at 100 $g$ at 24°C.

(n)     At the end of the Percoll centrifugation, cells appear at the top and as a pellet at the bottom. The top layer contains undigested gastric glands, parietal, and chief cells, and the bottom pellet is enriched in gastric mucous epithelial cells.

(o)     Wash the cells from the top layer three times at 20 $g$ for 3 min in serum-free cell culture media, then resuspend in 2 ml of PBSA for use later in immunofluorescence studies. Remove the rest of the Percoll, and resuspend the bottom pellet (containing the enriched population of gastric mucous epithelial cells); wash three times at 20 $g$ for 3 min in serum-free cell culture media. The surgical gastric specimens yield ∼6–9 × $10^6$ of enriched gastric mucous epithelial cells.

(2)     *Cell Culture*

(p)     Resuspend the enriched pellets of gastric mucous epithelial cells in cell culture media with 10% serum.

(q)     Plate the cell suspensions at ∼50,000 cells/cm² in matrix-coated 12-well dishes or filter well inserts.

### Protocol 2.10.   Primary Culture of Human Gastric Mucous Epithelium Using Collagenase/Dispase Digestion and Plating on Collagen

This procedure is taken from the method of Smoot and colleagues [2000].

#### Reagents and Materials
*Sterile*
- ❏ Leibowitz's L-15 medium containing 1% penicillin/streptomycin stock
- ❏ Ham's F12 cell culture medium with 10% FBS
- ❏ PBSA
- ❏ Collagen, type I rat-tail

❑ Collagenase/Dispase (see Section 8.1.5)
❑ Multiwell plates, six-well, or 35-mm Petri dishes
❑ Scalpels: #5 blade, for mincing tissue
*Non-sterile*
❑ Trypan Blue viability stain
❑ Shaking incubator or water bath
*(1) Cell Isolation*
(a) Collect gastric biopsies (both antral and fundic) in Leibowitz's L-15 medium containing 1% penicillin/streptomycin.
(b) Dilute type 1 rat-tail collagen into cell culture medium at a concentration of 0.05%.
(c) Incubate dishes at 37°C for 1 to 2 h, then vacuum-aspirate off the collagen solution and fill the dish with culture medium.
(d) Isolate the gastric cells enzymatically after mechanically mincing the two to four biopsies (measuring approximately 8–10 mm$^2$ each) into pieces less than or equal to 1–mm$^2$ in size. It is important that the collagenase/dispase solution be made fresh, as the enzyme activity will diminish when stored in the freezer.
(e) Place the biopsy into a 10 cm culture dish with 5 ml of collagenase/Dispase solution within a laminar flow hood.
(f) Mince the tissue with a #5 blade scalpel into pieces 1 mm in size or less, which should take approximately 10 min.
(g) Place the minced tissue in collagenase/dispase solution into a small sterile bottle (100 ml) with a magnetic stirrer.
(h) Add another 20 ml of collagenase/Dispase solution (total of 25 ml) and stir (approx. 100 g) for 60 min at 37°C.
(i) Pellet the tissue by centrifugation at 100 g for 5 min at 4°C, discard the collagenase/Dispase, then wash the tissue once in 10 ml of PBSA and pellet again by centrifugation.
(j) Resuspend the cell pellet in 2 ml of Ham's F12 cell culture medium with 10% FBS, and place into a six-well tissue culture plate or 3.5-cm dish. Cells from two to four biopsies can be placed in each well, depending on the density of cell growth desired. If only two biopsies are placed into each well then 1 ml of medium is sufficient. For higher density cultures, a 12-well tissue culture cluster can be used.
*(2) Cell Culture*
(k) Before plating, add 4 μl of Trypan Blue solution to 46 μl of the cell suspension and visualize by using a hemocytometer in order to determine whether the cultures being set up consist of viable cells. A cell count may performed, but the count might not be accurate because many of the epithelial cells are still within fragmented glands.

(l)   The next day, rinse the cell cultures twice with PBSA to remove nonadherent cells and excess tissue, then add 1–2 ml of Ham's F12 cell culture medium to each dish. Change the cell culture medium every 3–4 days thereafter.

### Protocol 2.11.   Primary Culture of Human Gastric Mucous Epithelium Using Collagenase, Pronase, and Hyaluronidase

The following procedure is taken from the method of Wagner and colleagues [1998].

*Reagents and Materials*
*Sterile*
❑  Growth medium: RPMI 1640 medium (Gibco) supplemented with 10% heat-inactivated FBS, gentamycin (100 μg/ml), penicillin (100 IU/ml), and amphotericin B (2.5 μg/ml)
❑  Medium-I: 70 mM NaCl, 20 mM NaHCO$_3$, 1.5 mM Na$_2$HPO$_4$, 5 mM KCl, 1.5 mM MgCl$_2$, 1 mM CaCl$_2$, 11 mM glucose, 50 mM HEPES
❑  Collagenase solution: medium 1 containing 1 mg/ml collagenase (*Clostridium histolyticum*)
❑  PCH solution: medium 1 supplemented with 0.6 mg/ml pronase (from *Streptomyces griseus*) 1 mg/ml collagenase, 0.5 mg/ml hyaluronidase and 0.2% BSA
❑  Wash medium: medium-I supplemented with 0.2% BSA
❑  Collagen-coated 24-well plastic multiwell plates or 75 cm$^2$ plastic flasks
*Non-sterile*
❑  Formalin, 10%
❑  Trypan blue, 0.4%
❑  LDH viability assay
*(1)   Cell Isolation*
(a)   Isolate human gastric epithelial cells from gastric tissue specimens obtained during gastric surgery. Obtain surgical specimens (1–7 g mucosa wet mass) from the macroscopically unaffected part of the stomach immediately after surgical removal, and place in 50 ml medium 1.
(b)   Fix part of the tissue samples in 10% formalin for histological examination to prepare hematoxylin-eosin-stained and Giemsa-stained sections.
(c)   Scrape off gastric mucosa, cut into small pieces, and suspend in medium 1.
(d)   Centrifuge the mucosal material (200 g, 5 min), and resuspend in 40 ml collagenase solution.

(e) Preincubate the suspension for 10 min at 37°C, and gas with 95% $O_2$, 5% $CO_2$ while stirring.

(f) Sediment tissue fragments (200 $g$, 3 min), and incubate in 40 ml PCH solution.

(g) After 50 min, stop the enzyme digestion by filtering the cells through a nylon cloth (60 μm) and washing them three times in medium I supplemented with 0.2% BSA.

(h) Centrifuge and resuspend the isolated cells in RPMI 1640 medium.

(i) Determine the number of isolated gastric epithelial cells and their viability by using a hemocytometer, by Trypan Blue exclusion and phase-contrast microscopy.

(2) *Cell Culture*

(j) Incubate human gastric epithelial cells in collagen-coated, 24-well plastic multiwell plates or in 25 ml in 75 cm$^2$ plastic flasks. Spread approximately $3 \times 10^5$ cells per well of the plates and use $3 \times 10^6$ cells per 75 cm$^2$ flask containing 25 ml RPMI 1640 growth medium.

(k) Incubate the flasks or plates at 37°C in air enriched with 5% $CO_2$ in 100% humidity.

(l) Assess the viability of cultured cells by Trypan Blue exclusion and by photometric determination of released lactate dehydrogenase (LDH) enzyme activity by using a commercial test.

Our previous work has shown that short-term incubation (12 h) of isolated human gastric epithelial cells results in the selective attachment of gastric mucous cells, leaving the chief cells and parietal cells to be recovered from the culture medium [Adashi et al., 1993]. Thus, these incubation conditions enabled us to cultivate a partially enriched cell fraction of gastric mucous cells without prior use of specific cell-separation procedures. For the evaluation of mucin synthesis, we cultured $3 \times 10^6$ cells in 25-cm$^2$ plastic flasks in a total volume of 3 ml containing RPMI 1640 medium supplemented with 1% heat-inactivated FBS. Unless stated otherwise, the culture medium was changed every other day.

## 8.4. Human Fetal Gastric Cell Isolation and Culture

### Protocol 2.12. Culture of Human Fetal Gastric Cells

The following protocol is condensed from Basque et al. [1999]. Tissues from 48 fetuses, varying in age from 17 to 20 weeks of gestation, were obtained from normal elective pregnancy terminations. No tissue was collected from cases associated with known fetal abnormality or fetal death.

### Reagents and Materials

*Sterile*

❏ Dissection medium: Leibovitz L-15 plus gentamycin and nystatin (40 µg/ml each)

❏ DMEM/F12: DMEM and Ham's F12 formulation (1:1) supplemented with penicillin (50 U/ml), streptomycin (50 µg/ml), and 10% FBS

❏ Matrisperse

### Protocol

*(1) Cell Isolation*

(a) Bring the stomach to the culture room, immerse in collection medium, and proceed within 30 min at room temperature.

(b) Remove cardiac and pyloric segments from the stomach, leaving the body and fundic regions.

(c) Cut tissue specimens into explants (3 × 3 mm), and rinse with dissection medium. Dissociate the gastric epithelium by using ice-cold Matrisperse for 16–20 h and gently agitate for approximately 1 h at 4°C, allowing dissociation of gastric cells as intact epithelial sheets or large aggregates.

(d) Centrifuge the resulting material and resuspend in DMEM/F12.

(e) Disperse epithelial aggregates mechanically into multicellular clumps, seed in plastic six-well multiwell plates (5 × 10⁴ cells in 3 ml) or 24-well plates (1.5 × 10⁴ cells in 1 ml), and leave undisturbed for at least 24 h to allow attachment.

(f) Discard culture medium with nonattached material, and renew every 48 h.

8.5.  Human Gastric Antral Cell Isolation and Culture

The following protocol is condensed from [Richter-Dahlfors et al. (1998], and specific details can be found in this reference.

### Protocol 2.13.  Human Gastric Antral Cell Isolation and Culture

### Reagents and Materials

*Sterile*

❏ Dissection medium: HBSS containing 0.1% BSA

❏ DMEM/F12: DMEM and Ham's F12 formulation (1:1) supplemented with penicillin (50 U/ml), streptomycin (50 µg/ml), and 10% fetal bovine serum (ICN)

❏ Growth medium: 50:50 DMEM/Ham's F-10, 1.0 mM Ca²⁺, 8 µg/ml insulin, 50 µg/ml hydrocortisone, 5% FCS, 10 µg/ml penicillin, 10 µg/ml streptomycin, and 10 µg/ml gentamycin

- Collagenase, 300 U/ml (type I, Sigma) in BME (Gibco)
- Collagenase, 600 U/ml (type I, Sigma)
- Collagenase, 900 U/ml (type XI, Sigma)
- EDTA, 0.5 mM
- HBSS BSA-DTT: HBSS/BSA supplemented with 0.01% dithiothreitol
- HBSS-BSA-DTT containing 0.001% deoxyribonuclease (Sigma)
- Scalpels, #11 blade, for dissection
- Nitex gauze, 240 μm mesh
- Nitex gauze, 62 μm mesh
- Multiwell plates, 12-well either untreated or coated with rat-tail collagen, laminin, fibronectin, or 3-aminopropyltriethoxysilane (APES)
- Beckman elutriator centrifuge and rotor

### Protocol

*(1)* *Cell Isolation*

(a) Dissect human antral and the accompanying corpus and small bowel tissue carefully, and wash in HBSS containing 0.1% BSA.

(b) Before digestion, remove a small portion of antral tissue (1 cm$^2$), fix in Bouin's solution and process for immunocytochemistry.

(c) Dissect the mucosa carefully from the submucosa and mince into small fragments (about 4 mm$^3$) by using a pair of scalpel blades.

(d) Digest antral tissue (8–10 g/50 ml) in 5 ml 300 U/ml collagenase for 15 min in an agitating water bath at 37°C.

(e) Discard the resulting dispersed cells, which consist mainly of mucous cells.

(f) Digest the remaining tissue further by sequential collagenase treatment (5 ml 600 U/ml, Type I) for two 1-h periods, followed by 5 ml 900 U/ml (type XI, Sigma) for 1 h, with each step followed by the addition of 500 μl of 5 mM EDTA (final concentration of 0.5 mM) for 15 min.

(g) Filter the cell suspension resulting from each collagenase/EDTA digest through 240 μm Nitex mesh.

(h) Wash the filtrate with HBSS BSA-DTT.

(i) Wash the filtrate with HBSS-BSA-DTT containing 0.001% deoxyribonuclease, and resuspend in 20 ml of the latter medium. Deoxyribonuclease is used to degrade any DNA that may have been liberated during the digestion process, thus reducing the viscosity of the cell pellet.

(j) Before elutriation, pool the filtrates from the second, third, and fourth digests, and filter through finer 62-μm Nitex mesh.

(k) Achieve further enrichment for G-cell content by centrifugal elutriation by using a Beckman elutriator rotor.

(i)   Sterilize the elutriation tubing and rotor with water, followed by 70% ethanol, and wash with sterile water.

(ii)  Collect cells under sterile conditions by using HBSS–BSA as eluant at a loading density of $1 \times 10^8$ cells per run.

(iii) Load cells at a rotor speed of 1500 g and a flow rate of 25 ml/min, and wash for 3 min under identical conditions.

(iv)  Adjust the rotor speed and flow rate to 1080 g and 40 ml/min, respectively, and collect a 100-ml fraction (fraction 1).

(v)   Obtain a second 100-ml fraction (fraction 2) at a rotor speed of 1080 g and a flow rate of 55 ml/min.

By immunocytochemical and radioimmunoassay, the majority of immunoreactive G-cells are eluted in fraction 1, and a greater proportion of immunoreactive D-cells are eluted in fraction 2.

(2)   *Cell Culture*

(l)   Resuspend cells in the F1 fraction at $0.5 \times 10^6$ cells/ml in growth medium. Use the DMEM/Ham's F-10 to reduce the extracellular calcium concentration to 1 mM rather than 1.8 mM in DMEM alone.

(m)   Plate the cells at 1 ml/well on 12-multiwell plates, either untreated or coated, and maintain in a 5% $CO_2$ incubator at 37°C for 2 days before experiments.

## 8.6.   Characterization of Gastric Mucous Epithelial Cell Cultures

### 8.6.1.   Cell Proliferation

**Protocol 2.14.   Measurement of Gastric Epithelial Cell Growth In Vitro**

*Reagents and Materials*
*Sterile*
❑   2× trypsin/EDTA solution
❑   Growth medium

*Protocol*
(a)   At various times after cell plating, add 0.25 ml of 2× trypsin/EDTA solution, to each culture well or porous filter.

(b)   Add 0.25 ml of growth medium with serum 20 min later to stop the trypsinization.

(c)   Remove the trypsinized cell suspension, centrifuge at 50 g for 3 min, and resuspend the pellet in 1 ml of PBSA.

(d)   Electronic cell counter: From a 1 ml suspension of cells, add 200 µl to 20 ml of Hematall for cell counts by using a cell counter with a multichannel particle size analyzer.

(e)    Use 200 μl to measure DNA fluorometrically with Hoechst-33258 on a Hoefer mini-fluorometer using calf thymus DNA for the standard curve.

(f)    MTT Assay: This procedure is for adherent cells in 96-well plates, if larger plates are used, then adjust volumes accordingly.
   (i)    Make a solution of 5 mg/ml MTT dissolved in PBSA.
   (ii)   Four hours before the end of the incubation, add 20 ml of MTT solution from step one to each well containing cells.
   (iii)  Incubate the plate in a $CO_2$ incubator at 37°C for 5 h.
   (iv)   Remove media with pipettor needle and syringe.
   (v)    Add 200 μl of DMSO to each well, and pipette up and down to dissolve crystals.
   (vi)   Put plate into the 37°C incubator for 5 min.
   (vii)  Transfer to plate reader, and measure absorbance at 550 nm.

BrdU Assay: see the following reference for details of this assay, as well as protocols that are available from commercially available kits (Gibco, Sigma, Promega) [Dolbeare and Selden, 1994].

### 8.6.2.    Mucin Production

## Protocol 2.15.    Identification of Mucin by PAS Cytochemistry

### Reagents and Materials
*Non-sterile*
❏    Formalin, 3.7%
❏    PBSA
❏    Paraffin embedding reagents
❏    Schiff's reagent
❏    Periodic acid, 1%
❏    $NaHSO_4$, 0.52%
❏    Light Green counterstain
❏    Microtome
❏    Coverslips
❏    Mountant

### Protocol
(a)    Fix human stomach tissues in 3.7% formalin, paraffin-embedded.
(b)    Cut 4–5 μm sections.
(c)    Deparaffinize the sections.
(d)    Rinse in PBSA.
(e)    Oxidize for 10 min in 1% periodic acid.
(f)    Rinse the sections once in PBSA, and add Schiff's reagent for 10 min.

(g)     Pass the sections directly through three successive baths of 2 min each of 0.52% $NaHSO_4$.

(h)     Rinse the sections then for 10 min in PBSA.

(i)     Lightly counterstain with Light Green.

(j)     Mount coverslip with mounting medium for microscopic visualization.

The above procedure is also used for human gastric mucous cell cultures, except for the deparaffinization step.

## Protocol 2.16.   Mucin Immunohistochemistry of Gastric Cells

### Reagents and Materials
*Non-sterile*
❑   Formalin, 3.7%
❑   Paraffin embedding and deparaffinizing reagents
❑   Microtome

### Protocol
(a)     Fix human gastric mucosae in 3.7% formalin.

(b)     Paraffin-embed.

(c)     Cut 4–5 μm sections.

(d)     Deparaffinize the sections and react with an anti-mucin antibody at dilutions of 1:50–:1200 for 2 h at 24°C.

(e)     Wash the sections, and add a peroxidase labeled goat-antirabbit antibody at a 1:100 dilution for 1 h at room temperature.

(f)     Wash the slides.

(g)     Lightly counterstain with Toluidine Blue.

(h)     Cover with mountant and cover-glass for microscopic visualization.

The above protocol was also used for the human gastric mucous cell cultures without the deparaffinization step.

### 8.6.3.   Epithelial Markers

## Protocol 2.17.   Identification of Gastric Cells by Cytokeratin Staining

### Reagents and Materials
*Sterile*
❑   24-multiwell dishes or porous filter inserts (Falcon or Costar)
❑   PBSA at 24°C

- ❏ Formalin, 3.7%
- ❏ Acetone at 4°C
- ❏ Anti-cytokeratin or anti-vimentin antibody

### Protocol

(a) Plate human gastric mucous epithelial cells on 24-multiwell dishes or porous filter inserts, and allow to grow for four days in culture.

(b) At the end of this period, wash the cultures once with PBSA at 24°C.

(c) Fix for 5 min with 3.7% formalin.

(d) Remove the fixative, and rinse the cultures three times with PBSA at 24°C.

(e) Expose the cultures to acetone at 4°C for 2.5 min.

(f) Rinse four times with PBSA at 24°C for 10 min.

(g) Treat with an anti-cytokeratin or anti-vimentin antibody at a dilution of 1:100 for 30 min at 24°C.

(h) Wash the cultures twice with PBSA, then visualize by the addition of a peroxidase labeled anti-mouse antibody.

## Protocol 2.18.  Identification of Gastric Cell Type Using Immunofluorescence Against Pepsinogen-II and Gastric H$^+$/K$^+$-ATPase

### Reagents and Materials

*Sterile*

- ❏ Glass coverslips or porous filters

*Non-sterile*

- ❏ Paraformaldehyde, 2.5%
- ❏ PBSA
- ❏ Methanol
- ❏ Triton-X, 0.1% in PBSA
- ❏ Polyclonal antibody against human pepsinogen-II

or

- ❏ Monoclonal antibody against the hog gastric H$^+$/K$^+$-ATPase
- ❏ FBS or goat serum
- ❏ Rhodamine-labeled goat anti-rabbit IgG

or

- ❏ Rhodamine-labeled goat anti-mouse total IgG
- ❏ Mountant: FluoroSave

### Protocol

(a) Grow human gastric mucous epithelial cells or glands on glass coverslips or porous filters.

(b) Fix with 2.5% paraformaldehyde for 10 min.

(c) Rinse the cultures and glands three times with PBSA at 24°C.

(d) Permeabilize with methanol for 1 min.

(e) Rinse three times with PBSA at 24°C.

(f) Treat with 0.1% Triton in PBSA for 3 min.

(g) Rinse the glands and cultures with PBSA.

(h) Incubate with either a polyclonal antibody against human pepsinogen-II or a monoclonal antibody against the hog gastric $H^+/K^+$-ATPase at a 1:75 dilution at 4°C overnight in a humidified chamber. Use FBS or goat serum in place of the primary antibodies for negative controls.

(i) After the overnight incubation, wash the cultures and glands three times with 24°C PBSA.

(j) Incubate with to following:
  (i) Rhodamine-labeled goat anti-rabbit IgG at a 1:40 dilution.
  or
  (ii) Rhodamine-labeled goat anti-mouse total IgG at a 1:100 dilutions for 1 h at 24°C in a humidified chamber.

(k) After the 1 h incubation, wash the cultures and glands three times with PBSA at 24°C.

(l) Treat with a drop of FluoroSave, and examine with a Nikon inverted microscope using epifluorescence.


### 8.6.4. Morphology

## Protocol 2.19.  Light Microscopy of Gastric Epithelial Cultures

### Reagents and Materials
*Sterile*
❑ Filter inserts precoated with fibronectin and type 1 collagen
*Non-sterile*
❑ Formaldehyde 3.0%, glutaraldehyde 3.0%, in cacodylate buffer (pH 7.4)
❑ $OsO_4$, 1.0%
❑ Ethanol, graded series, 20–100%
❑ LR-White resin
❑ Lee's methylene blue/basic fuchsin
❑ Photographic film: 35 mm, Kodak Tech-Pan, 50 ISO
❑ Scalpel and #11 blade

### Protocol
(a) Culture gastric cells on filter inserts precoated with fibronectin and type 1 collagen.

(b) Fix the gastric tissues and cultures for 24 h in 3.0% formaldehyde and 3.0% glutaraldehyde in cacodylate buffer (pH 7.4).

(c) After fixation, remove the porous filter with the adherent cell culture from its plastic holder by using a scalpel and #11 blade.

(d) Postfix both the filter and tissue in 1.0% $OsO_4$ for 1 h.

(e) Dehydrate in a graded series (20–100%) of ethanol solutions.

(f) Embed in LR-White resin.

(g) Cut thick sections (0.5–1.0 μm).

(h) Stain with Lee's methylene blue/basic fuchsin.

(i) Take photographs with a Nikon inverted microscope by using Tech-Pan 35-mm film at 50 ISO.

## Protocol 2.20. Electron Microscopy of Gastric Epithelial Cultures

### Reagents and Materials

*Sterile*

❏ Filter inserts precoated with fibronectin and type 1 collagen

*Non-sterile*

❏ Formaldehyde, 3.0%; glutaraldehyde, 3.0%, in cacodylate buffer (pH 7.4)

❏ $OsO_4$, 1.0% in cacadylate buffer (pH 7.4)

❏ Ethanol, graded series, 20–100%

❏ LR-White resin

❏ Uranyl acetate/lead citrate

❏ Glutaraldehyde, 2.0%, 1.0 mM $CaCl_2$, in cacodylate buffer (pH 7.4)

❏ NaCl, 150 mM

❏ Periodic acid ($H_5IO_6$), 1.0%

❏ Thiocarbohydrazide, 1.0% in 10% acetic acid

❏ Photographic film: 35 mm, Kodak Tech-Pan, 50 ISO

❏ Scalpel and #11 blade

❏ JEOL 100-CXII electron microscope

### Protocol

(a) For conventional electron microscopy, process cultures and tissues as described above for light microscopy (Protocol 2.19, a–f).

(b) Cut thin sections (0.1–0.2 μm).

(c) Post-stain with uranyl acetate/lead citrate.

(d) Photograph by using an electron microscope, e.g., JEOL 100-CXII.

(e) For electron cytochemical demonstration of mucoproteins:

    (i) Fix cell cultures and tissues for 10 min in 2.0% glutaraldehyde/1.0 mM $CaCl_2$ in cacodylate buffer (pH 7.4).

(ii) Rinse the specimens three times in 150 mM NaCl for 3 min.

(iii) Immerse in 1.0% periodic acid ($H_5IO_6$) for 5 min. Wash the specimens then once in 150 mM NaCl for 1 min.

(iv) Immerse in 1.0% thiocarbohydrazide in 10% acetic acid for 30 min at 37°C.

(v) Rinse the specimens twice in cacodylate buffer for 5 min.

(vi) Immerse in 1.0% $OsO_4$ cacodylate buffer for 30 min.

(vii) Wash the specimens are then in cacodylate buffer.

(viii) Embed in LR-White resin.

(ix) Cut thin sections.

(x) Examine directly or post-stain with lead citrate.

(xi) Check for electron-opaque deposits by using an electron microscope.

## 8.7. Main Applications of Normal Gastric Epithelial Culture

The isolation and growth of any human primary gastric cell is difficult, and even though there are no set guidelines for producing a successful cell culture, there are several things to consider when using any of the above protocols. The first consideration is getting the surgical specimen as quickly as possible from the hospital operating room into the appropriate media. In most cases, a serum-free medium with HEPES at pH 7.4 is usually best for transporting most tissue to the laboratory. In the cell isolation process, the viability of the cells should be monitored often (Trypan Blue procedure) to maximize the yield of cells. If poor viability or low plating efficiency persists, a reduction in the exposure time to the digestion enzymes should be tried. For good gastric attachment and cell growth, a high plating density and suitable matrix should be used. Also, an enriched media that is supplemented growth factors, vitamins, and a high-quality serum (e.g., Hyclone) is important for gastric cell attachment and growth within 48 h after plating. In some cases, the gastric cells may initially adhere to their substrate, but after 48 h a massive cell cytolysis occurs. This may be due to a low-level mycoplasma or other bacterial contamination. Finally, the success of any cell culture procedure depends not only a detailed protocol, but also on the expertise of the laboratory personnel involved. It is the authors' experience that it usually takes several months for a laboratory technician to acquire the necessary training that is needed to consistently produce good results in the isolation and the growth of gastric cells. In summary, the culturing of normal human gastric cells is still in its infancy, but the continuing progress being made in cell/molecular biology techniques as well as the new field of stem cell research is likely to produce some exciting advances in gastric epithelial cells.

## ACKNOWLEDGMENTS

The author is grateful to Dr. Ian Freshney and Dr. Roswitha Pfragner for the opportunity to write this chapter, as well as a special thanks to all of the investigators for their permission to use portions of their previously published works. The author also apologizes in the event of any oversight in the reporting or omission of certain methods or individuals in the writing of this chapter. The author would also like to give a special thanks to Dr. Brett Sheppard and Dr. Clifford Deveney for their input and support of this research over the years.

## SOURCES OF MATERIALS

| Item | Supplier |
|------|----------|
| Amino acids | Gibco (Invitrogen) |
| Antibiotics: penicillin/streptomycin, gentamycin, fungizone (amphotericin-B), mycostatin, | Gibco (Invitrogen) |
| Antibodies: anti-cytokeratin antibody; anti-vimentin antibody; goat-antirabbit antibody IgG; goat-antimouse IgG; polyclonal antibody against human pepsinogen-II; monoclonal antibody against the hog gastric $H^+/K^+$-ATPase | Dako |
| Bovine serum albumin powder (BSA), RIA-grade | Sigma, A7888 |
| BrdU DNA cell labeling kit | Gibco (Invitrogen); Promega; Sigma |
| Celect Gold FBS | ICN |
| Cell counter | Beckman Coulter |
| Cell counting solution: Hematall | Beckman Coulter |
| Cell culture inserts: | Becton Dickinson |
|     Falcon Cyclopore | Corning |
|     Transwell-COL | |
| Cell scraper | Nalge Nunc |
| Centrifuge tubes | Falcon (Becton Dickinson) |
| Collagen, Type-I rat tail | Sigma |
| Collagenase (from *Clostridium histolyticum*) | Boehringer Mannheim, Germany |
| Collagenase, type-I | Sigma |
| Coverslips, #1 | Corning |
| Deoxyribonuclease | Sigma |
| Dishes and flasks | Corning |
| Dispase | Boehringer (Roche) |
| DMEM | Gibco (Invitrogen) |
| Epon; LR-White | SPI-Pon™ from SPI Supplier |
| Erlenmeyer flasks, tissue culture grade | Kimble |
| FBS | Hyclone; ICN |
| Fibronectin, human | Boehringer Mannheim (Roche) |
| Ficoll/metrizoate | Amersham Pharmacia; ICN |

## SOURCES OF MATERIALS *(Continued)*

| Item | Supplier |
|------|----------|
| Fixatives: 37% formaldehyde solution; glutaraldehyde, 70% grade-1; osmium tetroxide, 4% solution; paraformaldehyde; methanol; Bouin's | Sigma |
| FluorSave, fluorescent mounting medium | Calbiochem |
| Growth factors: epidermal growth factor, (human recombinant EGF); transforming growth factor $\alpha$, (human recombinant TGF-$\alpha$) | Sigma |
| Ham's F-10 | Gibco (Invitrogen) |
| Histological stains | Sigma |
| Hyaluronidase | Sigma |
| Isotonic counting solution, Hematall | Beckman Coulter |
| Lactate dehydrogenase (LDH) kit viability assay | Boehringer Mannheim (Roche); Sigma |
| Laminin | Becton Dickinson; Gibco (Invitrogen) |
| Leibovitz's L-15 medium | Gibco (Invitrogen) |
| Lymphosep | ICN |
| Matrices: Type-I rat tail collagen | Vitrogen; Sigma |
| Matrigel | Becton Dickinson |
| Matrisperse | Collaborative Biomedicals (Becton Dickinson) |
| MTT cell viability kit | Sigma |
| Multiwell plates | Falcon (Becton Dickinson) |
| Nylon mesh, Nitex | Tekmar |
| Percoll cell separation media | Pharmacia Biotech |
| Petri dishes | Falcon (Becton Dickinson) |
| Petri dishes, 60 mm | Falcon (Becton Dickinson) |
| Pipettes | Falcon (Becton Dickinson) |
| Pronase (from Streptomyces griseus | Boehringer Mannheim |
| RPMI 1640 | Gibco (Invitrogen) |
| Siliconizing solution (Sigmacote) | Sigma |
| Stains | Sigma |
| Sylgard plastic | Dow Corning |
| TCM, serum-free supplement | Celox |
| Tissue histological mounting medium | SPI Supplies |
| Triton-X | Sigma |
| Trypan Blue | Gibco |
| Vitamins | Gibco (Invitrogen) |
| Wescodyne disinfectant | STERIS Corporation |

## REFERENCES

Adashi, E.Y., Resnick, C.E., Tedeschi, C., Rosenfeld, R.G. (1993) A kinase-mediated regulation of granulosa cell-derived insulin-like growth factor binding proteins (IGFBPs): disparate response sensitivities of distinct IGFBP species. *Endocrinology* 132:1463–1468.

Ayalon, A., Sanders, M.J., Thomas, L.P., Amirian, D.A., Soll, A.H. (1982) Electrical effects of histamine on monolayers formed in culture from enriched canine gastric chief cells. *Proc. Natl. Acad. Sci. USA* 79:7009–7013.

Basque, J.R., Chailler, P., Perreault, N., Beaulieu, J.F., Menard, D. (1999) A new primary culture system representative of the human gastric epithelium. *Exp. Cell Res.* 253:493–502.

Basque, J.R., and Menard, D. (2000) Establishment of culture systems of human gastric epithelium for the study of pepsinogen and gastric lipase synthesis and secretion. *Microsc. Res. Tech.* 48:293–302.

Beinborn, M., Giebel, J., Linck, M., Cetin, Y., Schwenk, M., Sewing, K.F. (1993) Isolation, identification and quantitative evaluation of specific cell types from the mammalian gastric mucosa. *Cell Tissue Res.* 274:229–240.

Chew, C.S. (1994) Parietal cell culture: new models and directions. *Ann. Rev. Physiol.* 56:445–461.

Chew, C.S., Ljungstrom, M., Smolka, A., Brown, M.R. (1989) Primary culture of secretagogue-responsive parietal cells from rabbit gastric mucosa. *Amer. J. Physiol.* 256:254–263.

Correa, P., and Chen, V.W. (1994) Gastric cancer. *Cancer Surv.* 19 20:55–76.

Defize, J., Arwert, F., Kortbeek, H., Frants, R.R., Meuwissen, S.G., Eriksson, A.W. (1985) Pepsinogen synthesis in monolayer culture of human and rabbit gastric mucosal cells. *Virchows Arch. B Cell Pathol. Incl. Mol. Pathol.* 49:225–230.

Dolbeare, F., and Selden, J.R. (1994) Immunochemical quantitation of bromodeoxyuridine: application to cell-cycle kinetics. *Methods Cell Biol.* 41:297–316.

Freshney, R.I., and Freshney, M.G. (2002) Culture of Epithelial Cells. 2nd edition, New York, NY: John Wiley and Sons.

Fuchs, C.S., and Mayer, R.J. (1995) Gastric carcinoma. *N. Eng. J. Med.* 333:32–41.

Fukamachi, H., et al. (1994) Fetal rat glandular stomach epithelial cells differentiate into surface mucous cells which express cathepsin E in the absence of mesenchymal cells in primary culture. *Differentiation* 56:83–89.

Giebel, J., Arends, H., Fanghanel, J., Cetin, Y., Thiedemann, K.U., Schwenk, M. (1995) Suitability of different staining methods for the identification of isolated and cultures cells from guinea pig (*Cavia aperea porcellus*) stomach. *Eur. J. Morphol.* 33:359–372.

Heim, H.K., Piller, M., An, X., Kilian, P., Netz, S., Sewing, K.F. (1997) Recovery of chalecystokinin response of porcine gastric chief cells during monolayer culture. *Digestion* 58:10–18.

Heim, H.K., Piller, M., Schwede, J., Kilian, P., Netz-Piepenbrink, S., Sewing, K.F. (1997) Pepsinogen synthesis during long-term culture of porcine chief cells. *Biochem. Biophys. Acta* 1359:35–47.

Landis, S.H., Murray, T., Bolden, S., Wingo, P.A. (1998) Cancer statistics. *CA Cancer J. Clin.* 48:6–29.

Neugut, A.I., Hayek, M., Howe, G. (1996) Epidemiology of gastric cancer. *Semin. Oncol.* 23:281–291.

Oie, H.K, Russell, E.K., Carney, D.N., Gazdar, A.F. (1996) Cell culture methods for the establishment of the NCI series of lung cancer cell lines. *J. Cell Biochem.* Suppl. 24:24–31.

Okoyama, N., Itoh, M., Joh, T., Miyamoto, T., Takeuchi. T., Moriyama, A., Kato, T., (1998) Regulatory mechanism of pepsinogen gene expression in monolayer cultures guinea pig gastric chief cells. *Dig Dis. Sci.* 43:2744–2749.

Okayama, N., Joh, T., Miyamoto, T., Kato, T., Itoh, M. (1994) Role of myosin light-chain kinase and protein kinase C in pepsinogen secretion from guinea pig gastric chief cells in monolayer culture. *Dig. Dis. Sci.* 39:2547–2557.

Olson, C.E. (1988) Glutathione modulates toxic oxygen metabolite injury of canine chief cell monolayers in primary culture. *Am. J. Physiol.* 254:49–56.

Ota, S., Terano, A., Hiraishi, H., Mutoh, H., Nakada, R., Hata, Y., Shiga, J., Sugimoto, T. (1990) A monolayer culture of gastric mucous cells from adult rabbits. *Gastroenterol. Jpn* 25:1–7.

Palli, D. (2000) Epidemiology of gastric cancer: an evaluation of available evidence. *J. Gastroenterol.* 35 Suppl 12:84–89.

Park, J.G., Frucht, H., LaRocca, R.V., Bliss, D.P. Jr., Kurita, Y., Chen, T.R., Henslee, J.G., Trepel, J.B., Jensen, R.T., Johnson, B.E., Bang, Y.J., Kim, J.P., Gazdar,

A.F. (1990) Characteristics of cell lines established from human gastric carcinoma. *Cancer Res.* 50:2773–2780.

Park, J.G., and Gazdar, A.F. (1996) Biology of colorectal and gastric cancer cell lines. *J. Cell Biochem.* Suppl 24:131–141.

Park, J.G., Yang, H.K., Kim, W.H., Chung, J.K., Kang, M.S., Lee, J.H., Oh, J.H., Park, H.S., Yeo, K.S., Kang, S.H., Song, S.Y., Kang, Y.K., Bangm Y.J., Kim, Y.H., Kim, J.P. (1997) Establishment and characterization of human gastric carcinoma cells. *Int. J. Cancer* 70:443–449.

Richter-Dahlfors, A., Heczko, U., Meloche, R.M., Finlay, B.B., Buchan, A.M. (1998) *Helicobacter pylori*-infected human antral primary cell cultures: effect on gastrin cell function. *Am. J. Physiol.* 275:393–401.

Roder, J.D., Bottcher, K., Siewert, J.R., Busch, R., Hermanek, P., Meyer, H.J. (1993) Prognostic factors in gastric carcinoma. Result of German Gastric Cancer Study 1992. *Cancer* 72:2089–2097.

Rutten, M.J., Campbell, D.R., Luttropp, C.A., Fowler, W.M., Hawkey, M.A., Boland, C.R., Krauss, E.R., Sheppard, B.C., Crass, R.A., Devenay, K.E., Devenay, C.W. (1996) A method for the isolation of human gastric mucous epithelial cells for primary cell culture; A comparison of biopsy vs. surgical tissue. *Methods in Cell Science* 18:269–281.

Rutzky, L.P., and Moyer, M.P. (1990) Human cell lines in colon cancer research. In Moyer MP (ed) *Colon Cancer Cells*. London: Academic Press, pp. 155–202.

Sanders, M.J., Amirian, D.A., Ayalon, A., Soll, A.H. (1983) Regulation of pepsinogen release from canine chief cells in primary monolayer culture. *Am. J. Physiol.* 245:641–646.

Scheiman, J.M., Kraus, E.R., Bonnville, L.A., Weinhold, P.A., Boland, C.R. (1991) Synthesis and prostaglandin E2-induced secretion of surfactant phospholipid by isolated gastric mucous cells. *Gastroenterology* 100:1232–1240.

Sekiguchi, M., and Suzuki, T. (1994) Gastric tumor cell lines. In Hay, R.J., Park, J.G., and Gazdar, A.F. (eds) *Atlas of Human Tumor Cell Lines*. 3rd Edition. San Diego, CA: Academic Press, pp. 287–316.

Smoot, D.T., Sewchand, J., Young, K., Desbordes, B.C., Allen, C.R., Naab, T. (2000) A merthod for establishing primary cultures of human gastric epithelial cells. *Cell Sci.* 22:133–136.

Tanaka, T., and Tani, S. (1995) Interaction among secretagogues on pepsinogen secretion from rat gastric chief cells. *Biol. Pharm. Bull* 18:859–865.

Tani, S., Kobayashi, S., Tanaka, T. (1994) Secretion of intrinsic factor from cultured rat gastric chief cells. *Biol. Pharm. Bull* 17:1333–1336.

Tani, S., Tanaka, T., Kudo, Y., Takahagi, M. (1989) Pepsinogen secretion from cultured rat gastric mucosal cells. *Chem. Pharm. Bull ( Tokyo )* 37:2188–2190.

Wagner, S., Enss, M.L., Cornberg, M., Mix, H., Schumann S., Kirchner, G., Jahne, J., Manns, M.P., Beil, W. (1998) Morphological and molecular characterization of human gastric mucous cells in long-term primary culture. *Pflugers Arch.* 436:871–881.

# 3

# Establishment of Cell Lines from Colon Carcinoma

Robert H. Whitehead

*Novel Cell Line Development Facility, 4148D, MRB III, Vanderbilt University, Nashville, Tennessee, robert.whitehead@vanderbilt.edu*

*Culture of Human Tumor Cells*, Edited by Roswitha Pfragner and R. Ian Freshney.
ISBN 0-471-43853-7  Copyright © 2004 Wiley-Liss, Inc.

## I. INTRODUCTION

Colon cancer is one of the most significant cancers in the Western world, with more than 130,000 cases diagnosed each year in the United States alone. As only 50% of these cancers will be cured by surgery and other treatment modalities, there is considerable need for further research into this disease. For in vitro studies to have clinical relevance, cell lines are needed that are representative of the disease. We therefore need cell lines derived from different stages of the disease progression as well as from sporadic colon cancer and the different familial forms of the disease. As colon cancer has been shown to follow a defined developmental pathway from normal to hyperplastic to adenoma to primary tumor to metastatic disease, it would also be advantageous to have cell lines that represent these different stages of disease progression. Many colon carcinoma cell lines have been published [Tompkins et al., 1974; Drewinko et al., 1976; Fogh et al., 1977; Brattain et al, 1983; McBain et al., 1984; Whitehead et al., 1985; Willson et al., 1987]. A number are deposited with the American Type Culture Collection (ATCC) and European Collection of Cell Cultures (ECACC) cell banks and are readily available. One problem with many of these cell lines is that they have been in culture for many years and are often at very high passage levels. The resemblance of the cultures available from these cells lines to the culture originally established is questionable. In addition, where tumor stage has been reported, most cell lines have been derived from later stages of the disease [Park et al., 1987; Willson et al., 1987]. In comparison, few cell lines have been established from early tumors (Dukes A or B) or from adenomas [Paraskeva et al., 1984; Willson et al., 1987; Whitehead et al., 1991].

Because most of the colon cancer cell lines have been in culture for many years or because they were derived from advanced undifferentiated tumors, very few of these cell lines express any of the differentiated features of the colonic mucosa [Chantret et al., 1988]. There is a need for new cell lines that can be used at low passage levels and can retain differentiated characteristics that identify them as originating in the colonic mucosa [Whitehead et al., 1992, Whitehead and Watson, 1994]. One of the problems facing anyone planning to establish cell lines is that many of the tumors are heterogeneous and may yield cell lines with mixed characteristics [Brattain et al., 1981; 1983].

## 2. PREPARATION OF MEDIA AND REAGENTS

### 2.1. Media

Many media have been used successfully to establish colon carcinoma cell lines. Dulbecco's modified Eagle's medium (DMEM), RPMI 1640 and Ham's F12, and combinations of DMEM and Ham's F12 have been the most common media used. Serum concentrations have varied from 2% to 20% without making any apparent difference in success rates. Supplement use has varied similarly, although a general consensus indicates that addition of insulin enhances epithelial cell growth. Little has been published describing the use of serum-free media for the establishment of colon carcinoma cell lines; however, the successful use of these media for culturing other normal epithelial tissues suggests that they may be useful especially for inhibiting fibroblast growth from the primary biopsy.

#### 2.1.1. Recommended Medium

500 ml RPMI 1640
25 ml fetal bovine serum (FBS; 5%)
5 ml of 500 μg/ml insulin (approx. $10^{-6}$ M).
5 ml of $10^{-4}$ M hydrocortisone
100 unit/ml penicillin
100 μg/ml streptomycin
2.5 μg/ml fungizone (optional)
15 mM $N$-(2-hydroxyethyl)piperazine-$N'$-ethanesulfonic acid (HEPES; sodium salt)
10 ml of 200 mM glutamine (if the medium does not already contain it)
24 mM $NaHCO_3$ (if not already included in medium)
The medium should be stored at 4°C after preparation.

#### 2.1.2. Medium Preparation

RPMI 1640 can be purchased as single-strength, or 10× concentrated liquid medium, or as powdered medium that has to be prepared in the laboratory. The single-strength medium sometimes requires the addition of glutamine (final concentration 2 mM) or sodium bicarbonate, depending on the formulation used by the manufacturer.

If powdered medium is used, it is normally available in packs to make either 1 L or 10 L of medium. It is better to use the smaller packs rather than to use part of the larger pack, as many of the media components are deliquescent and it is difficult to store the opened pack in a totally moisture-free environment. Also, there is no guarantee that all of the media components—there are 39 different salts, amino acids, and vitamins in RPMI 1640 and more in other media—are evenly distributed throughout the powder.

To make medium from powder:

(i)    Dissolve the powder in either reverse-osmosis/deionized $H_2O$ with a resistance of >18 $M\Omega cm$ or double-glass-distilled $H_2O$.

(ii)    Add the powder to 80% of the final volume of $H_2O$, and stir at 4°C until dissolved. Check the pH and, if needed, carefully adjust the level to that recommended by the manufacturer (normally pH 7.2) and add $H_2O$ to the final volume.

(iii)    Stir the medium at 4°C for at least 2 h to ensure complete solubilization of all of the ingredients.

(iv)    Filter through a 0.2-μm filter, aliquot into sterile, well-washed and rinsed bottles, and store at 4°C until used.

Note: Sufficient medium should be made for only 3–4 weeks of normal use. Longer storage can seriously degrade the glutamine in the solution, and old media should be supplemented with fresh glutamine.

(v)    Add 15 mM HEPES to the medium, which assists in buffering and is especially important in the early stages of culture when there are very few cells in each well and very little cell metabolism.

(vi)    Dissolve insulin in acidified $H_2O$ (pH < 2, made by adding acetic acid to $H_2O$). Filter the stock solution, aliquot in small volumes, and store at −20°C.

(vii)    Add insulin alone or in combination with transferrin and selenium. This mixture is available from most media suppliers as ITS solution and is normally either 100× or 1000×. Transferrin and selenium are not really necessary if serum has been added to the medium, but they are essential if serum-free medium is to be used.

(viii)    Dissolve hydrocortisone in ethanol. Alternatively, hydrocortisone hemisuccinate can be used, as this is water-soluble.

(ix)    Penicillin and streptomycin can be added separately or can be purchased as a mixture and added together. Other antibiotics have been used, and the most common is gentamycin, which is added at a final concentration of 25–50 μg/ml. The possibility of fungal contamination from luminal contents has led some workers to add amphotericin B (fungizone) at 2.5 μg/ml or mycostatin (nystatin) at 50–200 units per ml. (I have had good success without it.)

## 2.2.    Digestive Enzymes

An enzyme solution for disaggregating the tumor tissue is made by using collagenase Type 1 and neutral protease (Dispase). Sufficient collagenase and dispase are weighed to give final concentrations in solution of 300 units per ml of collagenase and 1 unit per ml of dispase. These are dissolved in growth medium and are sterilized by filtration. The solution is then aliquoted in 20 ml amounts in sterile screw-capped tubes and is stored at $-20°C$ until used. Make sufficient amounts for 1 month's use only.

### 2.3.  Sterilizing Solution

Dilute 1 ml 4% sodium hypoclorite (0.04% final concentration) to 100 ml in PBS (0.1 M phosphate buffered saline).

### 2.4.  Collagen Coating

### Protocol 3.1.  Coating with Collagen I for Colon Carcinoma Culture

#### Reagents and Materials
*Sterile*
- ❏  Collagen Type I. The collagen can be purchased as a sterile preparation from tissue culture suppliers
- ❏  0.1% Acetic acid in $H_2O$ sterilized by filtration
- ❏  Growth medium
- ❏  0.1 N NaOH
- ❏  24-Well tissue culture plates
- ❏  Tissue culture flasks

*Non-sterile*
- ❏  Plastic bags to contain culture plates

#### Protocol
(a)  Make a 3 mg/ml solution of collagen by dissolving collagen Type I in sterile 0.1% acetic acid in $H_2O$. When the solution is dissolved, store it at 4°C.

(b)  To coat growth surfaces, dilute the collagen 1:4 in growth medium and adjust the pH to 7.2 by carefully adding 0.1 N NaOH. This is done by adding the NaOH solution in small aliquots until the color of the medium has returned to normal.

(c)  Use this solution to coat the growth surface of the 24-well tissue culture plates and small flasks by adding the solution to each well and immediately removing most of the solution. This leaves a film of collagen solution in each well that will cross-link in a few minutes. Because the collagen solution will cross-link and form a gel once the pH is neutralized, this procedure must be done quickly.

(d)  The coated plates can be stored for a short period in the tissue culture hood or can be sealed in a plastic bag and stored at 4°C until used.

## 3. ESTABLISHING A CELL LINE FROM A TUMOR BIOPSY

The specimen should be obtained from the operating theater as soon as possible after removal. In most countries the specimen must first be taken to a regulated pathology department to be examined by a pathologist. Practitioners must comply with all local rules regarding the ethical use of human tissues. Normally, compliance includes an application to the local human ethics committee for permission to use human tissue in your studies.

It is helpful to have either a technician or a nurse in the operating theater to collect the specimen and to prevent its accidental placement in formalin.

### Safety

Once received in the laboratory, the specimen must be handled only by staff who are familiar with the risks involved in handling human tissue. It must be processed in a Class II vertical laminar flow cabinet, and gloves must be worn. All staff exposed to human tissue should be vaccinated against hepatitis virus.

### 3.1. Primary Culture

The following procedure can be followed for both a piece of tissue from a primary or secondary tumor or an involved lymph node.

### Protocol 3.2. Primary Culture of Colonic Tumors

#### Reagents and Materials

*Sterile*
- ❑ Growth medium (see Section 2.1)
- ❑ PBSA
- ❑ Collagenase/neutral protease solution (see Section 2.2)
- ❑ Trypsin/EDTA: trypsin, 0.25% in 1 mM EDTA
- ❑ Scalpels or sharp scissors
- ❑ Petri dish, 10 cm
- ❑ Collagen-coated flasks, plates, or dishes (see Protocol 3.1)
- ❑ Centrifuge tubes or universal containers
- ❑ Cloning discs, 3 mm diameter
- ❑ Cell scraper

*Non-sterile*
- ❑ Sodium hypochlorite, 4%

#### Protocol

(a) Obtain a piece of tissue approximately 1 cm$^3$ from the margin of the tumor, distant from any necrotic tissue.

(b)   Trim the tissue of any extraneous matter, such as fat or stroma.

(c)   Wash the specimen thoroughly with PBSA (3 to 5 times) to remove blood and any extraneous material.

(d)   Because it cannot be assumed that the specimen is sterile after being handled by the pathologist, sterilize the surface by incubation in 0.04% sodium hypochlorite for 20 min at room temperature. From this point in the procedure, the tissue can be considered sterile, and all subsequent procedures must be under sterile technique.

(e)   Rinse the tissue in PBSA, place in a Petri dish, and mince into approximately 1 mm$^3$ pieces with a scalpel or sharp scissors. After use, all of the instruments from this procedure must be disinfected and then washed before being sterilized.

(f)   Collect the chopped tissue in a centrifuge tube in PBSA and wash by centrifugation at 100 g for 5 min.

(g)   Remove the supernatant fluid and resuspend the small pieces of tissue in 20 ml of the collagenase/neutral protease solution for 90 min at 37°C with occasional shaking.

(h)   After 90 min, allow the remaining tissue pieces to settle, remove the supernatant to a new centrifuge tube centrifuge at 100 g for 5 min.

(i)   Resuspend the remaining tissue pieces in a new collagenase-neutral protease solution and incubate at 4°C overnight. The digestion step will liberate both single cells and small cell aggregates that, in the case of colon carcinoma, are crypt-like structures.

(j)   After centrifugation, wash the cell pellet once by resuspending in growth medium and repeating the centrifugation step.

(k)   Resuspend the cell pellet in 20 ml of growth medium and plate 1 ml into collagen-coated 25 cm$^2$ flasks. In addition, cells are plated in 0.2 ml volumes into the wells of 24-well dishes. Collagen-coated plates are not essential but have been found to aid the attachment of the tumor organoids and single cells liberated by the enzyme digestion process. Most tumor cells do adhere well to normal tissue culture flasks and dishes, especially if the plating volume is kept low so that contact with the growth surface is enhanced.

(l)   Once the cells are added to the flasks or plates, incubate these at 37°C in an incubator with an atmosphere of 5% $CO_2$ in air and 100% humidity.

(m)   The cultures should be left undisturbed for at least 48 h to allow organoid attachment. If the plates are handled earlier, any organoids that detach rarely re-attach. At this time the medium volume should be increased to 5 ml in 25cm$^2$ flasks and 1 ml in

24-well dishes by gentle addition of fresh medium, again taking care not to detach the organoids.

(n) The medium should be changed after 4 days and then twice a week. The medium removed from the plates and flasks should be cultured in fresh flasks as a number of colon carcinoma cell lines grow in suspension.

(o) The cultures are examined at least weekly by using an inverted microscope. Any epithelial cell patches are marked. The presence of fibroblasts should be noted. These contaminating cells often can be removed by scraping.

(p) Small patches of epithelial cells in wells with a mixture of cell types can be isolated by harvesting the epithelial colony.

    (i) Soak a small disk of filter paper (cloning disk, 3 mm) in trypsin/ ethylene diamine tetraacetate (EDTA) solution and place it on top of the epithelial colony by using sterile forceps.

    (ii) Remove the filter after 5 min and transfer to a new well containing growth medium.

## 3.2. Subculture

Once the epithelial cells are almost confluent, they are passaged by using trypsin/EDTA solution and standard procedure. The medium is removed, and 5 ml of trypsin/EDTA solution is added to the flask and then removed. The residual trypsin is sufficient to digest the cell–cell and cell–surface junctions. The culture is examined with the inverted microscope and, when the cells are rounding up, the flask is struck sharply on the base with the hand, which helps detach the cells. The cells are then resuspended in growth medium, and the cell suspension divided among three new flasks. The cells should not be split too much, as a high split ratio allows the most aggressive cells in the population to overgrow and the phenotype of the original tumor can be lost.

## 3.3. Cryopreservation

The cell population should be amplified until at least 2 or 3 × 75 cm$^2$ flasks are obtained. At this point, the cells should be stored in liquid $N_2$ or at $-80°C$ if liquid $N_2$ storage is not available, again using standard procedure. To do this, a solution of growth medium containing 20% FBS and 10% dimethyl sulfoxide (DMSO) is made up and is filter-sterilized. The cells are trypsinized and resuspended in the freezing medium at a final cell concentration of $2 × 10^6$/ml. A quantity of 1 ml cell suspension is added to liquid $N_2$ storage vials, and these are then frozen at $1°C$ per minute for 2 h (down to approximately $-80°C$) in an $-80°C$ freezer. The vials are then transferred to a liquid $N_2$ storage tank. The remaining cells should be cultured in growth medium to check for sterility. One vial should be removed from the $N_2$, thawed rapidly in the $37°C$ waterbath, resuspended in 15 ml of medium, and transferred to a 75 $cm^2$ flask to check the viability of the cells after freezing. The cells should be stored in liquid $N_2$ at the lowest passage possible. A series of freezings should be performed to obtain a sufficient stock of cells for future experiments.

## 3.4. Characterization

The cultured cells should then be characterized as well as possible for colon-specific differentiation properties and to prove that they are epithelial. To do this, the cells are grown on sterilized coverslips or LabTek chamber slides. When confluent, the cells are fixed and stained for the presence of brush border proteins, such as villin, brush border enzymes, such as dipeptidyl peptidase IV and sucrase, and mucin. The presence of endocrine cell properties can be determined by staining with an antibody to chromogranin A. Well-differentiated cultures of absorptive cells will often polarize in the culture vessel and form a brush border at the apical surface, even in monolayer cultures. This tendency can be demonstrated by scraping the monolayer from the growth surface and embedding it in a freezing compound, such as Tissue-Tek OCT. Frozen sections can then be cut and stained with an antibody to villin, which will show staining only on one surface of the monolayer [see Fig. 6, Whitehead et al, 1992]. Good tight junction formation can appear as "domes" in the culture [Kirkland, 1985; see Fig. 8, Paraskeva and Hague, 1991].

The cultured cells can also be examined by electron microscopy to determine the presence of a brush border on the apical surface of the cells, tight junctions at the apical surface, and the presence of mucin goblets. This can often be demonstrated best by using the scraped monolayer described above [Whitehead et al, 1985].

If possible, the genetic fingerprint of the cells should also be compared with that of the patient or of the tumor tissue from which the culture was derived [Stacey, 1991; MacLeod, 1992]. This comparison is important, as there have been many instances of contamination of primary cultures with previously established cell lines that were also being maintained in the same laboratory [Vermeulen et al, 1998]. An early study by Nelson-Rees and colleagues showed that more than 30% of 150 different cell lines derived from different types of cancer, which were being used in laboratories around the world, were all HeLa cells [Nelson-Rees et al, 1981]. More recent studies have confirmed that culture cross-contamination continues to be a problem

[Vermeulen et al, 1998; Markovic and Markovic, 1998]. To avoid contamination, it is essential that all work on primary cultures be performed separately from any work on established cell lines. If possible, *no established lines should be cultured in the laboratory where primary cell lines are being established.* If such conditions are not possible, separate media bottles must be kept for the each cell line and primary culture and the established cultures must never be in the tissue culture hood at the same time. This problem is not only restricted to colon carcinoma cell lines; it has recently been reported that widely used prostate carcinoma cell lines share common origins [van Bokhoven et al, 2001]. It is also important to know the initial phenotype of the cultured cells as the phenotype of the cells can change with continued culture [Williams et al, 1990].

The tumorigenicity of the newly established cell line can be established by injecting $10^6$ cells into nude mice and by observing the mice for tumor growth. To perform this task, cells are trypsinized from growing flasks and are resuspended at $1 \times 10^7$ per ml in PBSA. After which, 0.1 ml of the cell suspension is injected subcutaneously into the flank, back, or mammary line of nude mice. If a tumor grows, it should be harvested and processed for histology to compare the histology of the xenografts with the histology of the original tumor.

## 4. ESTABLISHING A CULTURE FROM ASCITIC FLUID

Patients with advanced tumors can also have ascitic fluid in their abdomen that contains tumor cells. This fluid can be a good source of cells for culture purposes.

### Protocol 3.3. Culture of Colonic Tumor Cells from Ascitic Fluid

#### Reagents and Materials
*Sterile*
- ❑ Growth medium
- ❑ Collection bag
- ❑ Centrifuge tubes, 50 ml
- ❑ Culture plates or flasks

#### Protocol
(a) Collect the fluid in a sterile bag, which can be transported to the laboratory.
(b) Place the fluid in 50 ml centrifuge tubes and centrifuged at 100 g for 5 min.

(c)   Resuspend the cell pellet in growth medium and seed into tissue culture plates or flasks. The majority of the inflammatory cells in the ascitic fluid will not adhere; however, the medium should be collected and centrifuged, as some tumor cells will not adhere but will grow in suspension. Because of this tendency, both the growth surface and the medium should be examined by using the inverted microscope.

## 5.   OTHER METHODS

### 5.1.   Feeder Layer

Some reports have described better success in establishing colon carcinoma cell lines if the tumor cells are seeded onto a feeder layer of fibroblasts [Brattain et al, 1982; Paraskeva et al, 1984]. The method normally uses 3T3 cells, which are a mouse fibroblast cell line available from the ATCC (www.atcc.org). The feeder layer must be growth-inhibited, either by irradiation (10–30 Gy) or by preincubation with mitomycin C, 20 µg/ml for 1 h [Stanley, 2002] or 0.25 µg/ml overnight (18$^L$) [Macpherson and Bryden 1971; Freshney, 2000]. The concentration of mitomycin C to be added to the medium and the period of incubation in this medium (30 min to 8 h) must be determined under the local growth conditions, ensuring the cells are in exponential growth. It is essential that all of the feeder layer cells be inactivated, as even one viable feeder cell can overgrow and contaminate the more slowly growing human tumor cells. It can prove to be very difficult to eliminate these cells from the new human colon carcinoma cell line.

In brief, the mouse fibroblast cell line or other feeder cell line is growth-inhibited by either irradiation or mitomycin C treatment. The cells are then plated at a density sufficient to give approximately 60% coverage of the growth surface and are allowed to attach. The tumor cell suspension is then added to the culture vessel and is allowed to attach. The culture is incubated. It may be necessary to passage the tumor cells onto a feeder layer for the first few passages; however, once the tumor cells are established, then they can be grown without the feeder layer. Also, if the culture is growing slowly and the feeder layer cells are dying, it may be necessary to add fresh growth-inhibited feeder cells to the culture.

### 5.2.   Xenograft

Another method that has been used especially with valuable specimens is the establishment of the tumor as a xenograft in nu/nu (nude) or SCID mice, as well as in tissue culture. Many of the original colon carcinoma cell lines were established in this way [Fogh et al, 1977].

## 5.3. Culture of Colon Carcinomas from Patients with Hereditary Syndromes

The methods described above are also suitable for the establishment of cell lines from tumors from patients with familial polyposis coli (FAP) and HNPCC. Only a few cell lines have been described from either FAP patients [Paraskeva et al, 1984; Miyazaki et al, 1996; Okada et al, 2000] or HNPCC patients [Whitehead et al, 1985].

## 6. SUPPLIERS

### 6.1. Tissue Culture Plastics

There are many brands of tissue culture plastics available. Any new brand should be tested for cell adhesion qualities before being ordered in bulk. Some suppliers that I have used are listed below.

### 6.2. Tissue Culture Media

There are also many suppliers of tissue culture media. Using local suppliers can save on shipping costs, but medium should be compared with the medium in current use before any changes are made.

## SOURCES OF MATERIALS

| Item | Supplier | Catalogue No. |
|---|---|---|
| Cloning disks, 3 mm | Sigma | Z37, 443-1 |
| Collagen Type I | Sigma | C8919 |
| Collagenase Type I | Sigma | C2674 |
| | Gibco | 17100-017 |
| Dipeptidyl peptidase IV (CD26) | Neomarkers | clone 202.36 |
| | Biosource International | AHS2611 |
| Dispase | Sigma | P3417 |
| | Gibco | 17105-032 |
| DMSO | Sigma | D1435 |
| Fetal bovine serum* | | |
| Flasks, plates and dishes | Becton Dickinson; Gibco; Sigma; TPP | |
| Fungizone | Gibco | 15290-018 |
| Gentamycin | Sigma | G1272 |
| | Gibco | 15710-064 |
| Ham's F12 liquid medium | Sigma | N8641 |
| | Gibco (Invitrogen) | *11765-054* |
| Ham's F12 powder medium | Sigma | N4388 |
| | Gibco | 21700-075 |
| Hydrocortisone hemisuccinate | Sigma | H4881 |
| Insulin–transferrin–selenium | Sigma | I 1884 |
| | Gibco | 41400-045 |
| Lab-Tek 8 Chamber Tissue Culture slides | Nalge Nunc International | 154534 |
| MUC-1 (mucin) | Chemicon | MAB4058 |
| Mucin, epithelial | Chemicon | MAB4074 |
| Neutral protease (Dispase) | Sigma | P3417 |
| | Gibco | 17105-032 |

## SOURCES OF MATERIALS *(Continued)*

| Item | Supplier | Catalogue No. |
|---|---|---|
| Nystatin | Sigma | N6261 |
| | Gibco | 15340-052 |
| OCT | ProSciTek | |
| Penicillin–streptomycin | Sigma | P0781 |
| | Gibco | 15140-122 |
| RPMI 1640 liquid medium | Sigma | R7388 |
| | Gibco | 11875-093 |
| RPMI 1640 powder medium | Sigma | R7755 |
| | Gibco | 31800-022 |
| Tissue-Tek OCT | ProSciTek | |
| Trypsin/EDTA | Sigma | T4049 |
| | Gibco | 25200-056 |
| Villin | Novacastra Laboratories | NC-L-Villin |
| | Chemicon | MAB1671 |

*Fetal bovine serum: This is available from many suppliers and should always be tested for its ability to support cell growth before use. (I have tested batches of FBS that were toxic to cells.) It is best not to take the suppliers test results on trust without testing the serum on your cells, as they normally use established cell lines in their testing procedures. It is best to find a supplier in your home country if possible.

## REFERENCES

Augeron, C., and Laboisse, C.L. (1984) Emergence of permanently differentiated cell clones in a human colonic cancer cell line in culture after treatment with sodium butyrate. *Cancer Res.* 44:3961–3969.

Brattain, M.G., Brattain, D.E., Sarrif, A.M., McRae, L.J., Fine, W.D., Hawkins, J.G. (1982) Enhancement of growth of human colon tumor cell lines by feeder layers of murine fibroblasts. *JNCI* 69:767–771.

Brattain, M.G., Fine, W.D., Khaled, M., Thompson, J., Brattain, D.E. (1981) Heterogeneity of malignant cells from a human colonic carcinoma. *Cancer Res.* 41:1751–1756.

Chantret, I., Barbat, A., Dussaulx, E., Brattain, M.G., Zweibaum, A. (1988) Epithelial polarity, villin expression, and enterocytic differentiation of cultured human colon carcinoma cells: a survey of twenty lines. *Cancer Res.* 48:1936–1942.

Drewinko, B., Romsdahl, M.M., Yang, L.Y., Ahern, M.J., Trujillo, J.M. (1976) Establishment of a human carcinoembryonic antigen producing colon adenocarcinoma cell line. *Cancer Res.* 36:467–475.

Fogh, J., Fogh, J.M., Orfeo, T. (1977) One hundred and twenty seven cultured human tumor cell lines producing tumors in nude mice. *J. Natl. Cancer Inst.* 59:221–225.

Freshney, R.I. (2000) Culture of Animal Cells 4th Ed. New York, Wiley-Liss, p. 200.

Kirkland, S.C. (1985) Dome formation by a human colonic adenocarcinoma cell line (HCA-7). *Cancer Res.* 45:3790–3795.

Kirkland, S.C., and Bailey, I.G. (1986) Establishment and characterization of six human colorectal adenocarcinoma cell lines. *Br. J. Cancer* 53:779–785.

Leibovitz, A., Stinson, J.C., McCombs, W.B., McCoy, C.E., Mazur, K.C., Mabry, N.D. (1976) Classification of human colorectal adenocarcinoma cell lines. *Cancer Res.* 36:4562–4569.

McBain, J.A., Weese, J.L., Meisner, L.F., Wolberg, W.H., Willson, J.V.K. (1984) Establishment and characterization of human colorectal cancer cell lines. *Cancer Res.* 44:5813–5821.

MacLeod, R.A., and Mitomucin, C. (1992) DNA fingerprints of cell lines. *Nature (London)* 359:681–682.

Macpherson, I., and Bryden, A. (1971) Mitomycinc treated cells as feeders. *Exp. Cell Res.* 69:240–241.

Markovic, O., and Markovic, N. (1998) Cell cross-contamination in cell cultures: the silent and neglected danger. *In Vitro Cell Dev. Biol. Anim.* 34:1–8.

Miyazaki, M., Tsuboi, S., Mihara, K., Kosaka, T., Fukaya, K., Kino, K., Mori, M., Namba, M. (1996) Establishment and characterization of a human colon cancer cell line, OUMS-23, from a patient with familial adenomatous polyposis. *J. Cancer Res. Clin. Oncol.* 122:95–101.

Nelson-Rees, W.A., Daniels, D.W., Flandermeyer, R.R. (1981) Cross-contamination of cells in culture. *Science* 212:446–452.

Okada, F., Kawaguchi, T., Habelhah, H., Kobayashi, T., Tazawa, H., Takeichi, N., Kitagawa, T., Hosokawa, M. (2000) Conversion of human colonic adenoma cells to adenocarcinomas cells through inflammation in nude mice. *Lab Invest.* 80:1617–1628.

Okazaki, K., Nakayama, Y., Shibao, K., Hirata, K., Sako, T., Nagata, N., Kuroda, Y., Itoh, H. (2000) Establishment of a human colon cancer cell line (PMF-ko14) displaying high metastatic activity. *Int. J. Oncol.* 17:39–45.

Paraskeva, C., Buckle, B.G., Sheer, D., Wigley, C.B. (1984) The isolation and characterization of colorectal epithelial cell lines at different stages in malignant transformation from familial polyposis coli patients. *Int. J. Cancer* 34:49–56.

Paraskeva, C., and Hague, A. (1991) Colorectum. In: Masters, J.W.R. (ed) *Human Cancer in Primary Culture. A Handbook.* Dordrecht, The Netherlands: Kluwer Academic Publishers, pp. 151–168.

Park, J.G., Kramer, B.S., Steinberg, S.M., Carmichael, J., Collins, J.M., Minna, J.D., Gazdar, A.F. (1987) Chemosensitivity testing of human colorectal carcinoma cell lines using a tetrazolium-based colorimetric assay. *Cancer Res.* 47:5875–5879.

Stacey, G. (1991) DNA fingerprinting and the characterization of cell lines. *Cytotechnology* 6:91–92.

Stanley, M.A. (2002) Culture of human cervical epithelial cells. In Freshney, R.I., and Freshney, M.G., Eds., Culture of Epithelial Cells, 2nd Ed., New York, Wiley-Liss.

Tompkins, W.A.F., Watrach, A.M., Schmale, J.D., Schultz, R.M., Harris, J.A. (1974) Cultural and antigenic properties of newly established cell strains derived from adenocarcinomas of the human colon and rectum. *J. Natl. Cancer Inst.* 52:1101–1110.

Van Bokhoven, A., Varella-Garcia, M., Korch, C., Hessels, D., Miller, G.J. (2001) Widely used prostate carcinoma cell lines share common origins. *Prostate* 47:36–51.

Vermeulen, S.J., Chen, T.R., Speleman, F., Nollet, F., Van Roy, F.M., Mareel, M.M. (1998) Did the four human cancer cell lines DLD-1, HCT-15, HCT-8, and HRT-18 originate from one and the same patient? *Cancer Genet. Cytogenet.* 107:76–79.

Whitehead, R.H., Jones, J.K., Gabriel, A., Lukeis, R.E. (1987) A new colon carcinoma cell line (LIM1863) that grows as organoids with spontaneous differentiation into crypt-like structures in vitro. *Cancer Res.* 47:2683–2689.

Whitehead, R.H., Macrae, F.A., St John, D.J.B., Ma, J. (1985) A colon carcinoma cell line (LIM1215) derived from a patient with inherited nonpolyposis colorectal cancer. *J. Natl. Cancer Inst.* 74:759–765.

Whitehead, R.H., Van Eeden, P., Lukeis, R.E. (1991) A cell line (LIM2463) derived from a tubulovillous adenoma of the rectum. *Int. J. Cancer* 48:693–696.

Whitehead, R.H., and Watson, N.K. (1994) Gastrointestinal cell lines as a model for differentiation. In: Halter, F., Winton, D., Wright, N.A. (eds) *The Gut as a Model in Cell and Molecular Biology.* Falk Symposium No. 94. pp. 275–290.

Whitehead, R.H., Zhang, H.H., Hayward, I.P. (1992) Retention of tissue-specific phenotype in a panel of colon carcinoma cell lines: relationship to clinical correlates. *Immunol. Cell Biol.* 70:227–236.

Williams, A.C., Harper, S.J., Paraskeva, C. (1990) Neoplastic transformation of a human colon epithelial cell line: In vitro evidence for the adenoma to carcinoma sequence. *Cancer Res.* 50:4724–4730.

Willson, J.V.K., Bittner, G.N., Oberley, T.D., Meisner, L.F., Weese, J.L. (1987) Cell culture of human colon adenomas and carcinomas. *Cancer Res.* 47:2704–2713.

# 4

# Pancreatic Cancer-Derived Cultured Cells: Genetic Alterations and Application to an Experimental Model of Pancreatic Cancer Metastasis

Haruo Iguchi[1], Kazuhiro Mizumoto[2], Masaki Shono[2], Akira Kono[3] and Soichi Takiguchi[1]

[1] Division of Biochemistry, National Kyusyu Cancer Center Research Institute, Notame, Fukuoka, Japan; [2] Department of Surgery and Oncology, Graduate School of Medical Sciences, Kyushu University, Maedashi, Fukuoka, Japan; and [3] Transgenic Inc., Chuougai, Kumamoto, Japan, higuchi@nk-cc.go.jp

*Culture of Human Tumor Cells*, Edited by Roswitha Pfragner and R. Ian Freshney.
ISBN 0-471-43853-7   Copyright © 2004 Wiley-Liss, Inc.

## 1. INTRODUCTION

Pancreatic cancer is one of the most aggressive cancers; it is responsible for 5% of all cancer-related deaths in the Western world [Parker et al., 1996]. This is, in part, due to difficulty in early diagnosis and also to the biological characteristics of pancreatic cancer. The use of cell lines may serve as a valuable tool in determining the characteristic features of pancreatic cancer cells. Genetic alterations of pancreatic cancer have been well described during the past 10 years. Mutations in the Ki-*ras* gene have been shown in pancreatic lesions with minimal atypia, suggesting that alterations of Ki-*ras* is an earlier event in the development of pancreatic neoplasia [Luttges et al., 1999; Yanagisawa et al., 1993]. Incidence of inactivation of the *p16* tumor suppressor gene has been shown to increase along with the increase in cytological atypia in pancreatic duct lesions [Wilentz et al, 1998]. Loss of the *p53* and *DPC4* tumor suppressor genes, however, appears to occur later in the development of pancreatic neoplasia [DiGuiseppe et al., 1994; Yamano et al., 2000; Wilentz et al., 2000]. On the basis of these findings, the progression of pancreatic cancer seems to be associated with multiple genetic alterations; that is, point mutations in the Ki-*ras* gene, and inactivation of the *p16*, *p53*, and *DPC4* tumor suppressor genes. Liver metastasis and/or peritoneal dissemination are frequently seen at the time of diagnosis in pancreatic cancer patients but little is known about the aggressive behavior of pancreatic cancer. It is important to elucidate the mechanism of such metastasis, to make it possible to develop drugs against molecular targets in the metastatic process.

In this chapter, we introduce pancreatic cancer-derived cell lines from ductal adenocarcinoma and describe genetic alterations and experimental models for liver metastasis and peritoneal dissemination of pancreatic cancer.

## 2. ESTABLISHMENT OF CELL LINES

### 2.1. Origin of Tumor Tissues

#### 2.1.1. KP-1N, KP-1NL

We obtained tumor tissues from a liver metastasis of pancreatic cancer (moderately differentiated tubular adenocarcinoma) at surgery. A cell suspension was prepared from the tumor tissues according to the procedure described (see Section 2.2. Primary Culture), after which, this cell suspension was injected into the spleen of a nude mouse (BALB/C-nu/nu, Nippon CLEA Inc., Tokyo, Japan). KP-1N was established from the resulting tumor in the spleen of the mouse, while KP-1NL was established from the resulting liver metastasis of the mouse.

#### 2.1.2. KP-2

We obtained tumor tissue from a biopsy specimen of a primary lesion of pancreatic cancer (moderately differentiated tubular adenocarcinoma).

#### 2.1.3. KP-3, KP-3L

KP-3 was established from a liver metastasis of pancreatic cancer (adenosquamous carcinoma) obtained at autopsy. KP-3 cells were injected into the spleen of a nude mouse, and the resulting liver metastasis in the mouse was subjected to preparation for KP-3L.

#### 2.1.4. KP-4

Ascites was obtained from a patient with pancreatic cancer (poorly differentiated adenocarcinoma), in whom elevation of PTHrP was found in both the serum and ascites. KP-4 was established from the ascites.

### 2.2. Primary Culture

**Protocol 4.1. Primary Culture from Pancreatic Tumor Tissue**

*Reagents and Materials*
*Sterile*
- ❑ Hanks' balanced salt solution (HBSS)
- ❑ HBSS containing penicillin (100 U/ml) and streptomycin (100 µg/ml)

- ❑ PBSA: Phosphate buffered saline without $Ca^{2+}$ and $Mg^{2+}$
- ❑ PBSA/EDTA: PBSA with 0.5mM EDTA
- ❑ Trypsin, 2.5%
- ❑ Trypsin/EDTA: 0.1% trypsin in PBSA/EDTA
- ❑ PBSA/EDTA2: PBSA with 2 mM EDTA
- ❑ Collagenase purified from Clostridium histolyticum
- ❑ Hyaluronidase
- ❑ DNase
- ❑ Daigo's T medium
- ❑ Daigo's T medium containing 0.2% collagenase
- ❑ Daigo's T medium containing 0.1% collagenase, 0.005% DNase, and 0.002% hyaluronidase
- ❑ Growth medium: (Daigo's T medium supplemented with 10% FBS, 100 U/ml of penicillin, and 100 µg/ml of streptomycin)
- ❑ RPMI 1640 medium
- ❑ IMDM: Iscove's modified Dulbecco's medium
- ❑ FBS: Fetal bovine serum
- ❑ Culture dishes and flasks
- ❑ Collagen-coated culture dishes
- ❑ Centrifuge tubes, 50 ml
- ❑ Stainless steel or Nylon sieve with mesh ～200 µm
- ❑ Cell scraper

*Non-sterile*
- ❑ Shaking water bath


### Protocol

(a) Keep the resected tumor specimens in chilled HBSS containing penicillin and streptomycin.

(b) Mince the tumor tissue into small pieces (approximately 1 $mm^3$) with ophthalmologic scissors in HBSS in a culture dish.

(c) Transfer the minced tissue into a centrifuge tube (50 ml) containing HBSS, and centrifuge at 200 *g* for 5 min under room temperature.

(d) Discard the supernatant by pipetting.

(e) Disperse the pellet and incubate with 20 ml of Daigo's T medium containing collagenase, DNase, and hyaluronidase in the centrifuge tube (50 ml) at 37°C for 30 min with shaking in a water bath. Seal the centrifuge tube with Parafilm during incubation.

(f) Centrifuge the incubation mixture at 200 *g* for 5 min and discard the supernatant by pipetting.

(g) Disperse the pellet and incubate with 5 ml of PBSA/EDTA2 at 37°C for 15 min.

(h)   Centrifuge the incubation mixture at 200 $g$ for 5 min and discard the supernate by pipetting.

(i)   Disperse the pellet and incubate with 20 ml of Daigo's T medium containing 0.2% collagenase at 37°C for 30 min with shaking in the water bath.

(j)   Transfer the incubation mixture, using a wide-bore pipette to prevent clogging, into a sieve and filter through the mesh to remove undigested tissue. Do this by pipetting an aliquot of the incubation mixture onto the mesh, filling the sieve with growth medium (see below) to dilute the incubation mixture, and collecting the filtrate in a culture dish.

(k)   Transfer the filtrate collected in the culture dish into a centrifuge tube (50 ml), and centrifuge at 200 $g$ for 5 min.

(l)   Discard the supernate by pipetting.

(m)   Suspend the pellet in an appropriate volume of growth medium and seed into collagen-coated culture dishes. The cell number is not counted when seeded. A culture flask (25 cm$^2$ or 75 cm$^2$) can also be used for the primary culture, however, a culture dish is easier to handle when subsequently eliminating fibroblasts.

(n)   Maintain the cultures in a humidified incubator at 37°C with a 95% air–5% $CO_2$ atmosphere, and observe daily.

(o)   Replace a half volume of the growth medium every 3 to 7 days.

(p)   To eliminate fibroblasts, expose the cultured cells to trypsin/EDTA, and scrape off the fibroblasts mechanically using a silicon rubber cell scraper when a shape change of the fibroblasts is noticed under a stereomicroscope. Repeat this manipulation at an appropriate interval until no further growth of fibroblasts can be observed.

(q)   A confluent state of the epithelioid tumor cells will be achieved in the culture dishes 1 to 2 months later, at which point the tumor cells may be transferred into culture flasks (25 cm$^2$), and maintained in growth medium (Daigo's T medium + 10% FBS).

(r)   Subsequent passage of the cells is performed by using trypsin/EDTA.

*Note:* Collagen-coated dishes are not necessary after the cells have once reached confluence. KP-1N and KP-3 cells grown on plastic surfaces of the culture flask are shown in Figure 4.1. These cells reveal typical epithelial morphologies. RPMI 1640 medium supplemented with 10% FBS, 100 U/ml of penicillin, and 100 µg/ml of streptomycin can also be used as a growth medium, instead of Daigo's T medium, to maintain these pancreatic cancer-derived cell lines after confluence has been reached.

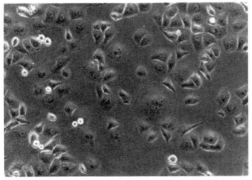
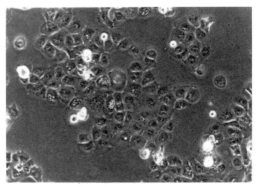

KP-1N                              KP-3

**Figure 4.1.** Photomicrographs of KP-1N (left) and KP-3 (right). Phase-contrast optics (×125).

## Protocol 4.2.   Primary Culture of Pancreatic Tumor Cells from Ascites

### Reagents and Materials
*Sterile*
- ❑ IMDM/20FB: IMDM supplemented with 20% FBS, 100 U/ml of penicillin, and 100 μg/ml of streptomycin
- ❑ 2 x IMDM/20FB: twofold concentrated IMDM supplemented with 20% FBS, 100 U/ml of penicillin, and 100 μg/ml of streptomycin
- ❑ IMDM/10FB: IMDM supplemented with 10% FBS, 100 U/ml penicillin, and 100 μg/ml streptomycin
- ❑ Heparinized tubes
- ❑ Collagen-coated culture dishes
- ❑ Cell scraper
- ❑ Trypsin/EDTA: 0.1% trypsin in PBSA with 0.5 mM EDTA

### Protocol
(a)  Aseptically collect an appropriate volume of ascites from a patient and transfer into a heparinized tube.

(b)  Centrifuge the ascites at 200 g for 5 min, and discard the supernate by pipetting.

(c)  Wash the pellet with PBSA three times, and suspend it in an appropriate volume of 2 x IMDM/20FB.

(d)  Seed the cells into collagen-coated culture dishes, and maintain the culture in a humidified incubator at 37°C with a 95% air–5% $CO_2$ atmosphere.

(e)  Replace a half volume of the growth medium every 3 to 7 days.

(f)  Eliminate fibroblasts by brief exposure to trypsin/EDTA followed by scraping with a cell scraper at an appropriate interval.

(g)   A confluent state of the epithelioid tumor cells will be achieved in
      1 to 2 months, at which point the cells may be transferred into
      culture flasks (25 cm$^2$), and maintained in IMDM/10FB.
(h)   Perform subsequent passages by using trypsin/EDTA


## 3.  LIVER METASTASIS ASSAY

Pancreatic cancer often metastasizes to the liver, which affects its
clinical course. Thus, it is important to control liver metastasis; its
mechanism, however, is still poorly understood to date. In order to
analyze the metastatic process, it may be useful to develop an experimental model by using cell lines. Therefore, we examined the metastatic potential of human pancreatic cancer-derived cell lines, including the series of KP-cells, to the liver according to the procedure
described by Morikawa et al. [1988].


### Protocol 4.3.   Assay of Metastasis by Xenografting in Nude Mice

*Note:* Great care must be taken when anesthetizing nude mice for
successful implantation, as nude mice can easily stop breathing under
anesthesia. The most stable anesthesia can be obtained when 80 mg/kg
of ketamine and 8 mg/kg of xylazine are injected intraperitoneally.
Under sufficient anesthesia, a left lateral laparotomy or mid-upper laparotomy is performed, and the distal pancreas and spleen are exposed.

*Reagents and Materials*
*Sterile*
❑   PBSA
❑   Scalpels, scissors, forceps
❑   Anesthetics: ketamine and xylazine
*Non-sterile*
❑   BALB/c nu/nu mice, 6 to 7 weeks old, 15 to 20 g
❑   75% ethanol,
❑   Cotton balls

*Protocol*
(a)   Anesthetize mice.
(b)   Wipe mouse skin with 75% ethanol.
(c)   Make a small incision in the skin and peritoneal wall at the left
      flank of a nude mouse.
(d)   Press out the spleen softly through the incision.
(e)   Inject tumor cells (1 × 10$^6$/0.05 ml PBSA) into the spleen.

(f) Return the spleen to the peritoneal cavity followed by suture of the abdominal wall.

(g) Perform autopsy 30–100 days, as appropriate, after inoculation of the cells, and assess metastasis to the liver by counting the tumor colonies.

**Table 4.1. Human pancreatic cancer-derived cell lines.**

| Cell line | Histology | Origin |
|---|---|---|
| AsPC-1 | duct cell (mo adeno*) | ascites |
| BxPC-3 | duct cell (mo adeno) | pancreas |
| SUIT-2 | duct cell (mo adeno) | liver metastasis |
| H-48N | duct cell (mo adeno) | pancreas |
| KP-1N | duct cell (mo adeno) | liver metastasis |
| KP-1N-L | duct cell (mo adeno) | liver metastasis of KP-1N in nude mice |
| KP-2 | duct cell (mo adeno) | pancreas |
| KP-3 | duct cell (adenosquamous) | liver metastasis |
| KP-3–L | duct cell (adenosquamous) | liver metastasis of KP-3 in nude mice |
| MIA-Paca-2 | undifferentiated | pancreas |
| PANC-1 | undifferentiated | pancreas |

*mo adeno, moderately differentiated adeno carcinoma

### 3.1. Results from Metastasis Assay

Pancreatic cancer-derived cell lines were tested, and their metastatic potential to the liver is shown in Tables 4.1 and 4.2. Liver metastasis was graded as follows: grade 0, tumor-free; grade I, histological evidence of tumor growth; grade II, <10 tumor colonies; grade III, ≥10 or uncountable tumor colonies. Metastatic potential of each cell line was assessed based on the grade and frequency of the liver metastasis, and the cell lines were divided into three groups according to the metastatic potential as follows: high metastatic potential, all mice bore tumor colonies and more than a half of the mice revealed grade III; low metastatic potential, all mice bore tumor colonies and the mice revealed grade I or II, or all mice did not bear tumor colonies and some of them revealed grade II or III; no metastatic potential, tumor colonies were not observed more than 10 weeks post-inoculation of the cells. The cell lines (KP-1NL, KP-3L) established from the liver metastasis of the nude mice revealed higher metastatic potential compared with their paternal cell lines (KP-1N, KP-3).

## 4. ORTHOTOPIC IMPLANTATION MODEL OF HUMAN PANCREATIC CANCER

Peritoneal dissemination often accompanies pancreatic cancer, and it is clearly important to try to prevent such dissemination, in addition to liver metastasis, in patients with pancreatic cancer. We developed an in vivo model for peritoneal dissemination of pancreatic cancer in

**Table 4.2.** Metastatic potential of pancreatic cancer-derived cell lines assessed by a liver metastasis assay. Human pancreatic cancer-derived cell lines.

| Cell-line | Days* | Grade | Frequency | Cell-line | Days | Grade | Frequency |
|---|---|---|---|---|---|---|---|
| High metastatic group | | | | Low metastatic group | | | |
| KP-1N-L | 47 | III | | KP-1N | 78 | 0 | |
| | 47 | III | 3/3 | | 78 | 0 | 1/3 |
| | 47 | III | | | 78 | III | |
| KP-3–L | 56 | III | 2/2 | KP-3 | 91 | II | 2/2 |
| | 105 | III | | | 91 | II | |
| | | | | PANC-1 | 71 | 0 | |
| | | | | | 83 | 0 | 1/3 |
| | | | | | 127 | II | |
| AsPC-1 | 43 | II | | MIA-Paca-2 | 78 | 0 | |
| | 45 | III | 4/4 | | 78 | 0 | 1/4 |
| | 51 | III | | | 84 | II | |
| | 55 | III | | | 131 | II | |
| SUIT-2 | 28 | I | | BxPC-3 | 45 | I | 2/2 |
| | 34 | II | 4/4 | | 51 | I | |
| | 50 | III | | | | | |
| | 50 | III | | No metastatic group | | | |
| | | | | H-48N | 101 | 0 | 0/3 |
| | | | | | 102 | 0 | |
| | | | | KP-2 | 90 | 0 | |
| | | | | | 90 | 0 | 0/3 |
| | | | | | 90 | 0 | |

*Days, days after inoculation

order to understand its mechanism. Here, we introduce an orthotopic implantation model by using human pancreatic cancer-derived cell line, SUIT-2 [Iwamura et al., 1987], which is not only simple and reproducible but also has characteristics that resemble those of human pancreatic cancer. This model could be a valuable tool for investigating the therapeutic effects of anti-cancer drugs for pancreatic cancer.

## 4.1. Surgical Procedure

## Protocol 4.4. Orthotopic Implantation of Pancreatic Cancer Cell Lines

### Reagents and Materials
*Sterile*
❏ SUIT-2 cells
❏ DMEM/10FB: DMEM + 10% FBS, 100 U penicillin, and 100 µg streptomycin/ml medium

- ❏ Trypsin/EDTA: 0.1% trypsin in PBSA with 0.5 mM EDTA
- ❏ Anesthetics: ketamine and xylazine

*Non-sterile*
- ❏ BALB/c nu/nu mice, 6 to 7 weeks old, preweighed, 15 to 20 g

**Protocol**

(a)   Culture SUIT-2 cells in DMEM/10FB.

(b)   Prepare a cell suspension by treatment with trypsin/EDTA.

(c)   Anesthesize mice with ketamine and xylazine: ketamine, 80 mg/ kg, and xylazine, 8 mg/kg.

(d)   Inject a total of 0.2 ml of cell suspension ($1 \times 10^6$ cells) directly into the distal pancreas (see Note in Protocol 4.3, above).

## 4.2.   Natural Course of Nude Mice with Orthotopic Implantation of Pancreatic Cancer Cells

After implantation of the SUIT-2 cells into the pancreas of nude mice, abdominal tumors are palpable within 3 weeks and apparent, massive ascites is observed through their dried skin. Cancer cachexia, characterized by decrease in food and water intakes, body weight loss, and tissue wasting, is observed in the nude mice within 4 to 5 weeks after the implantation. This cachexia increases in parallel with tumor growth, and is also seen in patients with advanced pancreatic cancer.

## 4.3.   Macroscopic Appearances of the Implanted Tumor

At laparotomy of the moribund mice (Fig. 4.2), xenografted primary tumors, measuring 2 to 3 cm in diameter, are noted at the implantation site of the pancreas. The tumors invade neighboring organs, including the spleen, stomach, and small or large bowels. In all tumor-bearing mice, small disseminated nodules are recognized throughout the peritoneal cavity, including the mesentery of bowels, retroperitoneum, and diaphragm. Some mice have bloody ascites in the abdomen. Liver metastasis is observed as white nodules on the surface of the liver. Incidence of liver metastasis is more than 70% of the nude mice when SUIT-2 cells are implanted. Lung or mediastinal metastases are rarely observed. The mice die or become moribund due to the huge primary implanted and metastatic lesions.

## 4.4.   Microscopic Appearances of the Implanted Tumor

Histologically, xenografted SUIT-2 cells proliferate in a medullary fashion (Fig. 4.3). The neoplastic cells show nuclear pleomorphism, loss of polarity, and some aberrant mitotic figures. The carcinoma cells invade normal pancreatic tissue of the host mice without any capsular

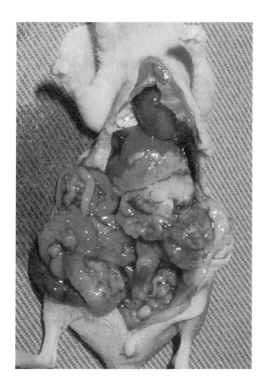

**Figure 4.2.** Macroscopic appearance of the abdomen in a nude mouse with orthotopic implantation of human pancreatic cancer-derived cell line, SUIT-2. An implanted tumor replaced the entire pancreas and invaded the spleen. A liver tumor is recognized as a large white nodule. Multiple disseminated nodules are observed throughout the peritoneal cavity.

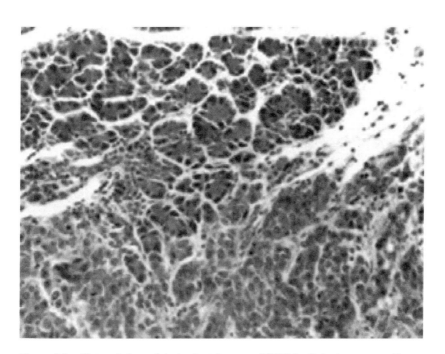

**Figure 4.3.** Histopathology of the implanted tumor of SUIT-2 cells in the pancreas. Cancer cells invade the pancreatic tissue of a host nude mouse without capsular formation.

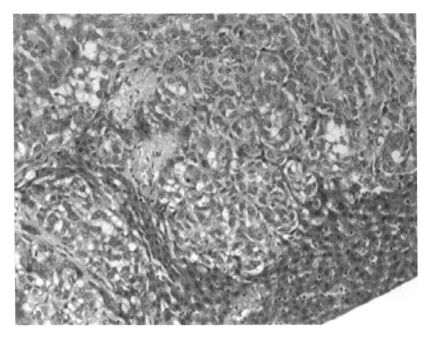

**Figure 4.4.** Histopathology of the liver metastasis from the implanted tumor in the pancreas.

formation. The metastatic liver tumors or peritoneal tumors similarly show invasive growth in nodular fashion, which resemble human liver metastases from pancreatic cancer (Fig. 4.4). When anti-pancreatic cancer drugs are given to mice with the orthotopic implantation of the cancer cells, therapeutic effects may be evaluated by tumor weight, number of the metastatic nodules, and survival rates of the animals. Anti-angiogenic drugs or matrix metalloproteinase (MMP) inhibitors are currently the focus among the various new therapeutic drugs being developed for pancreatic cancer. The orthotopic implantation model of SUIT-2 cells described here could be useful to examine the effects of these new drugs on the primary lesion, liver metastasis, and/or peritoneal dissemination of pancreatic cancer.

## 5. GENETIC ALTERATIONS

Genetic alterations are involved in the oncogenic process of various cancers. Metastasis is a characteristic feature of most malignant cells, and genetic changes have been observed during acquisition of metastatic potential. Hence, we decided to examine alterations of Ki-*ras*, *p53*, and *CDKN* 2A in human pancreatic cancer-derived cell lines with different metastatic potential.

## 5.1. Ki-*ras* Gene Mutation (Codon 12)

The Ki-*ras* sequence of exon 1 was amplified by polymerase chain reaction (PCR) with primers, (forward) GACTGAATATAAACT-TGTGG and (reverse) CTATTGTTGGATCATATTCG. The amplified DNA was sequenced by using a T7-sequencing kit.

## 5.2. *p53* Gene Mutations

The *p53* sequences of exon 5 to 8 were amplified by PCR. PCR products were purified by ethanol precipitation, digested with EcoRI, and repurified with phenol–chloroform extraction and ethanol precipitation. The purified PCR products were ligated into EcoRI site of pBluescript KS plasmids and were introduced to *E. coli* DH5-α cells. Transformed *E. coli* cells were cultured overnight at 37°C in 3 ml of L-broth and were selected by ampicillin. Double-strand plasmids were isolated and sequenced by using T7-sequencing kit.

## 5.3. Alterations of *CDKN* 2A Gene

Deletion of the *CDKN* 2A gene was determined by a comparative PCR method using Cy5 fluorescent labeled primers. The $\beta$-globin gene was used as an internal reference to confirm amplification by PCR. PCR products were mixed with an equal volume of formamide and were then heated at 85°C for 5 min and immediately cooled on ice. Denatured samples (5 μl) were loaded onto 6% Long Ranger™/6M urea gel. Electrophoresis was performed at 1000 V, 35 mA, 30 W for 2.5 h by using ALFred autosequencer. Analysis of the electrograms was performed by using Fragment Manager™ software, according to the manufacturer's instructions, and the ratio of the fluorescent intensity of the *CDKN* 2A/$\beta$-globin gene was calculated.

Methylation of the 5' region (CpG island) of the *CDKN* 2A gene was analyzed by digestion with methylation-sensitive restriction enzyme (Sma I) followed by PCR amplification. PCR products were electrophoresed on 6% Long Ranger/6 M urea gel, and fluorescent intensity was analyzed by using ALFred.

Sequence abnormalities of exon 2 of the *CDKN* 2A gene were analyzed by PCR followed by sequencing. Point mutations were determined according to the orphan peek analysis (OPA) method [Hattori et al., 1993].

## 5.4. Results from Genetic Studies

Alterations of Ki-*ras*, *p53*, and *CDKN* 2A genes are shown in Table 4.3. Point mutations in the Ki-ras gene at codon 12 were found in 9 of 11 cell lines (82%). The mutations at codon 12 consisted of a G → A

**Table 4.3.** Genetic alterations in pancreatic cancer-derived cell lines with different metastatic potential.

| | Ki-*ras* (codon 12) | *p53* (exon 5–8) | CDKN 2A deletion | CDKN 2A sequence abnormalities | methylation |
|---|---|---|---|---|---|
| High metastatic | | | | | |
| KP-1N-L | GAT | CGT → CAT (codon 273) | homo[a] | | − |
| KP-3–L | GTT | none | hemi[b] | | + |
| AsPC-1 | GAT | 1 bp deletion (codon 134–135) | hemi | 2 bp deletion (codon 69–70) | − |
| SUIT-2 | GAT | CGT → CAT (codon 273) | R[c] | GAG → TAG (codon 61) | − |
| Low metastatic | | | | | |
| KP-1N | GAT | CGT → CAT (codon 273) | homo | | − |
| KP-3 | GTT | none | R | | + |
| PANC-1 | GAT | CGT → CAT (codon 273) | homo | | − |
| MIA-Paca-2 | TGT | CGG → TGG (codon 248) | homo | | − |
| BxPC-3 | wild type | TAT → TGT (codon 220) | homo | | − |
| No metastatic | | | | | |
| H-48N | GTT | none | homo | | − |
| KP-2 | wild type | GCC → GAC (codon 161) | homo | | − |

[a] homo, homozygous deletion; [b] hemi, hemizygous deletion; [c] R, retention of its alleles

transversion at the second base (Asp) ($n = 5$); a G → T transversion at the second base (Val) ($n = 3$); and a G → T transvertion at the first base (Cys) ($n = 1$).

Sequence abnormalities of the p53 gene (exon 5 to 8) were found in 8 of 11 cell lines (73%). These abnormalities consisted of point mutations [CGT → CAT at codon 273 ($n = 4$), CGG → TGG at codon 248 ($n = 1$), TAT → TGT at codon 220 ($n = 1$), GCC → GAC at codon 161($n = 1$)] and 1bp deletion at codon 134–135(TTT TGC → TTTGC) ($n = 1$), which resulted in an appearance of a stop codon (TGA) between codon 169–170 (ATG ACG → A TGA CG).

Alterations of the *CDKN* 2A gene (deletion, sequence abnormalities, methylation) were found in all the cell lines tested. Deletion analyses revealed homozygous deletion in 7 of 11 cell lines (73%), hemizygous deletion in two cell lines (18%), and retention of its alleles in two cell lines (18%). In the two cell lines with hemizygous deletion, 2 bp deletion at codon 69–70 and methylation of the CpG island were found in AsPC1 and KP-3-L, respectively. In the remaining two cell lines with retention of the *CDKN* 2A gene alleles, point mutation of codon 61 and methylation of the CpG island were found in SUIT-2 and KP-3, respectively. As a result of such genetic changes in the *CDKN* 2A gene, a p16 protein is inactivated in all the pancreatic cancer-derived cell lines.

We describe various genetic changes in the pancreatic cancer-derived cell lines, however, such changes are not associated with the metastatic potential. Further studies, in particular on those events related to metastasis (adhesion, invasion, cell growth) using these cell lines, could contribute to elucidating characteristics of malignant behavior in pancreatic cancer.

## SOURCES OF MATERIALS

| Item | Supplier |
| --- | --- |
| ALFred autosequencer | Amersham Pharmacia |
| BALB/c nu/nu mice | Nippon CLEA |
| Cell scraper | Nalge Nunc |
| Collagenase purified from Clostridium histolyticum | Wako |
| Collagen-coated culture dish | Iwaki Glass |
| Culture dishes and flasks | Iwaki |
| Cy5 fluorescent labeled primers | Amersham Pharmacia |
| Daigo's T medium | Wako |
| DNase | Sigma |
| Ethylene diamine tetraacetate disodium salt [D16] (EDTA) | Wako |
| Fetal bovine serum (FBS) | JRH |
| Fragment Manager$^{TM}$ software | Amersham Pharmacia |
| Hyaluronidase | Sigma |

## SOURCES OF MATERIALS (Continued)

| Item | Supplier |
|---|---|
| Iscove's modified Dulbecco's medium (IMDM) | Invitrogen |
| Ketamine | Sigma |
| Long Ranger ™/6M urea gel | Applied Biosystems |
| Nude mice | Nippon CLEA |
| PBSA | Invitrogen |
| Penicillin–streptomycin | Invitrogen |
| RPMI 1640 medium | Invitrogen |
| Sieve with 200 μm mesh | Becton Dickinson; Tekmar |
| T7-sequencing kit | Amersham Pharmacia |
| Trypsin, 2.5% | Invitrogen |
| Xylazine | Sigma |

## REFERENCES

DiGiuseppe, J.A., Hruban, R.H., Goodman, S.N., Polak, M., van den Berg, F.M., Allison, D.C., Cameron, J.L., Offerhaus, G.J.A. (1994) Overexpression of p53 protein in adenocarcinoma of the pancreas. *Am. J. Clin. Pathol.* 101:684–688.

Hattori, M., Shibata, A., Yoshioka, K., Sasaki, Y. (1993) Orphan peak analysis: a novel method for detection of point mutations using an automated fluorescence DNA sequencer. *Genomics* 15:415–417.

Iguchi, H., Morita, R., Yasuda, D., Takayanagi, R., Ikeda, Y., Takada, Y., Shimazoe, T., Nawata, H., Kono, A. (1994) Alterations of the p53 tumor suppressor gene and Ki-ras oncogene in human pancreatic cancer-derived cell lines with different metastatic potantial. *Oncol. Rep.* 1:1223–1227.

Iwamura, T., Katsuki, I., Ide, K. (1987) Establishment and characterization of a human pancreatic cancer cell line (SUIT-2) producing carcinoembryonic antigen and carbohydrate antigen 19-9. *Jpn. J. Cancer Res.* 78:54–62.

Luttges, J., Schlehe, B., Menke, M.A., Vogel, I., Henne-Bruns, D., Kloppel, G. (1999) The K-ras mutation pattern in pancreatic ductal adenocarcinoma usually is identical to that in associated normal, hyperplastic, and metaplastic ductal epithelium. *Cancer* 85:1703–1710.

Morikawa, K., Walker, S.M., Jessup, M., Fidler, I.J. (1988) In vivo selection of highly metastatic cells from surgical specimens of different primary human colon carcinomas implanted into nude mice. *Cancer Res.* 48:1943–1948.

Parker, S.L., Tong, T., Bolden, S. (1996) Cancer statistics. *CA. Cancer J. Clin.* 46:5–27.

Sugimoto, Y., Morita, R., Hikiji, K., Imura, G., Ogata, Y., Yasuda, D., Kono, A., Igucho, H. (1998) Alteration of the CDKN2A gene in pancreatic cancers: Is it a late event in the progression of pancreatic cancer? *Int. J. Oncol.* 13:699–676.

Wilentz, R.E., Geradts, J., Maynard, R., Offerhaus, G.J.A., Kang, M., Goggins, M., Yeo, C.J., Kern, S.E., Hruban, R.H. (1998) Inactivation of the p16(INK4A) tumor suppressor gene in pancreatic duct lesions: loss of intranuclear expression. *Cancer Res.* 58:4740–4744.

Wilentz, R.E., Iacobuzio-Donahue, C.A., Argani, P. Mac Carthy, D.M., Parsons, J.L., Yeo, C.J., Kern, S.E., Hruban, R.H. (2000) Loss of expression of Dpc4 in pancreatic intraepithelial neoplasia: evidence that DPC4 inactivation occurs late in neoplastic progression. *Cancer Res.* 60:2002–2006.

Yamano, M., Fuji, H., Takagaki, T., Kadowaki, N., Watanabe, H., Shirai, T. (2000) Genetic progression and divergence in pancreatic carcinoma. *Am. J. Pathol.* 156:2123–2133.

Yanagisawa, A., Ohtake, K., Ohashi, K., Hori, M., Kitagawa, T., Sugano, H., Kato, Y. (1993) Frequent c-Ki-ras oncogene activation in mucous cell hyperplasias of pancreas suffering from chronic inflammation. *Cancer Res.* 53:953–956.

# 5

# The Establishment and Characterization of Bladder Cancer Cultures In Vitro

Vivian X. Fu, Steven R. Schwarze, Catherine A. Reznikoff and David F. Jarrard

*Department of Surgery, Section of Urology, University of Wisconsin-Madison Medical School, Madison, Wisconsin*
*jarrard@surgery.wisc.edu*

*Culture of Human Tumor Cells*, Edited by Roswitha Pfragner and R. Ian Freshney.
ISBN 0-471-43853-7 Copyright © 2004 Wiley-Liss, Inc.

## I.  INTRODUCTION

### 1.1.  Background

Bladder cancer is the sixth most common cancer in the United States and the third most prevalent cancer in men over sixty years of age. The American Cancer Society estimated 57,400 new cases of bladder cancer in 2003 resulting in 12,500 deaths [Jemal et al., 2003]. Bladder cancer is three to four times more common among men than women. The incidence of the disease increases with age. Numerous etiologic agents have been associated with the development of bladder cancer, including the arylamines 4-aminobiphenyl and 2-naphthylamine [Vineis, 1994]. These agents are encountered in industrial settings but are also present in cigarettes, the leading risk factor associated with bladder cancer in North America [Vineis and Pirastu, 1997]. These carcinogens are concentrated in the urine and, therefore, in contact with bladder urothelium. Chronic infections and parasites, notably *Schistosoma hematobium*, are also responsible for a large number of bladder cancer cases in Egypt and other countries when bladder cancer is endemic [Christie et al., 1986]. Thus, the vast majority of bladder cancers are thought to be due to exogenous causes, and characteristically these occur in a delayed fashion after the exposure. For bladder cancer diagnosis and treatment, see review articles [Droller, 1998; Gschwend, 2000; Metts et al., 2000].

## 1.2. Pathology

The bladder is a pelvic organ that contains several distinct tissue layers. The most superficial layers consist of the urothelium, the lamina propria, and the interstitium. Multiple layers of crossing muscle fibers constitute the deeper muscularis propria that supports these superficial layers and has important implications with regard to cancer progression. Those tumors that invade into the muscle layers (T2, T3) are generally of higher grade and have a much higher risk of metastatic spread. Approximately 75% of these tumors will develop distant metastases over the course of five years if left untreated. [For review, see Messing and Catalona, 1998]. In contrast, tumors that remain superficial (Ta, T1) have a significantly lower risk of metastatic spread and progression. Thus, a basic biological distinction is made between tumors containing these two phenotypes.

Common histologies for bladder tumors include transitional cell carcinomas (TCC), squamous cell carcinomas, and adenocarcinomas. TCC is the most common form of bladder cancer and accounts for 90% of bladder cancers in North America. Roughly 20% of tumors are muscle-invasive at diagnosis. The majority (80%) of TCCs are superficial, low-grade lesions. Bladder tumors typically have a high propensity for recurrence with 50% to 75% forming subsequent tumors. This characteristic of bladder cancer is indicative of a "field" defect or generalized alteration that has resulted in the initiation and progression of multiple cells throughout the bladder. A subset of superficial tumors with an aggressive phenotype is termed carcinoma *in situ* (CIS). The five-year survival for low-grade TCC approaches 90%, while survival rates for myoinvasive tumors with regional spread (N+) or distant metastases (M+) survival rates are less than 50% and 6%, respectively, [Bane et al., 1996; Epstein et al., 1998].

## 1.3. Principles of the Method

Several in vitro bladder cancer culture systems, in addition to our own, have been developed in recent years [Leighton et al., 1977; Abaza et al., 1978; Bulbul et al., 1986; [Reese et al., 1976; Leighton and Tchao, 1984; Leighton et al., 1985; Lawson et al., 1986; Leighton, 1991; Crook et al., 2000]. These systems provide a unique approach for defining key genetic and epigenetic events associated with the development of bladder cancer in pure populations of tumor cells. In addition, these alterations may then be associated with the acquisition of selected cellular phenotypes, including the ability to form immortal cell lines, grow in soft agar, develop drug resistance, and become established tumors in immunodeficient mice.

The principle aspects of our method include obtaining fresh bladder cancer tissue and, under sterile conditions, transporting these tissues to the culture facility. We utilize an explant technique in which the cells are grown on a collagen substrate in a semi-defined media consisting of supplemented Ham's F12 with minimal fetal bovine serum (FBS). This system allows the growth and expansion of pure TCC cells without contaminating muscle or fibroblast cells. Once these cells begin to proliferate, they undergo a rapid expansion that provides multiple cultures for testing.

In our studies, we have demonstrated that superficial cancers will have a limited lifespan and spontaneously undergo senescence after 4 to 7 passages in culture [Yeager et al., 1998]. However, myoinvasive TCC samples will spontaneously immortalize and form cell lines. The ability to correlate clinical outcomes, specifically the risk of metastatic disease, with phenotypic aspects in culture for individual tumors is one of the unique advantages of this method. This system can also be used not only to detect certain genetic or biological changes in a tumor, but to correlate these with cellular phenotypes as well. These cells are also useful for predicting clinical responses to chemotherapy and for screening new anti-tumor agents in biological research. In summary, this system provides an excellent model to study the biological, biochemical, genetic, and molecular mechanisms of carcinogenesis, as well as address relevant questions regarding the prognosis of individual patients in the clinical setting. Examples include determining which papillary tumors are likely to recur or progress based on their phenotypic or genetic makeup or establishing the best treatment strategy, either a type of chemotherapy or radiation therapy, for a tumor with a specific genotype or phenotype.

## 2. PREPARATION OF MEDIA AND REAGENTS

### 2.1. Media and Salt Solutions

#### 2.1.1. Ham's F12 Media

(i) Dissolve 11.76 g of $NaHCO_3$ in three liters of ultra-pure water (UPW).

(ii) Add a 10-L package of F12 power to the $NaHCO_3$ solution.

(iii) Bring pH to 7.3 with HCl/NaOH.

(iv) Stir until all components are dissolved, and bring the total volume up to 9 L with UPW.

(v) Filter-sterilize through a 0.22-μm bottle top filter.

(vi) Store at 4°C for up to two months.

### 2.1.2. Supplemented F12+ Medium

(i) For ~1 L of sterile F12+ medium, add the following amounts of sterile stock solutions to 900 ml of F12 media:

| | |
|---|---|
| Regular insulin (100 units/ml) | 2.5 ml |
| Hydrocortisone (1,000 μg/ml) | 1 ml |
| Human transferrin (500 μg/ml) | 10 ml |
| Dextrose (0.27 g/ml) | 10 ml |
| Non-essential amino acids (10 mM) | 10 ml |
| Penicillin (10,000 units/ml)/streptomycin (10,000 μg/ml) | 10 ml |
| L-glutamine (200 mM) | 10 ml |
| Fetal bovine serum | 10 ml (for 1%) or 50 ml (for 5%) |

(ii) Store at 4°C.

### 2.1.3. Transferrin Stock Solution

(i) Dissolve 1 g transferrin in 2 L of phosphate buffered saline (PBS) solution with 0.3 mM $CaCl_2$.

(ii) Filter-sterilize and aliquot into 10 ml tubes.

(iii) Store at −20°C.

### 2.1.4. Dextrose Stock Solution

(i) Add 540 g dextrose to 1300 ml of heated to nearly boiling UPW.

(ii) Bring final volume to 2000 ml.

(iii) Cool solution to room temperature, and filter-sterilize through 0.22 μm bottle top filter.

(iv) Aliquot to 10 ml and store at −20°C.

### 2.1.5. Collagenase Solution

(i) Dissolve 400 mg collagenase into 180 ml of Hanks' solution with $Mg^{2+}$ and $Ca^{2+}$ to make a final concentration of 400–600 U/ml.

(ii) Add 20 ml bovine album fraction V.

(iii) Filter-sterilize the solution, and dispense in 7.5 ml aliquots

(iv) Check the pH, which should be around 7.3.

(v) Store aliquots in −20°C.

### 2.2. Collagen-Coated Plates

We find that the growth of tumor cells is superior on plastic culture dishes coated with a thick (3–4 mm) collagen gel. Collagen is available commercially; however it may also be made with minimal difficulty in large quantities in the laboratory. We utilize a protocol for making Type I collagen in which the collagen is harvested from rat-tail tendons and reconstituted with ammonia to form a gel.

### 2.2.1. Extraction of Collagen from Rat Tails

This protocol yields one liter of collagen solution that is sufficient for coating two hundred 10 cm culture dishes. The collagen solution is stored at 4°C and is good for one year.

### Protocol 5.1. Extraction of Collagen from Rat Tails

***Materials and Reagents***

*Sterile*
- Scalpels and hemostat
- Four 250 ml centrifuge tubes
- Two 500 ml beakers
- PBSA: Dulbecco's phosphate-buffered saline solution A (i.e. lack $Ca^{2+}$ and $Mg^{2+}$)
- One liter 0.1% acetic acid

*Non-sterile*
- Rat tails, stored at −20°C
- Two liters 70% ethanol

***Protocol***

(a) Thaw 16–20 rat tails per liter of collagen by socking them in a 2:1 solution of UPW:EtOH.
(b) Make a longitudinal cut along the entire length of the tail with the scalpel.
(c) Peel the skin off the rat tails.
(d) Grasp the end of the tail with the hemostat, and break the joints about 1 cm away from the end and then pull out the tendons.
(e) Cut the exposed tendons free, and place them in a sterile beaker of PBSA.
(f) Repeat steps (d) and (e) as many times as necessary to remove all the tendons from the tail. The yield is about 0.6 g of wet tendons per tail.
(g) Wash the tendons in the beaker by repeated rinse with sterile PBSA until the PBS is clear.
(h) Place the washed tendons in a sterile beaker. At this point, proceed by using only sterile conditions and techniques.
(i) Determine the wet weight of the tendons by weighing them in the sterile beaker. Calculate and record the yield.

(j)  Dissolve 10 g wet tendons in 1 liter of 0.1% acetic acid.

(k)  Stir the above solution for 48 h on a magnetic stir plate at room temperature by using medium speed; then leave the solution for five days at 4°C without stirring.

(l)  Transfer the collagen solution to four sterile 250 ml centrifuge bottles.

(m)  Centrifuge at 12,000 g for 2 h at 4°C to bring down the fibers.

(n)  Pour the supernatant that contains the collagen solution into a sterile storage bottle, and store at 4°C.

(o)  Determine the protein concentration by the standard Lowry assay.

(p)  If necessary, dilute stock collagen solution with sterile 0.1% acetic acid to result in a protein concentration of 2 mg/ml.

### 2.2.2.  Ammonia-Reconstituted Collagen Gel Substrate

**Protocol 5.2.  Ammonia-Reconstituted Collagen Gel Substrate**

*Materials and Reagents*
*Sterile*
❑  Collagen stock solution, 2 mg protein/ml
❑  $NH_4OH$, 28%
❑  Ham's F12 medium with penicillin, 100 U/ml and streptomycin, 100 µg/ml (P/S) only
❑  Supplemented Ham's F12+ medium with 1% FBS
❑  Tissue culture dishes, 10 cm or 6 cm
❑  Plastic Petri dishes, 15 cm
❑  Cotton swabs

*Protocol*
(a)  Pipette 6 ml of stock collagen solution into 10 cm plastic tissue culture dishes.

(b)  Rotate plate to spread the solution evenly.

(c)  Place tissue culture dishes with collagen solution into 15 cm drying dishes (note: leave top off of culture dish when drying).

(d)  Dip three cotton swabs into ammonium hydroxide and place on opposite sides of the drying dishes.

(e)  Cover the drying dishes and leave at room temperature in the culture hood or chemical hood overnight.
    *Note:* Generally, one night is long enough to solidify the collagen gel, although increased humidity can make the drying time longer. Over-drying the collagen makes it difficult to rehydrate.

(f)  Add 10 ml of penicillin/streptomycin-supplemented medium to each solidified collagen dish, and incubate in humidified $CO_2$ incubator at 37°C overnight.

(g)  Change with P/S-supplemented medium two more times for the next two days.

(h)  Change one last time with appropriate, serum supplemented F12 medium and the plates will then be ready to use the next day.

(i)  Change the media once every week with the fully supplemented media that will be used for the cells.

## 3.  SAFETY PRECAUTIONS

Handling human samples poses a risk for HIV, hepatitis, and other blood-borne pathogens. Eye protection, lab coats, and gloves are necessary whenever working with these cells. Use care when using blades and needles for tissue isolation. If any cuts or needle sticks occur, contact the biosafety department and/or an emergency facility immediately.

## 4.  CULTIVATION

### 4.1.  Acquisition of Tissue

TCC samples are procured from the operating room utilizing one of several different approaches. Typically, a transurethral bladder tumor resection is performed by using a cystoscope and a wire loop containing electrocautery. Large papillary tumors may often be debulked by using this wire loop in the absence of electricity ("cold looping"). Samples may also be obtained by using cup biopsy forceps to prevent thermal injury to the tumor tissue. Avoiding the use of electrocautery during sample procurement improves the viability of the specimen in culture. The tumor base and sufficient sample must remain to allow for complete pathologic staging of the patient; thus smaller lesions (less than 1 cm tumor) are not recommended for analysis unless there are multiple samples. It has been our practice to also flash-freeze in liquid nitrogen a small piece of tissue isolated for culture in order to provide reference tissue for any subsequent molecular studies, as well as providing additional diagnostic tissue should the pathologist require it. Control tissue for molecular studies may be obtained from peripheral blood by harvesting normal blood lymphocytes.

An additional source of tissue is obtained at the time of cystectomy or during removal of the whole bladder for bladder cancer. This is generally performed for large tumors that are muscle-invasive and provides an excellent source for tissue. After the specimen is removed, the bladder is opened up under sterile conditions, typically on the operative field, and tumor tissue is identified and resected for culture. The harvested tumor tissue is immediately placed into cold, sterile supplemented-F12+ media (see Section 2.1.2) and transported to the basic laboratory facility for culture. Rapidly processing the tissue (<2 h) is important for tissue viability and subsequently establishing the culture. Bladder specimens also provide the opportunity to culture non-tumor urothelial tissue and these urothelial cells can serve as a control.

*Note:* All tissue acquisition from patients should be handled through internal review board approved protocols and strict patient confidentiality is required.

## 4.2. Primary Culture

### Protocol 5.3.  Primary Culture of Bladder Tumors

*Materials and Reagents*

*Sterile*
- ❏ Petri dishes, 10 cm
- ❏ Forceps
- ❏ Scissors
- ❏ Surgical knives and blades, #9 or #10
- ❏ Supplemented-F12+ medium (see Section 2.1.2) with 1% FBS
- ❏ Supplemented-F12+ medium with 5% FBS
- ❏ Collagen-coated culture dishes(see Protocol 5.2)

*Protocol*
(a)  Mince tissue to about 1-mm³ pieces in a sterile dish inside the hood.
(b)  Resuspend pieces in supplemented-F12+ medium with 5% FBS.
(c)  Transfer tissue onto collagen-coated plates, and spread evenly. Roughly 10–30 explants are placed on each 10 cm dish.
(d)  Adjust the amount of medium so that it just covers the explants (3–4 ml for 10 cm dish). This will prevent the explants from floating and will allow them to attach to the collagen gel in the shortest time.
(e)  Place dishes in 37°C, 5% $CO_2$ humidified incubator.
(f)  After 12–24 h, remove medium carefully and replace with fresh supplemented 1% FBS F12+ media. Outgrowth from the explants can be detected as early as 24 h after the initial plating, depending on the viability of the tumor cells.

(g) Change the media every two days until the culture reaches late log phase ($2-3 \times 10^6$ cells/10 cm dish) or the plate is 70–80% confluent. Generally this process takes about one week after the initiation of the culture.

(h) Disperse the culture by using trypsinization for passage or cryo-preservation.

### 4.3.  Subculture of TCC Cultures

### Protocol 5.4.  Subculture of Primary Cultures from Bladder Tumors

***Materials and Reagents***
*Sterile*
- ❏ Trypsin stock: 0.5% trypsin in $Ca^{2+}$- and $Mg^{2+}$-free Hanks' balanced salt solution (HBSS) with 5.3 mM EDTA·4Na, $10\times$ stock
- ❏ Trypsin/EDTA: 0.05% trypsin in $Ca^{2+}$- and $Mg^{2+}$-free HBSS with 0.53 mM EDTA
- ❏ Collagen-coated culture dishes, 6 cm
- ❏ Supplemented-F12+ with 5% FBS
- ❏ 15 ml conical centrifuge tubes
- ❏ Rotating platform incubator at 37°C

***Protocol***
(a) Remove the growth medium from the culture dishes.

(b) Rinse the plate (5 ml/10 cm dish) with Trypsin/EDTA for 2–3 min.

(c) Add same amount of fresh Trypsin/EDTA to disperse cells.

(d) Incubate at 37°C on a rotary platform for 10–20 min, and peri-odically examine the cells under a microscope. When the cells begin to round up and a small percentage (5%) is floating, they are ready to be removed.
*Note:* Leaving cells in Trypsin-EDTA solution for long periods results in cell damage and reduced viability.

(e) Loosen cells from the collagen substrate with a pipette by gently lavaging the plate with medium. The cells and medium may then be transferred to a centrifuge tube. Rinse the dish with an equal volume of growth medium to collect any residual cells and add this to the centrifuge tube as well. Check under the microscope to make certain all cells have been detached.

(f) Add a couple of drops of FBS to the Trypsin/EDTA-cell suspension to a final concentration of approximately 5%, and gently mix. This step helps to inactivate the trypsin.

(g) Spin down the cells at 250 *g* for 5 min.

(h) Discard the supernate, and resuspend the cell pellet in a small volume of supplemented F12+ media with 5% FBS by using gentle pipetting to break up any clumped cells.

(i) Take an aliquot for cell counting using 0.1% Trypan Blue to assess viable cells.

(j) Add desired volume of F12+ with 5% FBS, and dispense cells (about $5 \times 10^5$ cells in each 6 cm plate) for secondary culture. Generally, a 1:2 or 1:3 split is utilized.

(k) Change the medium with 1% FBS supplemented-F12+ the next day. The majority of cells should be adhering to the collagen at this point.

### 4.4. Cryopreservation and Thawing Cells

Because of the limited lifespan of the TCC culture, multiple cultures should be frozen down for future use to avoid losing the cell line. To do this, 10% dimethylsulfoxide (DMSO) or glycerol and 10% FBS are added to the growth medium. About $1 \times 10^6$ cells are placed in each freezing vial. The best results are obtained when cells are slowly cooled (2 h freezing time) to the final freezing temperature of liquid nitrogen or an automated cell freezer is used. To regenerate cell lines, cells are thawed rapidly in a 37°C water bath. The cells are transferred into 4 ml of 5% FBS supplemented-F12+ and spun for 5 min at 250 $g$. The supernate is discarded, and cells are resuspended in 3 ml of medium and seeded into 6 cm dishes.

*Safety note:* Wear transparent facemask, cryoprotective gloves, and lab coat when handling liquid nitrogen. Use a covered container when thawing vials because they may explode when they are warmed if they have absorbed any liquid nitrogen. Alternatively, store vials in the vapor phase to avoid possible explosion on thawing.

## 5. CRITERIA FOR ESTABLISHING TRANSFORMED STATUS

The lifespan of TCC in vitro cultured under the conditions described above varies from 2 to 7 passages using a 1:2 or 1:3 split. At this point, cells enter a phase in which they generally develop a senescence-associated morphology and cease proliferating. This senescence-associated morphology is identifiable by its characteristic light microscopy changes, which consist of an enlarged cytoplasm and nucleus ("fried egg" appearance; Fig. 5.1). In addition, they demonstrate senescence-associated $\beta$-galactosidase staining [Dimri et al., 1995].

However, a subset of cells, typically myoinvasive TCCs, may enter a crisis period in which they maintain the morphology of proliferating

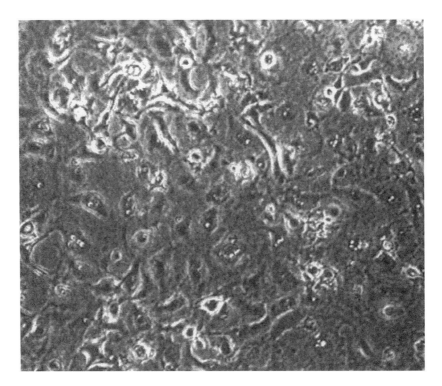

**Figure 5.1.** TCC cells, in this case from a superficial bladder cancer, demonstrate a typical senescent morphology after prolonged passaging in culture. This consists of an enlarged, rounded cytoplasm and nucleus. These cultures positively stain for SA-$\beta$ galactosidase. Note the absence of fibroblasts (100×).

cells but do not noticeably expand their numbers. An effort is generally made to passage the cells to smaller dishes (e.g., 6 cm) in order to maintain sufficient cell density to support growth. Often during this period, the cell numbers may remain stable for many weeks. Eventually, a clonal outgrowth of cells occurs and can be propagated indefinitely. An alternate pattern seen in cultures from some myoinvasive tumors is continued proliferation with no crisis period, which suggests the presence of a large number of immortal cells in vivo.

The criterion, generally used for immortalization, is the ability to be cultured continuously for over one year without any loss in growth rate. In our system we have noted that culture beyond 10 to 15 passages is typically associated with the development of an immortal phenotype. In normal human urothelial cell cultures, cells are never passaged beyond 3 to 5 times representing 30–50 population doublings [Reznikoff et al., 1994]. In addition, in more than 30 samples to date, superficial bladder tumors have never immortalized spontaneously, whereas myoinvasive tumors routinely do so [Yeager et al., 1998].

Chapter 5. Fu et al

# 6. TROUBLESHOOTING

## 6.1. Bacterial and Fungal Contamination

Extreme caution should be observed with regard to sterile technique during the entire procedure. Human tissue may be contaminated with bacteria or yeast at a number of stages before arriving at the basic science laboratory. Generally, patients are given antibiotics before surgery, which helps with preharvest sterility. If gross contamination occurs within one to two days after the sample procurement, this is often due to contamination at some point during the procurement. Gross bacterial contamination is identifiable by a cloudy culture dish, but minimal involvement requires microscopy at high power. In our experience extensive bacterial infections are difficult to treat and are easily propagated to other plates. Therefore, grossly contaminated plates should be discarded immediately. Using media supplemented with antibiotics generally will eradicate any low level of bacterial contamination. Attention should be paid to the sterility of the media and collagen. We routinely develop test plates containing new media (without antibiotics) and collagen for 7 days before using the batch with cultured tissue samples. Routine sterilization and care of the incubator and water bath are also performed. If fungal contamination occurs, amphotericin B is often effective in eliminating it without killing epithelial cells. Add 1 μg/ml of amphotericin B to the media for three days, with a fresh change of media daily. If possible, isolate the culture to reduce the risk of contaminating healthy cells. Note that amphotericin B is typically toxic and will slow the growth of the involved cultures. If fungal growth resumes, it is considered drug-resistant and the culture should be discarded.

## 6.2. Fibroblast Contamination/Poor Growth

The use of collagen and the low serum (1%) media was originally optimized for urothelial cell growth [Reznikoff et al., 1987]. Both the medium and collagen select against the growth of fibroblasts and other contaminating cells. Care must be taken to change the medium promptly after plating to the lower 1% FBS, because higher levels of FBS may encourage fibroblast growth. Normal urothelial cells may also grow in this setting, which makes the harvest and plating of pure tumor tissue critical. Because most tumors grow as papillary or as flat "sessile" lesions that are easily isolated from normal urothelium, this is typically done without too much difficulty. TCCs often exhibit a significantly altered morphology by forming multilayered structures in vitro reminiscent of papillary growths in vivo [Yeager et al., 1998]. Rapid plating of the tumor after procurement may also minimize fibroblast growth, as we have observed that these cells are more resistant to anoxia than urothelial and TCC cells are.

Successful culture is also dependent on collagen quality. Generally, thinner and harder collagen gels are less amenable to cell attachment and growth. The collagen plates must be rinsed fully to discard the acetic acid and any undefined components, which might decrease cell growth. Also, the plates must be fully equilibrated in the culture medium to provide optimal conditions for cell attachment and growth. Instead of providing essential growth requirement, the collagen gel functions in a physical manner providing a natural substrate for cell attachment and spreading.

A final element for success is careful passaging, with particular attention paid to explant density. Initial cultures must include a relatively high density of explants in order to obtain robust outgrowth. Typically, we recommend 30–60 explants per 10 cm dish. Do not create more dishes by using fewer explants per dish. Low density dishes will take longer to grow and will develop heterogeneous areas of viability and differentiation. The same principles apply to serial passages. Passaging confluent plates at 1:2 or 1:3 will permit the cell–cell interactions necessary for growth stimulation. We also leave as many tumors explants in the culture as possible when passage because they will continue to be a source of new cell growth.

### 6.3. Differentiation

Differentiation is a common phenomenon in non-immortalized cultured cells and is defined by alterations in morphology and an irreversible cell cycle arrest. Senescence can be distinguished from differentiated cells by their phenotype. Differentiated cells become isolated in tight clumps, have a keratinizing appearance, and lose their individual morphology. We find that this often occurs by culturing the cells at low densities. In addition, differentiated cells may result if the proliferating cell culture becomes confluent and is not replated. Density between 50% and 90% should be maintained to prevent these phenomena.

## 7. APPLICATIONS OF THE METHOD

One of the powerful aspects of this model includes the ability to assess biological characteristics of individual tumors in vitro. Other molecular and cytogenetic studies may also be performed and are correlated with these features. The ability to amplify large numbers of cells from a primary tumor for testing allows a number of features to be examined from a relatively small amount of starting material. The in vitro system complements and has advantages over studies performed on in vivo samples in a number of ways. First, the cultured material contains few normal contaminating cells. Second, a more malignant, aggressive phenotype may be selected in culture that would in time be selected in vivo. Third, numerous analyses, including immunoprecipitation and karyotyping, are more easily performed on cultured cells than on in vivo tissues.

Cytogenetic and molecular analyses may be performed on TCC samples in culture. Metaphase cells are generated, which enable karyotyping, fluorescent *in situ* hybridization (FISH), and spectral karyotyping (SKY) analyses to be performed [Bubendorf et al., 2001]. Changes in ploidy, modal chromosome number, and clonal gains and losses of chromosomal material can be detected. Finer mapping of genetic alterations can be performed by using comparative genomic hybridization (CGH) and specific alterations analyzed by using Southern analysis, restriction fragment length polymorphisms (RFLP), and polymerase chain reaction (PCR) technologies. These alterations may be compared directly to the expression of RNA and protein in culture. Immunoprecipitation and Western blotting are straightforward in these cultured cells in contrast to tumor samples. These analyses can frequently be done with cells from one proliferating 6- or 10-cm dish.

Biological endpoints in vitro may be assessed in order to judge tumor aggressiveness in vivo. Morphologic parameters have been assessed for low- and high-grade TCCs [Yeager et al., 1998]. Low-grade TCCs often demonstrate morphology similar to normal human urothelial cells; however, they can form multilayered structures reminiscent of their in vivo morphology (Fig. 5.2). Higher grade samples typically exhibit significantly altered morphology forming loosely adherent monolayers with cellular pleomorphism. Another useful endpoint is the acquisition of anchorage-independent growth [Reznikoff et al., 1983]. Typically $1 \times 10^6$ viable cells are used in this assay representing one confluent 10-cm TCC culture. Tumorigenicity in mice is another important parameter that may be tested with cultured cells. Normally $5 \times 10^6$ viable cells are injected subcutaneously into athymic nude mice at two flank or abdominal sites per mouse. To perform this assay in duplicate, two to four 10 cm dishes are required. One potential advantage of this technique is that culture may select for a subpopulation of aggressive cells that will lead to a higher rate of "take" in mouse xenografts.

The ability to demonstrate extended lifespan beyond 30–50 population doublings is another phenotype with clinical importance. We

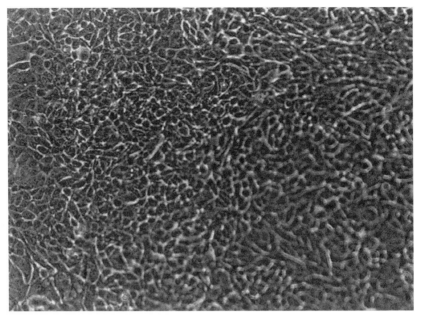

**A**

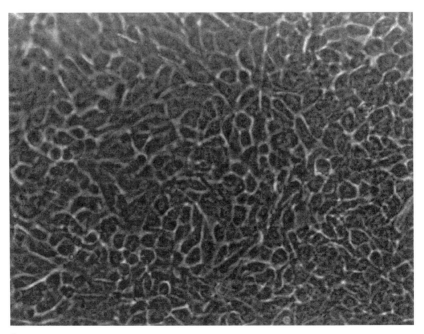

**B**

**Figure 5.2.** Comparative morphology of cultured normal human urothelial cells (HUC) and TCC bladder cancer cells. HUC cultures demonstrate an epithelial morphology with tight cell–cell adhesion, 100× (A) and 200× (B). High-grade TCCs display a unique morphology with cellular pleomorphism and prominent nucleoli. There is less cellular adhesion, 100× (C) and 200× (D).

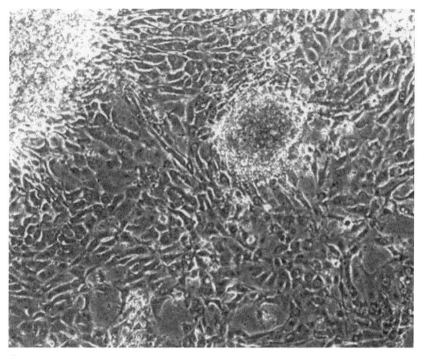

C

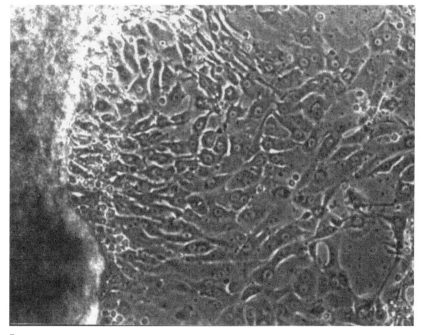

D

**Figure 5.2.** *(Continued)*

have demonstrated that only those bladder cancers that demonstrate myoinvasion—an aggressive clinical phenotype—will spontaneously immortalize in our culture system [Yeager et al., 1998]. In contrast, superficial tumors retain the ability to senescence. Immortal TCC phenotypes have been correlated with alterations in the p16-pRb cell cycle block and elevated rates of +20q and +8q gain by CGH analysis. We have used this model further to demonstrate other associations between immortalization and alterations in the p53 pathways for myoinvasive cancers [Sarkar et al., 2000]. Thus, the phenotype of bypassing senescence is important in bladder cancer; it is one of the applications of this model that may be studied in depth.

We have utilized our cell culture model to assess the response to radiotherapy in tumors with different genotypes, specifically those with *p53* alterations [Sarkar et al., 2000]. The in vitro toxicity of novel drug therapies may be also tested against tumors with different phenotypes and genotypes, thus yielding valuable information to be used in the clinical setting.

## 8.  ALTERNATE METHODOLOGIES

An alternative method for culturing bladder tumor samples is collagenase digestion before plating. Collagenase digestion is not always necessary because bladder tumor tissue typically lacks an acinar structure. However, in high-grade tumors invading deeply into the muscle, collagenase digestion may be of use. Other media and techniques have been described for establishing and propagating bladder cancer cells.

### 8.1. Collagenase Digestion

### Protocol 5.5.   Collagenase Digestion of Bladder Tumors

***Reagents and Materials***
*Sterile*
❑  Collagenase solution (see Section 2.1.5)
❑  Supplemented F12+ with 1% and 5% FBS (see Section 2.1.2)
❑  Collagen-coated dishes
*Non-sterile*
❑  Rocking platform

***Protocol***
(a)  Incubate minced bladder tumor tissue in a collagenase solution at 37°C on a rocking platform for 30 min–2 h.

(b) Spin down the digested tissue at 4,500 g for 5 min, and eliminate the supernatant by aspiration.

(c) Add 10 ml of supplemented F12+ medium with 5% FBS to the tube, and mix by inverting the tube for several times.

(d) Incubate at 37°C with shaking for 15 min.

(e) Spin at 4,500 g for 5 min and discard supernatant.

(f) Repeat steps (c) through (e) for a total of three times.
*Note:* Thorough rinsing with culture media is required after collagenase digestion because the left over collagenase inside the tissue will digest the collagen substrate to which the cells need to attach.

(g) Resuspend tissue pellet in an appropriate amount of F12+ medium with 5% FBS, and plate on collagen-coated culture dishes.

(h) Change medium to F12+ with 1% FBS the next day.

## 8.2. Single-Cell Suspension Culture and Other Culture Techniques

In addition to the explant culture technique, a single-cell suspension culture can be used as well. The single-cell suspension culture requires a thorough digestion and excludes the use of tissue explants as cell source [Lindert and Terris, 2000; Reznikoff et al., 1983]. Other culture systems, such as organ culture, gradient culture, and matrix culture, have been used with bladder cancers to test cytotoxic agents at clinically relevant concentrations [Leighton et al., 1977; Abaza et al., 1978; Bulbul et al., 1986; Reese et al., 1976; Lawson et al., 1986; Crook et al., 2000].

## 8.3. Other Requirements

Other media formulations have been used for culturing bladder cancer cells. For example, RPMI 1640, modified Eagle's medium (MEM), McCoy's 5A, and Waymouth's MB752/1 media can all be used effectively [Dangles et al., 1997]. The concentrations of FBS can vary from 5% to 20%. Most laboratories use 5% $CO_2$ for their incubators with some as high as 7.5–10% depending on the bicarbonate concentration. Dual gas, nitrogen and $CO_2$ incubators can be used to reduce the oxygen content to physiologic levels (4% $O_2$) or to hypoxic levels (<4%) to mimic the oxygen content within tumors. Besides tissue culture dishes, using a flask may eliminate contamination by taking advantage of a relatively closed environment. However, it is challenging to coat flasks with collagen [Kanno et al., 1998]. Bovine pituitary extract (BPE) and cholera toxin are two supplements that can be added to the F12 medium for culturing cells. BPE is mitogenic and stimulates cell growth especially in low-density culture. Cholera

**Table 5.1.** Continuous human bladder cancer cell lines.

| Cell line | Grade | Stage | Differentiation | Reference |
|---|---|---|---|---|
| 5637 | | | Carcinoma | (Fogh, 1978; Hader, Stach-Machado et al., 1987; Yeager and Reznikoff, 1998) |
| 253J | G4 | T4 | Moderately differentiated TCC | (Elliott, Cleveland et al., 1974; Elliott, Bronson et al., 1976) |
| 647V | G2 | | TCC | (Elliott, Bronson et al., 1976; Elliott, Bronson et al., 1977; Kunugi, Vazquez-Padua et al., 1990; Dangles, Femenia et al., 1997) |
| COLO232 | G3 | T2b | Poorly Differentiated TCC | (Moore, Morgan et al., 1978) |
| EJ (MGH-U1) | G3 | T2 | Anaplastic TCC | (O'Toole, Povey et al., 1983). (Kato, 1976; Kato, Irwin, Jr. et al., 1977; Evans, Irwin et al., 1977; Marshall, Franks et al., 1977; Kato, Ishikawa et al., 1978; Heaney, Ornellas et al., 1978; Knuchel, Hofstadter et al., 1989; Wang, Zhang et al., 2000) |
| EJ28 | | | Anaplastic, poorly differentiated | |
| EJ138 | | | | (Datta, Allman et al., 1997) |
| HOK-1 | G3 | T3 | Squamous & adeno-carcinoma | (Offner, Ott et al., 1991) |
| HT1197 | G4 | T2 | Mixture differentiated & anaplastic TCC with adenocarcinoma change | (Rasheed, Gardner et al., 1977; Masters, Hepburn et al., 1986) |
| HT1376 | G3 | T2 | Mostly differentiated TCC | (Rasheed, Gardner et al., 1977; Masters, Hepburn et al., 1986) |
| HU456 | G2 | T2 | Differentiated TCC | (Masters, Hepburn et al., 1986) |
| HU961T | G2 | T1 | Differentiated TCC | (Masters, Hepburn et al., 1986) |
| J82 | G3 | T3 | Poorly differentiated TCC | (O'Toole, Price et al., 1978; O'Toole, Povey et al., 1983) |
| J82COT | G3 | T3 | | (O'Toole, Povey et al., 1983) |
| JMSU1 | G3 | | Poorly differentiated | (Morita, Shinohara et al., 1995) |
| JTC-32 | G3 | | | (Kakuya, Yamada et al., 1983; Kugoh, Fujiwara et al., 2000) |
| KK-47 | G1 | | | (Hisazumi, Kanokogi et al., 1979; Nakajima and Hisazumi, 1987) |
| KU-1 | G3 | | | (Yajima, 1970; Kanno, Nonomura et al., 1998) |
| MGH-U2 | G3 | | Differentiated TCC | (Plotkin, Wides et al., 1979; Chi Wei Lin, Julia C.Lin et al., 1985; Masters, Hepburn et al., 1986) |
| NBT-2 | G3/4 | T3 | | (Yamamoto, 1979) |
| PS1 | G3 | T4 | Poorly Differentiated TCC | (Sanford, Geder et al., 1978) |
| RT112 | G1-2 | N/A | Moderately/Well differentiated TCC | (O'Toole, Povey et al., 1983; Walker, Povey et al., 1990) |
| RT4 | G1 | T2 | Well differentiated TCC | (O'Toole, Povey et al., 1983) |
| SCaBER | | T3 | Squamous cell carcinoma | (O'Toole, Nayak et al., 1976; O'Toole, Povey et al., 1983) |

**Table 5.1.** *(Continued)*

| Cell line | Grade | Stage | Differentiation | Reference |
|---|---|---|---|---|
| SW780 | | | TCC | (Fogh, 1978; Williams, 1980) |
| T24 | G3 | | Poorly Differentiated TCC | (Bubenik et al., 1973; Malkovsky and Bubenik, 1977; Kato et al., 1977; Kato, Ishikawa et al., 1978; Heaney, Ornellas et al., 1978; Kunugi, Vazquez-Padua et al., 1990; Dangles et al., 1997; Perabo et al., 2001) |
| TBC-1 | G3 | | | (Chang et al., 1995) |
| TCC-10 | | | | (Torino et al., 2001) |
| TCCSUP | G4 | T4 | Differentiated TCC | (Nayak et al., 1977; O'Toole et al., 1983; Perabo et al., 2001) |
| TPC-1 | G2 | | | (Chang et al., 1995) |
| UCRU-BL17 | G3 | T4b | TCC, SCC & adeno-carcinoma | (Brown et al., 1990) |
| UCRU-BL-17CL | G3 | T4b | TCC, SCC & adeno-carcinoma | (Russell et al., 1988) |
| UCRU-BL-28 | G3/G2 | T4 | | (Russell et al., 1993) |
| UMK-1 | G2 | | SCC & adenocarcinoma | (Muraoka et al., 1995) |
| UM-UC-1 | G2 | | | (Grossman et al., 1984) |
| UM-UC-2 | | | | (Grossman et al., 1984) |
| UM-UC-3 | | | TCC | |
| VM-CUB-1 | | | TCC | (Fogh, 1978) |
| VM-CUB-2 | | T4 | TCC | (Fogh, 1978) |
| VM-CUB-3 | G3 | | | |
| | | | Differentiated TCC | (Fogh, 1978; Masters et al., 1986) |
| TCC 92-1 | G3 | T4 | TCC | |
| TCC 96-2 | G2/3 | T3 | | (Sarkar et al., 2000) |
| TCC 97-6 | G3 | T3 | | |
| TCC 97-24 | G3 | T3 | | |
| TCC 97-7 | G2/3 | T1 | | |
| TCC 96-1 | G2/3 | T2 | | |
| TCC 97-18-I | G3 | T2 | | |
| TCC 97-21-M | G3 | T4 | | |
| TCC 94-10 | G2/3 | T3 | | |
| TCC 97-1 | G1/2 | T2 | | |
| TCC 97-15 | G3 | T1 | | |
| TCC 97-29 | G1/2 | T1 | | |

toxin may promote differentiation and has been useful in other culture systems.

## 9. LIST OF AVAILABLE CONTINUOUS CELL LINES

Cell lines derived from human carcinomas have been used for many years to study different aspects of tumor cell behavior. Numerous cell lines have been established and characterized from human TCC representing a broad range of grades and stages [O'Toole et al., 1983;

Masters et al., 1986; Masters et al., 1988]. Some of the widely used continuous bladder cancer cell lines are 5637, SCaBER, UM-UC-3, SW780, J82, T24, TCCSUP, and RT4. These cell lines are all available commercially from the American Type Culture Collection. Other cell lines that have been established by different research groups, including 12 TCC cell lines that were generated in our laboratory, are listed below. Our strains were originally generated from biopsies of muscle-invasive primary bladder tumors, which then spontaneously bypassed senescence and became immortalized cell lines (Table 5.1).

## ACKNOWLEDGMENTS

The work was supported by the University of Wisconsin Howard Hughes Medical Research Resources Program and National Institutes of Health CA76184-01.

## SOURCES OF MATERIALS

| Material | Supplier |
| --- | --- |
| Collagenase | Invitrogen |
| Hanks' solution | Cellgro |
| Bovine album fraction V | Invitrogen |
| Regular insulin | Eli Lilly & Company |
| Hydrocortisone | Pharmacia & Upiohn Co. |
| Human transferrin | Sigma |
| Dextrose | Fisher |
| Non-essential amino acids | BioWhittaker, Inc. |
| Penicillin/streptomycin | Pfizer |
| L-glutamine | Cellgro |
| Fetal bovine serum | Sigma |
| Bovine pituitary extract | Invitrogen |
| Ammonium hydroxide | Fisher |
| NaHCO$_3$ | Sigma |
| Ham's F12 power | Cellgro |
| 0.22 μm bottle top filter | Corning Inc. |
| Trypsin-EDTA | Invitrogen |
| Mg$^{2+}$ and Ca$^{2+}$-free Hanks balanced salt solution (HBSS) | Cellgro |
| DMSO | Sigma |
| Glycerol | Sigma |
| Rat Tail | Biotrol |

# REFERENCES

Abaza, N.A., Leighton, J., Zajac, B.A. (1978) Clinical bladder cancer in sponge matrix tissue culture: procedures for collection, cultivation, and assessment of viability. *Cancer* 42:1364–1374.

Bane, B.L., Rao, J.Y., Hemstreet, G.P. (1996) Pathology and staging of bladder cancer. *Seminars Oncol.* 23:546–570.

Brown, J.L., Russell, P.J., Philips, J., Wotherspoon, J., Raghavan, D., (1990) Clonal analysis of a bladder cancer cell line: an experimental model of tumour heterogeneity. *Br. J. Cancer* 61:369–376.

Bubendorf, L., Grilli, B., Sauter, G., Mihatsch, M.J., Gasser, T.C., Dalquen, P., (2001) Multiprobe FISH for enhanced detection of bladder cancer in voided urine specimens and bladder washings. *Am. J. Clin. Pathol.* 116:7986.

Bubenik, J., Baresova, M., Viklicky, V., Jakoubkova, J., Sainerova, H., Donner, J., (1973) Established cell line of urinary bladder carcinoma (T24) containing tumour-specific antigen. *Int. J. Cancer* 11:765–773.

Bulbul, M.A., Pavelic, K., Slocum, H.K., Frankfurt, O.S., Rustum, Y.M., Huben, R.P., Bernacki, R.J., (1986) Growth of human urologic tumors on extracellular matrix. *J. Urology*, 136:512–516.

Chang, J., Sui, Z., Ma, T., Ma, K., Zhang, X., Wang, J., Dong, K., Yao, Q., (1995) Establishment and characterization of two cell lines derived from human transitional cell carcinoma. *Chin. Med. J.* 108:522–527.

Chi, W.L., Lin, J.C., Prout, G.R.J. (1985) Establishment and characterization of four human bladder tumor cell lines and sublines with different degrees of malignancy. *Cancer Res.* 45:5070–5079.

Christie, J.D., Crouse, D., Kelada, A.S., Anis-Ishak, E., Smith, J.H., Kamel, I.A., (1986) Patterns of Schistosoma haematobium egg distribution in the human lower urinary tract. III. Cancerous lower urinary tracts. *Am. J. Tropical Med. & Hyg.* 35:759–764.

Crook, T.J., Hall, I.S., Solomon, L.Z., Birch, B.R., Cooper, A.J. (2000) A model of superficial bladder cancer using fluorescent tumour cells in an organ-culture system. *BJU International* 86:886–893.

Dangles, V., Femenia, F., Laine, V., Berthelemy, M., Le Rhun, D., Poupon, M.F., Levy, D., Schwartz-Cornil, I., (1997) Two- and three-dimensional cell structures govern epidermal growth factor survival function in human bladder carcinoma cell lines. *Cancer Res.* 57:3360–3364.

Datta, S.N., Allman, R., Loh, C.S., Mason, M., Matthews, P.N. (1997) Photodynamic therapy of bladder cancer cell lines. *Br. J. Urol.* 80:421–426.

Dimri, G.P., Lee, X., Basile, G., Acosta, M., Scott, G., Roskelley, C., Medrano, E.E., Linskens, M., Rubelj, I., Pereira-Smith, O., (1995) A biomarker that identifies senescent human cells in culture and in aging skin in vivo. *Proc. Natl. Acad. Sci. USA* 92:9363–9367.

Droller, M.J., (1998) Bladder cancer: state-of-the-art care. [see comments]. [Review] [73 refs]. *CA Cancer J. Clin.*, 48:269–284.

Elliott, A.Y., Bronson, D.L., Cervenka, J., Stein, N., Fraley, E.E. (1977) Properties of cell lines established from transitional cell cancers of the human urinary tract. *Cancer Res.* 37:1279–1289.

Elliott, A.Y., Bronson, D.L., Stein, N., Fraley, E.E. (1976) In vitro cultivation of epithelial cells derived from tumors of the human urinary tract. *Cancer Res.* 36:365–369.

Elliott, A.Y., Cleveland, P., Cervenka, J., Castro, A.E., Stein, N., Hakala, T.R., Fraley, E.E., (1974) Characterization of a cell line from human transitional cell cancer of the urinary tract. *J. Natl. Cancer Inst.* 53:1341–1349.

Epstein, J.I., Amin, M.B., Reuter, V.R., Mostofi, F.K. (1998) The World Health Organization/International Society of Urological Pathology consensus classification of urothelial (transitional cell) neoplasms of the urinary bladder. Bladder Consensus Conference Committee. [see comments]. [Review] [44 refs]. *Am. J. Surg. Pathol.* 22:1435–1448.

Evans, D.R., Irwin, R.J., Havre, P.A., Bouchard, J.G., Kato, T., Prout, G.R., Jr., (1977) The activity of the pyrimidine biosynthetic pathway in MGH-U1 transitional carcinoma cells grown in tissue culture. *J. Urol.* 117:712–719.

Fogh, J. (1978) Cultivation, characterization, and identification of human tumor cells with emphasis on kidney, testis, and bladder tumors. *Natl. Cancer Inst. Monogr.* 5–9.

Grossman, H.B., Wedemeyer, G., Ren, L., (1984) UM-UC-1 and UM-UC-2: characterization of two new human transitional cell carcinoma lines. *J. Urol.* 132:834–837.

Gschwend, J.E., Fair, W.R., Vieweg, J. (2000) Radical cystectomy for invasive bladder cancer: contemporary results and remaining controversies. [Review] [71 refs]. *Eur. Urol.* 38:121–130.

Hader, M., Stach-Machado, D., Pfluger, K.H., Rotsch, M., Heimann, B., Moelling, K., Havemann, K., (1987) Epidermal growth factor receptor expression, proliferation, and colony stimulating activity production in the urinary bladder carcinoma cell line 5637. *J. Cancer Res. Clin. Oncol.* 113:579–585.

Heaney, J.A., Ornellas, E.P., Daly, J.J., Lin, J.C., Prout, G.R., Jr. (1978) In vivo growth of human bladder cancer cell lines. *Invest. Urol.* 15:380–384.

Hisazumi, H., Kanokogi, M., Nakajima, K., Kobayashi, T., Tsukahara, K., Naito, K., Kuroda, K., Matsubara, F., (1979) [Established cell line of urinary bladder carcinoma (KK-47): growth, heterotransplantation, microscopic structure and chromosome pattern (author's transl)]. [Japanese]. *Nippon Hinyokika Gakkai Zasshi–Jpn. J. Urol.* 70:485–494.

Jemal, A., Murray, T., Samuels, A., Ghafoor, A., Ward, E., and Thun, M.J., (2003) Cancer Statistics, 2003. *CA Cancer J. Clin.* 53:5–26.

Kakuya, T., Yamada, T., Yokokawa, M., Ueda, T., (1983) Establishment of cell strains from human urothelial carcinoma and their morphological characterization. *In Vitro* 19:591–599.

Kanno, N., Nonomura, N., Miki, T., Kojima, Y., Takahara, S., Nozaki, M., Okuyama, A., (1998) Effects of epidermal growth factor on the invasion activity of the bladder cancer cell line. *J. Urol.* 159:586–590.

Kato,T., (1976) [Cell kinetics of the bladder carcinoma (author's transl)]. [Japanese]. *Nippon Hinyokika Gakkai Zasshi–Jpn. J. Urol.* 67:491–496.

Kato, T., Irwin, R.J., Jr., Prout, G.R., Jr., (1977) Cell cycles in two cell lines of human bladder carcinoma. *Tohoku J. Exper. Med.* 121:157–164.

Kato, T., Ishikawa, K., Nemoto, R., Senoo, A., Amano, Y. (1978) Morphological characterization of two established cell lines, T24 and MGH-U1, derived from human urinary bladder carcinoma. *Tohoku J. Exp. Med.* 124:339–349.

Knuchel, R., Hofstadter, F., Jenkins, W.E., Masters, J.R., (1989) Sensitivities of monolayers and spheroids of the human bladder cancer cell line MGH-U1 to the drugs used for intravesical chemotherapy. *Cancer Res.* 49:1397–1401.

Kugoh, H., Fujiwara, M., Kihara, K., Fukui, I., Horikawa, I., Schulz, T.C., Oshimura, M., (2000) Cellular senescence of a human bladder carcinoma cell line (JTC-32) induced by a normal chromosome 11. *Cancer Genet. & Cytogen.* 116:158–163.

Kunugi, K.A., Vazquez-Padua, M.A., Miller, E.M., Kinsella, T.J. (1990) Modulation of IdUrd-DNA incorporation and radiosensitization in human bladder carcinoma cells. *Cancer Res.* 50:4962–4967.

Lawson, A.H., Riches, A.C., Weaver, J.P., (1986) Human bladder tumors in organ culture. J. Urol. 135:1061–1065.

Leighton, J. (1991) Radial histophysiologic gradient culture chamber: rationale and preparation. *In vitro Cell. & Dev. Biol.*, 27A:786–790.

Leighton, J., Abaza, N., Tchao, R., Geisinger, K., Valentich, J. (1977) Development of tissue culture procedures for predicting the individual risk of recurrence in bladder cancer. *Cancer Res.* 37:t–9.

Leighton, J., and Tchao, R. (1984) The propagation of cancer, a process of tissue remodeling. Studies in histophysiologic gradient culture. *Cancer Metast. Rev.* 3:81–97.

Leighton, J., Tchao, R., Nichols, J. (1985) Radial gradient culture on the inner surface of collagen tubes: organoid growth of normal rat bladder and rat bladder cancer cell line NBT-II. *In vitro Cell. & Dev. Biol.* 21:713–715.

Lindert, K.A., and Terris, M.K., (2000) Effect of contrast material on transitional cell carcinoma viability. *Urology* 56:876–879.

Malkovsky, M., and Bubenik, J. (1977) Human urinary bladder carcinoma cell line (T24) in long-term culture: chromosomal studies on a wild population and derived sublines. *Neoplasma* 24:319–326.

Marshall, C.J., Franks, L.M., Carbonell, A.W. (1977) Markers of neoplastic transformation in epithelial cell lines derived from human carcinomas. *J. Natl. Cancer Inst.* 58:1743–1751.

Masters, J.R., Bedford, P., Kearney, A., Povey, S., Franks, L.M., (1988) Bladder cancer cell line cross-contamination: identification using a locus-specific minisatellite probe. *Br. J. Cancer* 57:284–286.

Masters, J.R., Hepburn, P.J., Walker, L., Highman, W.J., Trejdosiewicz, L.K., Povey, S., Parkar, M., Hill, B.T., Riddle, P.R., Franks, L.M., (1986) Tissue culture model of transitional cell carcinoma: characterization of twenty-two human urothelial cell lines. *Cancer Res.* 46:3630–3636.

Messing, E.M., and Catalona W.J. (1998) Urothelial tumors of the urinary tract. In Walsh, P.C., Retik, A.B., Vaughn, E.D., Wein, A.J. (eds), Vol. 3. W.B. Sanders: Philadelphia, PA, 2327–2410.

Metts, M.C., Metts, J.C., Milito, S.J., Thomas, C.R., Jr., (2000) Bladder cancer: a review of diagnosis and management.. *J. Natl. Med. Assoc.* 92:285–294.

Moore, G.E., Morgan, R.T., Quinn, L.A., Woods, L.K. (1978) A transitional cell carcinoma cell line. *In vitro* 14:301–306.

Morita, T., Shinohara, N., Honma, M., Tokue, A. (1995) Establishment and characterization of a new cell line from human bladder cancer (JMSU1). *Urol. Res.* 23:143–149.

Muraoka, K., Nabeshima, K., Kataoka, H., Kishi, J., Koono, M. (1995) Establishment of a new human urinary bladder mixed carcinoma cell line UMK-1 and its cell-density-dependent secretion of gelatinases and tissue inhibitor of metalloproteinase-1. *Urol. Inter.* 54:184–190.

Nakajima, K., and Hisazumi, H., (1987) Enhanced radioinduced cytotoxicity of cultured human bladder cancer cells using 43 degrees C hyperthermia or anticancer drugs. *Urol. Res.* 15:255–260.

Nayak, S.K., O'Toole, C., Price, Z.H. (1977) A cell line from an anaplastic transitional cell carcinoma of human urinary bladder. *Br. J. Cancer* 35:142–151.

O'Toole, C., Nayak, S., Price, Z., Gilbert, W.H., Waisman, J. (1976) A cell line (SCABER) derived from squamous cell carcinoma of the human urinary bladder. *Int. J. Cancer* 17:707–714.

O'Toole, C., Price, Z.H., Ohnuki, Y., Unsgaard, B. (1978) Ultrastructure, karyology and immunology of a cell line originated from a human transitional-cell carcinoma. *Br. J. Cancer* 38:64–76.

O'Toole, C.M., Povey, S., Hepburn, P., Franks, L.M., (1983) Identity of some human bladder cancer cell lines. *Nature (London)* 301:429–430.

Offner, F.A., Ott, G., Povey, S., Knuechel, R., Preisler, V., Fuezesi, L. Klosterhalfen, B. Ruebben, H., Hofstaedter, F., Kirkpatrick, C.J., (1991) Characterization of the new bladder cancer cell line HOK-1: expression of transitional, squamous and glandular differentiation patterns. *Intl. J. Cancer* 49:122–128.

Perabo, F.G., Kamp, S., Schmidt, D., Lindner, H., Steiner, G., Mattes, R.H., Wirger, A. Pegelow, K. Albers, P. Kohn, E.C. Ruecker, A. Mueller, S.C., (2001) Bladder cancer cells acquire competent mechanisms to escape Fas- mediated apoptosis and immune surveillance in the course of malignant transformation. *Br. J. Cancer* 84:1330–1338.

Plotkin, G.M., Wides, R.J., Gilbert, S.L., Wolf, G., Hagen, I.K., Prout, G.R., Jr., (1979) Galactosyl transferase activity in human transitional cell carcinoma lines and in benign and neoplastic human bladder epithelium. *Cancer Res.* 39:3856–3860.

Rasheed, S., Gardner, M.B., Rongey, R.W., Nelson-Rees, W.A., Arnstein, P. (1977) Human bladder carcinoma: characterization of two new tumor cell lines and search for tumor viruses. *J. Natl. Cancer Inst.*, 58:881–890.

Reese, D.H., Friedman, R.D., Smith, J.M., Sporn, M.B. (1976) Organ culture of normal and carcinogen-treated rat bladder. *Cancer Res.* 36:T-7.

Reznikoff, C.A., Belair, C., Savelieva, E., Zhai, Y., Pfeifer, K., Yeager, T., Thompson, K.J., DeVries, S., Bindley, C., Newton, M.A., (1994) Long-term genome stability and minimal genotypic and phenotypic alterations in HPV16 E7-, but not E6-, immortalized human uroepithelial cells. *Genes Dev.* 8:2227–2240.

Reznikoff, C.A., Gilchrist, K.W., Norback, D.H., Cummings, K.B., Erturk, E., Bryan, G.T., (1983) Altered growth patterns in vitro of human papillary transitional carcinoma cells. *Am. J. Pathol.* 111:263–272.

Reznikoff, C.A., Loretz, L.J., Pesciotta, D.M., Oberley, T.D., Ignjatovic, M.M. (1987) Growth kinetics and differentiation in vitro of normal human uroepithelial cells on collagen gel substrates in defined medium. *J. Cell Physiol.* 131:285–301.

Russell, P.J., Jelbart, M., Wills, E., Singh, S., Wass, J., Wotherspoon, J., Raghavan, D., (1988) Establishment and characterization of a new human bladder cancer cell line showing features of squamous and glandular differentiation. *Int. J. Cancer* 41:74–82.

Russell, P.J., Palavidis, Z., Rozinova, E., Philips, J., Wills, E.J., Lukeis, R., Wass, J., Raghavan, D., (1993) Characterization of a new human bladder cancer cell line, UCRU-BL-28. *J. Urol.* 150:1038–1044.

Sanford, E.J., Geder, L., Dagen, J.E., Laychock, A.M., Ladda, R., Rohner, T.J., Jr., (1978) Establishment and characterization of a new human urinary bladder carcinoma cell line (PS-1). *Invest. Urol.* 16:246–252.

Sarkar, S., Julicher, K.P., Burger, M.S., Della, V., V Larsen, C.J. Yeager, T.R. Grossman, T.B. Nickells, R.W. Protzel, C. Jarrard, D.F. Reznikoff, C.A., (2000) Different combinations of genetic/epigenetic alterations inactivate the p53 and pRb pathways in invasive human bladder cancers. *Cancer Res.* 60:3862–3871.

Torino, J.L., Burger, M.S., Reznikoff, C.A., Swaminathan, S., (2001) Role of TP53 in repair of N-(deoxyguanosin-8-yl)-4-aminobiphenyl adducts in human transitional cell carcinoma of the urinary bladder. *Carcinogenesis* 22:147–154.

Vineis, P. (1994) Epidemiology of cancer from exposure to arylamines. [Review] [19 refs]. *Environ. Health Persp.* 102:Suppl–10.

Vineis, P., and Pirastu, R., (1997) Aromatic amines and cancer. [Review] [71 refs]. *Cancer Causes & Control* 8:346–355.

Walker, M.C., Povey, S., Parrington, J.M., Riddle, P.N., Knuechel, R., Masters, J.R., (1990) Development and characterization of cisplatin-resistant human testicular and bladder tumour cell lines. *Eur. J. Cancer* 26:742–747.

Wang, Z., Zhang, Z., Liu, Y., Chen, Y., Li, Q., Duanguolan Qin, D., Liu, G., Wang, L., (2000) Effect of retinoic acid and its complexes with transition metals on human bladder cancer cell line EJ in vitro. *Urol. Res.* 28:191–195.

Williams, R.D. (1980) Human urologic cancer cell lines. *Invest. Urol.* 17:359–363.

Yajima,T. (1970) Monolayer culture of human urinary bladder tumors. II. [Japanese]. *Nippon Hinyokika Gakkai Zasshi–Jpn. J. Urol.* 61:805–821.

Yamamoto, T. (1979) Establishment of a new cell line (NBT-2) derived from a human urinary bladder carcinoma and its characteristics (author's transl). [Japanese]. *Nippon Hinyokika Gakkai Zasshi–Jpn. J. Urol.* 70:351–357.

Yeager, T.R., DeVries, S., Jarrard, D.F., Kao, C., Nakada, S.Y., Moon, T.D.,

Bruskewitz, R., Stadler, W.M., Meisner, L.F., Gilchrist, K.W., Newton, M.A. Waldman, F.M., Reznikoff, C.A., (1998) Overcoming cellular senescence in human cancer pathogenesis. *Genes Dev.* 12:163–174.

Yeager, T.R., Jarrard, J.F., Reznikoff, C.A., (1998) An in vitro model for human bladder cancer pathogenesis studies. In Kenneth W. Adolph (ed), CRC Press: Boca Raton, FL 159–178.

Yeager, T.R., and Reznikoff, C.A., (1998) Methotrexate resistance in human uroepithelial cells with p53 alterations. *J. Urol.* 159:581–585.

# 6

# Long-term Culture of Normal and Malignant Human Prostate Epithelial Cells

Robert K. Bright and Jennifer D. Lewis

*Department of Microbiology and Immunology, Southwest Cancer Center, Texas Tech University Health Sciences Center, 3601 4th Street, STOP 6591, Lubcock, Texas, robert.bright@ttuhsc.edu*

*Culture of Human Tumor Cells*, Edited by Roswitha Pfragner and R. Ian Freshney.
ISBN 0-471-43853-7   Copyright © 2004 Wiley-Liss, Inc.

## 1. INTRODUCTION

Prostate cancer is the most common cancer affecting men and the second most common cause of male cancer-related death in the Western world [Lalani et al., 1997; Lee et al., 1994]. As men survive longer, the likelihood of developing prostate cancer increases. For example, nearly 80% of 80-year-old men have prostate epithelial cells that could be considered cancerous. Whether clinical disease actually develops appears to be patient-specific. Prostate-intraepithelial neoplasia (PIN), a precursor for prostate cancer, has been seen in men as young as 30 years of age [Lalani et al., 1997]. More cases of prostate cancer are being diagnosed as a direct result of increased awareness and improved screening methods, such as digital rectal exams and serum PSA (prostate specific antigen) levels, thus attributing to increased incidence of disease.

Despite prostate cancer prominence, an understanding of prostate tumorigenesis and advancements in new therapies are lacking. Since the introduction of androgen ablation in the early 1940s, no major changes in mortality have been observed [Lee et al., 1994]. Lack of knowledge into the development of prostate cancer and limited advancement in therapy are due in part to the absence of suitable in vitro systems to study the biological events that cause initiation and progression of prostate cancer. An in vitro prostate cancer cell culture system would provide many benefits. First, it is easier to study growth effects of certain agents in an isolated system. Second, it is less difficult to determine specific changes involved in the early steps of carcinogenesis. Third, an in vitro system is more efficient when screening for potential carcinogens or anti-cancer agents, and the use of a cell system is more cost-effective, less time-consuming, and highly reproducible [Webber et al., 1984]. Fourth, long-term cultures of paired normal and tumor cells make possible quantitative analyses of differential gene expression for the discovery of new prostate cancer markers.

Finally, long-term cultures of paired normal and tumor cells are critical for studies focused on cellular immune responses to prostate cancer and the development of immunologic therapies or anti-cancer vaccines.

However, development of long-term prostate epithelial cell lines has proven difficult in the past. In previous attempts to culture prostate cells, researchers have faced obstacles, such as overgrowth of fibroblasts, inability to accomplish serial passage, and undefined culture conditions [Peehl, 1992]. Until 1993, only three long-term prostate cancer cell lines generated from metastatic lesions were available for use in in vitro experiments [Horoszewicz et al., 1983; Kaighn et al., 1979; Stone et al., 1978], and no long-term cultures of primary cancer cells existed. However, short-term cell cultures from primary (non-metastatic) prostate cancers were available for experimental use. Short-term cultures generally produce too few cells to support detailed and repeated molecular and immunologic studies. The only instances in which long-term immortalized prostate cells had been described were few and were limited to normal epithelial cells. Despite advances in culture media development, improvements in fresh tissue preparation, and epithelial cell culture techniques, generation of long-term prostate epithelial cell lines from primary adenocarcinomas, without immortalization in vitro, remains a seemingly insurmountable obstacle.

To address this, we developed methods for the successful generation of several distinct sets of matched, immortal normal prostate (NPTX) and tumor prostate-derived (CPTX) epithelial cell lines [Bright et al., 1997]. Briefly, fresh prostatectomy specimens obtained directly from the operating room are dissected under sterile conditions by an experienced pathologist, with neighboring sections from malignant or benign areas allocated for tissue culture, as well as for paraffin and frozen sections for subsequent pathologic analysis and genetic studies. Primary cell lines are initiated by mechanical disruption of the prostate tissue specimens, followed by culture in collagen-coated dishes containing defined medium supplemented with growth factors and fetal bovine serum (FBS). Early passages of adherent proliferating tumor-derived cell cultures are immortalized by transduction with a recombinant amphitropic retrovirus encoding the E6 and E7 transforming proteins of human papilloma virus serotype 16 (HPV16) [Halbert et al., 1991]. Similar techniques are used to establish autologous, immortal cell lines of normal prostate epithelium, normal seminal vesicle epithelium, and normal prostate-derived stromal cells or fibroblasts. To date, our efforts have yielded nearly 100% success rate in establishing more than three-dozen, long-term cell lines from patients with advanced prostate cancer.

It has been demonstrated that the E6 and E7 genes of HPV16 are sufficient for immortalization of primary human keratinocytes. The E7 protein binds to the tumor suppressor pRb protein, while E6 binds to the tumor suppressor p53 and targets it for degradation. As a result of destroying p53 and pRb tumor suppressors, the cell cycle is disregulated, which results in uncontrollable cell growth [Demers et al., 1994; Halbert et al., 1992; Munger et al., 1989].

Aside from the generation of long-term primary cancer lines, the second biggest challenge is the definitive characterization of the cell lines; that is, distinguishing cultivated prostate cancer cells from normal epithelial cells. Past cytogenetic evaluation of multiple short-term prostate epithelial cell cultures has revealed that the majority of lines generated from localized prostate cancers exhibit a normal male karyotype [Brothman et al., 1990; Brothman et al., 1991; Brothman et al., 1992]. This, combined with the unremarkable microscopic morphology of short-term cultures and a pervasive lack of success with xenotransplantation, has rendered accurate identification and characterization of human primary prostate cancer cell lines extremely difficult.

The initiation of prostate cancer is believed to occur as a result of multiple genetic changes within the cell, including the inactivation of potential tumor suppressor genes as manifested by allelic chromosomal deletions [reviewed in Isaacs et al., 1994]. Early studies examining chromosomal deletions in fresh (non-cultured) primary prostate cancer specimens exhibited allelic loss of heterozygosity (LOH) on several chromosomes to include 8p, 10q, and 16q [Carter et al., 1990; Bergerheim et al., 1991; Sakr et al., 1994]. Subsequent studies confirmed a remarkably high percentage of allelic loss on the short arm of chromosome 8, thus moving chromosome 8p to the forefront of the list of potential sites for prostate cancer-associated tumor suppressor genes [Bova et al., 1993; Trapman et al., 1994; Macoska et al., 1995]. Moreover, examination of 99 microdissected tumors [Vocke et al., 1996] and 54 microdissected PIN lesions [Emmert-Buck et al., 1995] for LOH on the short arm of chromosome 8 demonstrated strong evidence for the inactivation of a tumor suppressor gene(s) on chromosome 8p12–21 when compared with matched normal epithelial cells. Thus, examination of LOH within this minimal deletion region on chromosome 8p12–21 represents a potentially powerful alternative method for the identification and characterization of human prostate epithelial cell lines derived from primary tumors. In addition, we recently described the generation of an iso-chromosome 8q in prostate cancer, further supporting 8p loss as a possible mechanism for tumorigenesis, perhaps through the inactivation of a tumor suppressor gene [Virgin et al., 1999].

Phenotyping of the individual cell lines by immunocytochemistry and flow cytometry confirms the epithelial cell origin of all prostate-derived epithelial cell lines (all express cytokeratins) and demonstrates that all the established long-term tumor and normal cell lines express significant amounts of MHC class I molecules. In addition, incubation of the tumor cell lines in the presence of interferon (IFN)-gamma increases the expression of MHC class I molecules and induces expression of MHC class II molecules [Bright et al., 1997], as well as CD40, a T cell co-stimulation molecule of the tumor necrosis factor (TNF) receptor family (unpublished observations). Thus, we have systematically developed methods for successfully establishing paired prostate tumor and prostate normal cell lines from prostate cancer patients that have proven useful for our tumor-associated antigen identification and tumor immunity studies.

The following text will describe reliable methods for generating and characterizing continuously proliferating prostate cancer cell lines from primary tumors, as well as from patient matched normal tissue, as developed in our laboratory.

## 2. PREPARATION OF MEDIA

In a laminar flow hood (sterile), add each of the following reagents to one 500 ml bottle of keratinocyte-serum-free medium (SFM) and mix well. The supplemented media is referred to as keratinocyte growth medium (KGM).

(i)    1.5 ml bovine pituitary extract (BPE; 25 µg/ml final concentration)

(ii)    1 vial of epidermal growth factor (EGF; 5 ng/ml final concentration)

(iii)    5 ml L-Glutamine (2 mM final concentration)

(iv)    5 ml    N-(2-hydroxyethyl)piperazine-N'-ethanesulfonic    acid (HEPES) (10 mM final concentration)

(v)    2.5 ml Penicillin/streptomycin antibiotics (50 IU/ml and 50 µg/ml final concentrations, respectively)

(vi)    500 µl Gentamycin sulfate (50 µg/ml final concentration)

(vii)    1 ml Fungizone (250 ng/ml final concentration, antimycotic)

(viii) 25 ml of heat-inactivated FBS, which can be added directly to the Nalgene filtering apparatus or the keratinocyte-SFM bottle of base medium.

*Note:* The BPE and EGF supplements come with the Keratinocyte-Base Medium, but must be specifically requested at the time of order. Add all of the EGF by transferring 1 ml of medium into the EGF vial and adding back to keratinocyte medium. Be sure the vial of BPE is mixed well by vortexing before adding it to the base medium. The vial of BPE contains a total of 2 ml, but only 1.5 ml is used to make KGM. DO NOT reuse the remaining BPE. If prepared in this manner the final concentrations of EGF and BPE are 5 ng/ml and 25 μg/ml, respectively.

### Sterilization

(i) Place two pre-filters in the Nalgene filtering apparatus (0.45 μm) to prevent clogging due to the BPE.

(ii) Apply vacuum source.

(iii) Slowly pour enough keratinocyte-SFM (after all supplements are added) into the filter apparatus to wet and prevent floating of the pre-filters.

(iv) Add the rest of keratinocyte-SFM until entire solution is in filter reservoir.

(v) Add 5% heat-inactivated FBS, (25 ml) to Nalgene filtering apparatus if not already added to the base medium.

## 3. TISSUE PROCUREMENT

Tissue specimens used for generating cell lines are obtained from patients undergoing radical prostatectomies for treatment of intermediate to high-grade localized prostate cancer (Gleason grades 6–8, tumor stages T2C to T3C). Fresh prostatectomy specimens obtained directly from the operating room are dissected under sterile conditions by an experienced pathologist. Tissues designated as normal prostate, prostate cancer, or normal seminal vesicle, on gross inspection are dissected separately for the purpose of generating cell cultures, as well as for preserving as frozen and fixed sections (Fig. 6.1). When areas of tumor are less than obvious, sterile 6 mm punches are taken and confirmed as normal or tumor by microscopic evaluation of fixed specimens.

## 4. TISSUE PROCESSING

Special care must be taken when isolating tumor specimens, as there is often a mixture of benign prostate epithelium, benign prostatic hyperplasia (BPH), prostatic intra-epithelial hyperplasia (PIN), and invasive tumor cells identified by microscopic analysis. To establish a tumor cell line, pure tumor tissue is desired. Taking smaller tissue fragments (≤1 mm) increases the chance of obtaining pure tumor tissue from patients. The tissue fragments are divided randomly and equally for tissue cultures, and for frozen and paraffin sections (Fig.

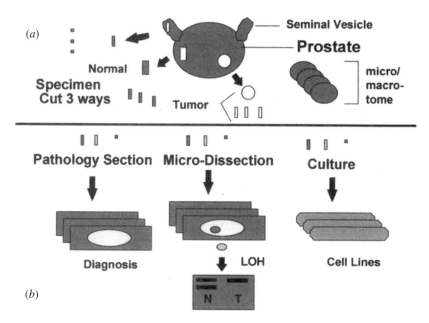

**Figure 6.1.** Diagram depicting the process of tissue procurement and distribution following radical prostatectomy. (*a*) Illustration of the dissection of tissue from a prostate gland following surgical removal. Tumor tissue is shown as a circle, and the normal prostate and normal seminal vesicle tissues as large and small rectangles, respectively. (*b*) Illustration summarizing the allocation of the tissue fragments following sterile dissection. From left to right, paraffin sections for microscopic confirmation of tissue type; that is, tumor or normal and percentages of mixed cell types, frozen sections for subsequent micro-dissection and LOH studies to confirm tumor cells genetically, and culture of tissue fragments for the establishment of long-term cell lines.

6.1). Whenever possible, multiple, distinct tumor tissue fragments should be selected for culture initiation from the individual specimens. If these guidelines are followed, it is possible to obtain tissue specimens containing at least 95% neoplastic cells.

## 5. ESTABLISHMENT OF PRIMARY CELL CULTURES

Cultures are initiated routinely by mechanical dissection (<1 cm diameter fragments). Tissue fragments are carefully minced into 2–3 mm cubes in a small volume of growth medium, and the resultant slurry of tissue and cells is dispensed into collagen-coated six-well plates. All cultures are initiated in a volume of 1 ml per well and incubated at 37°C, 5% $CO_2$. It is crucial that the cultures are left undisturbed for 2–3 days to allow viable cells and tissue chunks to settle and attach to the plates. After 2–3 days, the unattached debris is carefully aspirated, and 3–5 ml of fresh KGM is added to each well.

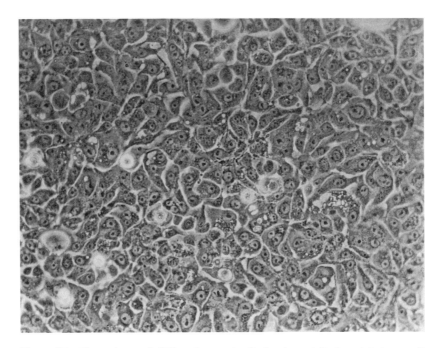

**Figure 6.2.** Photomicrograph (200×, phase contrast) of an immortalized prostate tumor cell line following 10–12 weeks in culture. The cobblestone appearance and general morphology are typical of all prostate epithelial cell lines generated, whether benign or malignant. Reproduced with permission (Bright et al., 1997).

*Note:* Cell proliferation is relatively slow with short-term cell cultures. In addition, the cultures survive only approximately 5–6 weeks unless immortalized. Though the cultures should have a cobblestone appearance indicative of epithelial cells (Fig. 6.2), it is impossible to distinguish between normal and tumor cell cultures by visual inspection by using an inverted light-microscope.

### Protocol 6.1.   Primary Culture of Normal, Benign, and Malignant Prostate

*Reagents and Materials*
- ❏  KGM growth medium (see Section 2)
- ❏  Six-well tissue culture plates coated with Type I rat tail collagen
- ❏  Petri dishes, 10 cm
- ❏  Two 18-gauge needles

*Protocol*
(a)   Place tissue sample in a sterile Petri dish with a small volume of KGM, and carefully pull apart by using 18-guage needles

(b)  Equally divide the resulting slurry between wells of six-well Type I rat tail collagen-coated tissue culture plates.

*Note:* It is critical to start cultures in 1 ml per well.

(c)  Incubate plates at 37°C, with 5% $CO_2$.
(d)  Do not disturb for 2–3 days to allow viable cells and tissue fragments to attach to the plates.
(e)  Carefully aspirate medium, and remove unattached debris after 2–3 days.
(f)  Carefully re-feed cells with 3–5 ml per well of fresh KGM.
(g)  Check culture daily, and replace medium every 2–3 days.

## 6.  PRIMARY CULTURE MAINTENANCE

Growth medium for prostate and seminal vesicle epithelial cell lines is as described in Section 2. To guard against the outgrowth of fibroblasts while initiating primary epithelial cell cultures from fresh tissue specimens, the concentration of FBS may be reduced to 1–2% and/or cholera toxin added at 10–20 ng/ml. In the rare event that fibroblasts persist in the epithelial cell cultures, differential trypsinization (incubation for 1–2 min at room temperature, followed by washing away of detached fibroblasts leaving the more adherent epithelial cells) is extremely successful in achieving pure epithelial cell cultures. The fibroblasts typically detach with trypsin in 1–2 min., whereas the epithelial cells remain attached. To help ensure the establishment of viable and actively growing cultures, it is necessary to replace the medium every 2–3 days. When culture wells appear to contain pure epithelial cells (i.e., no obvious contaminating fibroblasts) it is beneficial to use KGM with 5% FBS.

## 7.  IMMORTALIZATION OF PRIMARY CELL CULTURES

In order for primary cultures to survive beyond 5–6 weeks, in vitro immortalization is necessary. After the first couple of passages, the adherent monolayer of prostate epithelial cells can be readily transduced with recombinant retrovirus. The immortalizing retrovirus encodes the E6 and E7 transforming proteins of human papilloma virus serotype 16. Transduction with this retrovirus results in the establishment of long-term prostate epithelial cells (Fig. 6.3). Confirmation of successful transduction is noted when cells survive in geneticin (G418) at a concentration of 1 mg/ml and continue to proliferate.

Cell culture immortalization is accomplished by transduction of actively proliferating cells with a recombinant retrovirus encoding the E6 and E7 transforming proteins of human papilloma virus serotype

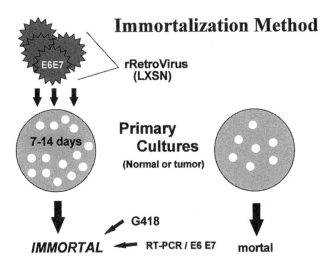

**Figure 6.3.** Diagram depicting the process of immortalization of primary prostate epithelial cell cultures. High-titer supernates containing the E6E7 expressing recombinant retrovirus (rRetrovirus) are used to infect proliferating primary epithelial cells after 1 to 2 passages at a split of 1:2. The right side of the illustration depicts a replica culture well that was not immortalized with the rRetrovirus. Immortal cell cultures will actively grow in 1 mg/ml G418. RT-PCR is used routinely to demonstrate expression of the E6 and E7 viral transgenes.

16 (HPV16) and the eukaryotic selection marker neomycin phosphotransferase, designated LXSN16E6E7 (generously provided by Dr. Denise Galloway, Fred Hutchinson Cancer Research Center, Seattle, WA) [Halbert et al., 1991]. In preparation for immortalization, short-term epithelial cell cultures (culture passages 1–2) are split and allowed to re-attach in six-well plates for at least 48 h, yielding cultures that are 50–60% confluent. Because the immortalizing vector is a recombinant retrovirus, it is imperative that the cell cultures are actively proliferating at the time of transduction. We have found it helpful to split each well 1:2, which allows them to reach confluency a second time, and then split them again 1:2 and immortalize 2–4 days later to guarantee that the cultures are proliferating. Transduction with the LXSN16E6E7 retrovirus is accomplished by replacing the culture medium with culture supernatant collected from the retrovirus producer line PA317, in the presence of 10 µg/ml DEAE-dextran (Sigma) for 24 h.

### Protocol 6.2. Immortalization of Prostatic Epithelial Cultures

#### Reagents and Materials
- PA317 producer cells
- Dulbecco's modified Eagle's medium (DMEM) with 4.5 g/L glucose, 10% FBS

- [ ] Trypsin (50 μg/ml)
- [ ] Filter apparatus, 0.45 μM
- [ ] DEAE/dextran stock solution, 10 mg/ml

### Protocol

(a) Grow PA 317 producer cells in DMEM with a high concentration of glucose (4.5 g/L) and 10% FBS at 37°C with 10% $CO_2$.

(b) Passage PA317 cells approximately 1:10 every 6–8 days, inoculating approximately $2 \times 10^6$ cells/75 $cm^2$.

(c) Detach cells from flask by using trypsin for 1–2 min.

Note: Be sure to keep trypsinization time to minimum needed to detach cells since trypsin can be toxic to these cells if left on too long.

(d) Quench trypsin with 3–4 volume of medium containing FBS.

(e) Split PA317 to $2 \times 10^6$ cells/10 cm dish in 10 ml media.

(f) Change medium after 24 h.

(g) Wait an additional 24 h, and collect supernatant for transduction of normal and malignant prostate epithelial cells.

(h) Filter supernate through 0.45 μm sterilization filter, and combine with DEAE/dextran at 10 μg/ml (final concentration).

Note: Use a 10 mg/ml stock solution of DEAE/dextran.

(i) Let supernate and DEAE/Dextran sit at room temperature for 10 min before adding to prostatic cells.

Note: Viral supernatants may be stored at −80°C for several weeks, but transduction efficiency may be reduced by 50% over time.

(j) Discard PA317 producer cells after 3–4 weeks in culture, and thaw a new vial of cells.

### 7.1. Safety Precautions

Culturing of the viral packaging line by this method results in high titer viral supernates, thus extreme caution should be taken handling the packaging line and the supernates. Biosafety level 2 practice and equipment with universal precautions should be used when handling the supernate and the packaging line. Although the stable immortalized prostate cell lines do not produce virus, it is recommended that stringent safety precautions be used for the first two to three passages following transduction.

# 8. CHARACTERIZATION OF LONG-TERM CELL LINES

## 8.1. Detection of Loss of Heterozygosity

Examination of LOH within a defined minimal deletion region on chromosome 8p12–21 represents a potentially powerful alternative method for the identification and characterization of human prostate epithelial cell lines derived from primary tumors.

## Protocol 6.3. Detection of Loss of Heterozygosity in Prostate Cell Lines

### Reagents and Materials

☐ The polymorphic DNA markers used for the detection of LOH on chromosome 8p12–21 include SFTP-2, D8S133, D8S136, NEFL, D8S137, D8S131, D8S339, and ANK.

### Protocol

(a) The polymerase chain reaction (PCR) is performed as previously described [Vocke et al., 1996]. Briefly, 12.5 ul PCR reaction mixtures contain 200 $\mu$M dATP, dGTP, and dTTP; 40 $\mu$M dCTP; 0.8 mM primers; 2 $\mu$Ci [$\alpha^{32}$P] dCTP; 16 $\mu$M tetramethylammonium chloride [Hung et al., 1990]; 1 PCR reaction buffer (containing 1.5 mM MgCl$_2$); and 1 unit of Taq polymerase. Dimethyl sulfoxide (DMSO) (5%) is added to reactions for the markers D8S133 and D8S137 to improve amplification and resolution of the products.

(b) Perform reactions with all markers as follows: 2 min at 95°C, followed by 28 to 40 cycles (depending on the marker) of annealing and extension (95°C for 30 s, annealing temperature for 30 s, and 72°C for 30 s), and a 2 min incubation at 72°C. Determine annealing temperatures for each marker empirically after an initial estimate based on primer length and composition.

(c) Denature the labeled amplified DNA samples for 5–10 min at 90°C.

(d) Load onto a gel consisting of 7% acrylamide (30:0.8 acrylamide: bisacrylamide), 5.6 M urea, 32% formamide, and 1 TBE [0.089 M Tris, pH 8.3; 0.089 M borate; 0.002 M ethylene diamine tetra-acetate (EDTA)] [Litt et al., 1993].

(e) Samples are electrophoresed at 95 W for 2–4 h.

(f) Transfer gels to sequencing gel filter paper.

(g) Perform autoradiography with Kodak X-OMAT film (Fig. 6.4).

Our criterion for LOH is at least 75% loss of one allele compared with an autologous fresh peripheral blood lymphocyte (PBL) control,

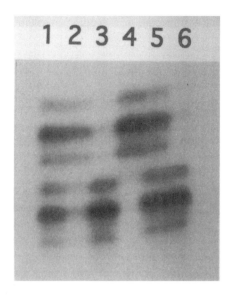

**Figure 6.4.** PCR analysis of microsatellite D8S136 on microdissected prostate tumor and normal cells compared with immortalized cell lines from a single patient. Lane 1: NPTX passage 26. Lane 2: microdissected tumor >75% loss of lower allele. Lane 3: uncloned CPTX passage 21 loss of upper allele. Lanes 4–6 single cell clones from CPTX passage 8. Reproduced with permission (Bright et al., 1997).

as determined by direct visualization by three independent investigators. When sufficient DNA is available, LOH should be verified with at least two independent experiments.

## 9. MAINTENANCE OF LONG-TERM CELL LINES

**Protocol 6.4. Maintenance of Long-term Cell Lines from Prostate**

### Reagents and Materials
- Trypsin, 5 µg/ml, EDTA, 0.3 mM
- Phosphate buffered saline (PBSA) or Hanks' balanced salt solution (HBSS)
- Growth medium (see Section 2)
- 75 cm² tissue culture flasks

### Protocol
(a) Routinely maintain immortalized, actively growing cultures [see Bright et al. 1997] in 75 cm² tissue culture flasks in KGM. A confluent 75 cm² flask will routinely yield 8–10 × 10⁶ cells.

(b) Split proliferating cells with trypsin/EDTA mixture when approximately 90% confluent.

*Note:* It is wise to freeze a couple of vials at 2–5 × 10⁶ cells/ml of early passage cells by using freeze media made of FBS and 10% DMSO (see Section 10) to ensure adequate numbers for studies involving cellular alterations over time.

(c) Warm trypsin to 37°C before adding to cells.
(d) Aspirate the old medium from the flask.
(e) Wash the monolayer one time with sterile PBSA or HBSS.
(f) Aspirate the PBSA or HBSS.
(g) Add appropriate amount of trypsin to the monolayer (based on the flask size).

*Note:* Use 2 ml of PBSA and trypsin for a 25 cm$^2$ flask or 5 ml for a 75 cm$^2$. To speed up trypsinization, place flasks in incubator for 2–3 min, and then tap flasks to loosen cells. Use 1-min intervals until cells are detached.

(h) Split cells 1:10 based on the amount of trypsin added into appropriate flask (i.e., if using 2 ml of trypsin, take 200 $\mu$l and add to 10 ml of media in a new 25 cm$^2$ flask).
(i) Incubate cells in 37°C incubator with 5% $CO_2$ until confluent.

*Note:* Be sure to loosen lids of flasks to allow exchange of gases.

(j) Passage adherent cells with trypsin when approximately 90–100% confluent.

## 10. CRYOPRESERVATION OF LONG-TERM CELL LINES

It is recommended that cells be cryopreserved whenever they are split for the first 10 passages, which equates to approximately 9 weeks for an immortalized cell line.

### Protocol 6.5. Cryopreservation of Immortalized Prostatic Cell Lines

***Reagents and Materials***
- PBSA
- Trypsin, 5 $\mu$g/ml
- KGM medium with 5% FBS
- 15 ml centrifuge tubes
- Hemocytometer
- Freeze medium: FBS and 10% DMSO
- Cryo-vials
- Liquid nitrogen freezer

*Protocol*

(a) Wash the cells with the appropriate volume of PBSA (2–5 ml depending on size of flask).

(b) Aspirate PBSA.

(c) Add a suitable amount of trypsin and incubate for 2–3 min at 37°C in an incubator with 5% $CO_2$.

(d) When cells are no longer attached to the bottom of the flask, pull up entire volume (i.e., 2 ml) of trypsin/cell mixture into a sterile 10 ml pipette containing 8 ml of KGM media with 5% FBS, to a total volume of 10 ml.

(e) Dispense into a sterile 15 ml centrifuge tube.

(f) Count cells on hemocytometer.

(g) Centrifuge for 10 min at 250 g.

(h) Aspirate media.

(i) Resuspend in freeze media (FBS and 10% DMSO) at $2 \times 10^6$ cells/ml.

(j) Aliquot 1 ml of cells/freeze media mixture per 1.8 ml cryo-vial.

(k) Cap vials firmly, and label vial with pencil or alcohol resistant marker.

(l) Place vials into an ethanol bath to ensure the cells freeze slowly, and place in −70°C for 24–72 h.

(m) Transfer vials to gas phase of liquid nitrogen freezer (−180°C) for long-term storage.

(see safety notes in Protocol 15.7)

*Note:* Cells may not survive at −70°C long-term (longer than 6 months), therefore it is recommended that frozen cells be transferred to liquid nitrogen within three days of freezing.

## Protocol 6.6. Thawing Cryopreserved Prostatic Cell Lines

*Reagents and Materials*
- Dry ice
- 37°C water bath
- KGM medium with 5% FBS
- 25 cm$^2$ or 75 cm$^2$ flask

*Protocol*

(a) Locate cells in liquid nitrogen log, remove vial from freezer, and place on dry ice.

(b) Thaw contents of vial by placing in 37°C water bath until a small piece of "ice" remains.

(see safety notes in Protocol 15.7)

*Note:* It is very important for cell viability that a small piece of "ice" remain in the vial and that the contents in the vial are diluted with KGM immediately after thawing. Immediate dilution is necessary because the concentration of DMSO used in the freeze medium is harmful to the cells unless frozen or quickly diluted.

(c) Carefully dilute the contents of the cryo-vial by pulling 9 ml of KGM medium into a pipette and slowly dispensing a small volume into the vial.

(d) After quickly but carefully diluting the contents of the cryo-vial, remove the mixture of medium and cells and place the total of 10 ml into a sterile 15 ml centrifuge tube.

(e) Centrifuge at 250 g for 10 min.

(f) Aspirate medium, and resuspend again in 10 ml of KGM .

(g) Spin again at 250 g for 10 min to dilute out DMSO and thoroughly wash cells.

(h) Aspirate medium and resuspend in 10 ml of KGM medium; transfer to a 25 cm$^2$ or 75 cm$^2$ (final of 10 ml or 20 ml of medium, respectively) flask.

(i) Place in 37°C incubator with 5% $CO_2$, and allow cells to adhere for approximately 48 h.

## 11. APPLICATIONS

The high mortality rate for many forms of cancer and the expected overall increase in the number of individuals with cancer in our aging population over the next few decades necessitates the need for more effective methods for treatment and prevention. Immunotherapeutic strategies for cancer represent attractive alternatives to standard treatment for the following reasons: i) The immune system has the ability to recognize changes as small as a single amino acid in a complex protein, as well as distinguish normal cells from diseased cells, which thereby eliminates the overwhelming and sometimes fatal side effects associated with many conventional therapies. ii) The immune system is by nature systemic; that is, the cells and products of an immune response possess the ability to reach tissue sites harboring microscopic metastatic deposits of tumor cells not detectable or resectable by using standard procedures. In this capacity, tumor antigen-specific immunotherapies hold great promise as an adjuvant therapy following surgical resection of primary tumors for prevention of recurrent disease and/or metastasis.

To attain this goal, the identification and immunologic characterization of novel prostate tumor associated antigens is critical. Very few candidate tumor-associated antigens have been described for human prostate cancer, and little is known about the human immune response to prostate cancer. This is due, in part, to the absence of unique, human prostate-derived cell lines necessary for the effective in vitro study of anti-tumor immunity in prostate cancer and for the identification and cloning of novel tumor associated antigens.

We studied differential gene expression in one of our novel immortalized prostate tumor cell lines compared with the paired normal prostate cell line by using a suppression subtractive hybridization technique to generate subtracted cDNA libraries. Differential expression of over 1000 randomly picked clones was confirmed by Northern hybridization. Sequence analysis identified a cDNA clone from the tumor up-regulated library that was identical to a novel protein previously cloned from human colon cancer cells, designated NY-CO-25 [Scanlan et al., 1998]. In addition, we have applied the technique of differential message display reverse transcriptase PCR (RT-PCR) to compare our paired normal and tumor cell lines for differential gene expression by using *RNAimage* kits (GenHunter). Using this method of differential analysis of gene expression, we have identified several novel candidate tumor-associated antigens that are either over-expressed in the tumor cell lines compared with the normal cell lines or are uniquely expressed in the tumor cell lines. Cloned cDNAs that exhibited a marked over or unique expression in the tumor compared with its autologous paired normal cell line by Northern blot were subcloned and sequenced, and the sequence was analyzed by using the GenBank. Clones that matched available sequences from GenBank were chosen for immediate study to evaluate the T cell immune response against the antigen, including the prostate, breast, and lung cancer associated antigen, *hD52* [Byrne et al., 1995]. Studies such as these illustrate the usefulness of these paired cell lines for tumor associated antigen discovery and immunologic studies.

In addition to immunologic studies, these cell lines have proven, and continue to prove, their usefulness for research into multiple areas of prostate tumor biology and therapy. For example, we recently published studies describing a possible mechanism for chromosome 8p loss and 8q gain frequently observed in high-grade localized prostate cancer [Virgin et al., 1999], studies on mechanisms of prostate tumor metastasis involving TIMP and MMP expression [Dong et al., 2001] and IL-13 receptor targeted gene therapy [Kawakami et al., 2001]. Prostate tumor tropism to the bone studies have also been aided by the use of these paired normal and tumor-derived cell lines.

## ACKNOWLEDGMENTS

This work was supported in part by funds from the Providence Research Foundation, a grant from the Chiles Foundation funds from the Southwest Cancer Center, and a grant from the American Cancer Society (RPG-00-207-01-CCE to R.K.B.).

## SOURCES OF MATERIALS

| Item | Company |
| --- | --- |
| Bovine pituitary extract (BPE) | Gibco/Invitrogen |
| cDNA libraries | Clontech |
| Cholera toxin | Sigma |
| Cryo-vials | Nalge Nunc |
| Falcon tissue culture flasks | Becton Dickinson |
| Fetal bovine serum (FBS) | Biowhittaker/Cambrex |
| Fungizone | Biowhittaker/Cambrex |
| Gentamycin sulfate | Biowhittaker/Cambrex |
| HEPES buffer | Biowhittaker/Cambrex |
| Keratinocyte- SFM | Gibco/Invitrogen |
| Kodak X-OMAT film | Eastman Kodak |
| L-Glutamine | Biowhittaker/Cambrex |
| Nalgene cryovials | Fisher |
| Nalgene filters | Fisher |
| Penicillin/streptomycin | Biowhittaker/Cambrex |
| Primers | Research Genetics |
|  | Or synthesized on an Applied Biosystems DNA synthesizer |
| Rat tail collagen I-coated tissue culture plates | Collaborative Biomedical Products |
| rEpidermal growth factor (rEGF) | Gibco/Invitrogen |
| *RNAimage* kits | GenHunter |
| Sequencing gel filter paper | Bio-Rad |
| Sterile 15 ml centrifuge tubes | Fisher |
| Taq polymerase | Boehringer Mannheim |
| Tissue culture flasks, 75 cm$^2$ | Falcon, Becton Dickinson |
| Trypsin | Fisher |
| Trypsin-EDTA | Fisher |

## REFERENCES

Bergerheim, U.S.R., Kunimi, K., Collins, V.P., Ekman, P. (1991) Deletion mapping of chromosomes 8, 10, and 16 in human prostatic carcinoma. *Genes Chrom. Cancer* 3:215–220.

Bova, G.S., Carter, B.S., Bussemakers, M.J.G., Emi, M., Fujiwara, Y., Kypriano, N., Jacobs, S.C., Robinson, J.C., Epstein, J.I., Walsh, P.C., Osaacs, W.B. (1993) Homozygous deletion and frequent allelic loss of chromosome 8p22 loci in human prostate cancer. *Cancer Res.* 53:3869–3873.

Bright, R.K., Vocke, C.D., Emmert-Buck, M.R., Duray, P.H., Solomon, D., Fetsch, P., Rhim, J.S., Linehan, W.M., Topalian, S.L. (1997) Generation and genetic characterization of immortal human prostate epithelial cell lines derived from primary cancer specimens. *Cancer Res.* 57:995–1002.

Brothman, A.R., Peehl, D.M., Patel, A.M., McNeal, J.E. (1990) Frequency and pattern of karyotypic abnormalities in human prostate cancer. *Cancer Res.* 50:3795–3803.

Brothman, A.R., Peehl, D.M., Patel, A.M., MacDonald, G.R., McNeal, J.E., Ladaga, L.E., Schellhammer, P.F., (1991) Cytogenetic evaluation of 20 cultured prostatic tumors. *Cancer Genet. Cytogenet.* 55:79–84.

Brothman, A.R., Patel, A.M., Peehl, D.M., Schellhammer, P.F. (1992) Analysis of prostatic tumor cultures using fluorescence in-situ hybridization (FISH). *Cancer Genet. Cytogenet.* 62:180–185.

Byrne, J.A., Tomasetto, C., Gasrnier, J.M. Rouyer, N., Mattei, M.G., Bellocq, J.P., Rio M.C., Basset, P. (1995) A screening method to identify genes commonly over-expressed in carcinomas and the identification of a novel complementary DNA sequence. *Cancer Res.* 55:2896–2903.

Carter, B.S., Ewing, C.M., Ward, W.S., Treiger, B.F., Aalders, T.W., Schalken, J.A., Epstein, J.I., Isaacs, W.B. (1990) Allelic loss of chromosomes 16q and 10q in human prostate cancer. *Proc. Natl. Acad. Sci. USA* 87:8751–8755.

Demers, G.W., Foster, S.A., Halbert, C.L., Galloway, D.A. (1994) Growth arrest by induction of p53 in DNA damaged keratinocytes is bypassed by human papillomavirus 16 E7. *PNAS* 91:4382–4386.

Dong, Z., Nemeth, J.A., Cher, M.L., Palmer, K.C., Bright, R.K., Fridman, R. (2001) Different regulation of matrix metalloproteinase-9, tissue inhibitor of metalloproteinase-1 (TIMP-1) and TIMP-2 expression in co-culture of prostate cancer and stronmal cells. *Int. J. Cancer* 93:507–515.

Emmert-Buck, M.R., Vocke, C.D., Pozzatti, R.O., Duray, P.H., Jennings, S.B., Florence, C.D., Zhuang, T.Z., Bostwick, D.G., Liotta, L.A., Linehan, W.M. (1995) Allelic loss on chromosome 8p12–21 in microdissected prostatic intraepithelial neoplasia. *Cancer Res.* 55:2959–2962.

Halbert, C.L., Demers, G.W., Galloway, D.A. (1991) The E7 gene of human papillomavirus type 16 is sufficient for immortalization of human epithelial cells. *J. Virol.* 65:473–478.

Halbert, C.L., Demers, G.W., Galloway, D.A. (1992) The E6 and E7 genes of human papillomavirus Type 6 have weak immortalizing activity in human epithelial cells. *J. Virol.* 66:2125–2134.

Horoszewicz, J.S., Leong, S.S., Kawinski, E. Karr., J.P., Rosenthal, H., Chu, T.M., Mirand, E.A., Murphy, G.P. (1983) LNCaP model of prostatic carcinoma. *Cancer Res.* 43:1809–1818.

Hung, T., Mak, K., Fong, K.A. (1990) A specificity enhancer for polymerase chain reaction. *Nucleic Acid Res.* 18:4953.

Isaacs, W.B., Bova, G.S., Morton, R.A., Bussemakers, J.D., Ewing, C.M. (1994): Molecular biology of prostate cancer. *Sem. Oncol.* 21:514–521.

Kaighn, M.E., Narayan, K.S., Ohnuki, Y., Lechner, J.F., Jones, L.W. (1979): Establishment and characterization of a human prostatic carcinoma cell line (PC3). *Invest. Urol.* 17:16–23.

Kawakami, K., Husain, S.R., Bright, R.K., Puri, R.K. (2001) Gene transfer of interleukin 13 receptor a2 chain dramatically enhances the antitumor effect of IL-13 receptor-targeted cytotoxin in human prostate cancer xenografts. *Cancer. Gene Ther.* 8:861–868.

Lalani, el-N., Laniado, M.E., Abel, P.D. (1997) Molecular and cellular biology of prostate cancer. *Cancer Metas. Rev.* 16:29–66.

Lee, M.S., Garkovenko, E., Yun, J.S., Weijerman, P.C., Peehl, D.M., Chen, L.S., Rhim, J.S. (1994) Characterization of adult human prostatic epithelial cells immortalized by polybrene-induced DNA transfection with a plasmid containing an origin-defective SV40 genome. *Int. J. Oncol.* 4:821–830.

Litt, M., Hauge, X., Sharma, V. (1993) Shadow bands seen when typing polymorphic dinucleotide repeats: some causes and cures. *Biotechniques* 15:280–284.

Macoska, J.A., Trybus, T.M., Benson, P.D., Sakr., W.A., Grignon, D.J., Wojno, K.D., Pietruk, T., Powell, I.J., (1995) Evidence for three tumor suppressor gene loci on chromosome 8p in human prostate cancer. *Cancer Res.* 55:5390–5395.

Munger, K., Phelps, W.C., Bubb, V., Howley, P.M., Schlegel, R. (1989) The E6 and E7 genes of the human papillomavirus type 16 together are necessary and sufficient for transformation of primary human keratinocytes. *J. Virol.* 63:4417–4421.

Peehl, D.M. (2002) Human prostatic epithelial cells. In Freshney, R.I., Freshney, M.G., ed., Culture of Epithelial Cells. Wiley-Liss, Inc.: New York, NY, pp. 172–194.

Sakr, W.A., Macoska, J.A., Benson, P., Grignon, D.J., Wolman, S.R., Pontes, J.E., Crissman, J.D. (1994) Allelic loss in locally metastatic, multisampled prostate cancer. *Cancer Res.* 54:3273–3277.

Scanlan, M.J., Chen, Y.T., Williamson, B., Gure, A.O., Stockert, E., Gordan, J.D., Tureci, O., Sahin, U., Pfreundschuh, M., Old, L.J. (1995) Characterization of human colon cancer antigens recognized by autologous antibodies. *Int. J. Cancer* 76:652–658.

Stone, K.R., Mickey, D.D., Wunderli, H., Mickey, G.H., Paulson, D.F. (1978) Isolation of human prostate carcinoma cell line (DU 145). *Int. J. Cancer* 21:274–281.

Trapman, J., Sleddens, H.F.B.M., van der Weiden, M.M., Dinjens, W.N.M., Konig, J.J., Schroder, F.H., Faber, P.W., Bosman, F.T. (1994) Loss of heterozygosity of chromosome 8 microsatellite loci implicates a candidate tumor suppressor gene between the loci D8S87 and D8S133 in human prostate cancer. *Cancer Res.* 54:6061–6064.

Virgin, J.B., Hurley, P.M., Cher, M.P., Nahhas, F., Bebchuck, K.G., Mohamed, A.N., Sakr, W.A., Bright, R.K., Cher, M.L. (1999) Isochromosome 8q is associated with 8p loss of heterozygosity in a prostate cancer cell line. *The Prostate* 41:49–57.

Vocke, C.D., Pozzatti, R.O., Bostwick, D.G., Florence, C.D., Jennings, S.B., Strup, S.E., Duray, P.H., Liotta, L.A., Emmert-Buck, M.R., Linehan, W.M. (1996) Analysis of 9 microdissected prostate carcinomas reveals a high frequency of allelic loss on chromosome 8p12–21. *Cancer Res.* 56:2411–2416.

Webber, M.M., Chaproniere-Rickenberg, D.M., Donohue, R.E. (1984): Isolation and growth of adulthuman prostatic epithelium in serum-free, defined media. In *Methods for Serum-Free Culture of Cells of the Endocrine System.* Wiley Liss-Inc.: New York, NY, pp. 47–61.

# 7

# The Development of Human Ovarian Epithelial Tumor Cell Lines from Solid Tumors and Ascites

**Anne P. Wilson**

*Woodbine Terrace, Stanton, Near Ashbourne, Derbyshire, DE6 2DA, UK*
*annienoble2003@yahoo.co.uk*

*Culture of Human Tumor Cells*, Edited by Roswitha Pfragner and R. Ian Freshney.
ISBN 0-471-43853-7 Copyright © 2004 Wiley-Liss, Inc.

## I. BACKGROUND

The following sections provide a brief overview of the pathology and clinical features of ovarian cancer. For a more detailed account of the pathology of ovarian tumors refer to "Tumors of the Ovary" by Fox and Langley [Fox and Langley, 1976]. All aspects of ovarian

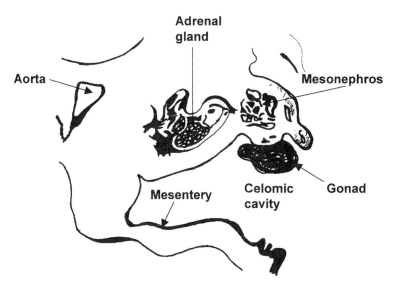

**Figure 7.1** Embryonic development of the gonad and the mesenteric lining of the celomic cavity.

cancer, including aetiology, genetics, biology and treatment are comprehensively covered in a series of volumes arising from the Helene Harris Memorial Trust Meetings [Sharp et al. (eds), 1990, 1992, 1995, 1996].

### 1.1. Embryology and Histology

A working knowledge of the embryonic development of the female reproductive tract and the histology of the normal human ovary illuminates the pathology and aetiology of the common epithelial tumors of the ovary. Embryonic development of the ovary takes place in four main stages. Firstly, undifferentiated germ cells migrate to the genital ridges, these being thickenings of the celomic epithelium located ventrally to the developing kidneys (Fig. 7.1). Proliferation of the celomic epithelium and underlying mesenchyme follows, giving rise ultimately to a peripheral cortex and inner medulla in the primitive gonad. In the male fetus the production of Müllerian Inhibiting Substance (MIS) leads to suppression of the development of the female phenotype, favouring development of the medulla, involution of the cortex and differentiation along the Wolffian pathway into the male genitourinary system. In the female the absence of MIS favours preferential development of the cortex. The reproductive tract develops along the Müllerian pathway with downward migration of the celomic epithelium and differentiation to produce the endocervix, the epithelial lining of the Fallopian tubes and the endometrial lining of the uterus.

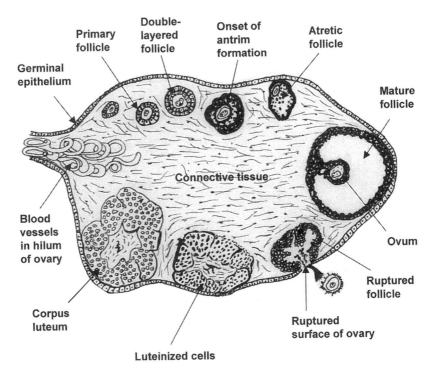

Germinal epithelium

Primary follicle

Double-layered follicle

Onset of antrim formation

Atretic follicle

Mature follicle

Connective tissue

Blood vessels in hilum of ovary

Ovum

Ruptured follicle

Corpus luteum

Ruptured surface of ovary

Luteinized cells

**Figure 7.2**  The cellular structure of the human ovary, showing different stages of follicular development.

The celomic epithelium forms the serosal surfaces of the abdominal cavity and pleural cavities and is composed of a squamous epithelium of mesothelial cells covering both the parietal and visceral surfaces of these cavities. The surface of the ovary is also covered by celomic epithelium, which is cuboidal rather than squamous. The term *germinal epithelium* has been commonly used to name this surface covering because the primordial germ cells were once believed to originate from this layer. Although this is now known to be incorrect the original term has remained in common usage.

The human ovary is a complex structure containing a variety of different cell types. The basic structure is shown schematically in Figure 7.2 with the various stages of ovulation shown around the perimeter of the cortex. Prior to puberty the ovary contains ∼300,000 primordial follicles in the cortical region. These are the functional units of the ovary, containing the female gametes and the steroid-producing cells. The two-layered theca folliculi surrounding the follicles consists of an outer layer of connective tissue cells (theca externa) and an inner layer of secretory epithelial cells (theca interna). The inner layer of granulosa cells surround the ovum and are separated from the theca interna by a basement membrane.

The hormonal control of the ovulation cycle is a complex subject but the main events can be summarized briefly to indicate the relevant hormones and their sites of production. Maturation of the follicles begins at puberty with proliferation of granulosa cells and secretion of follicular fluid into the central cavity under the influence of follicle stimulating hormone (FSH) and luteinizing hormone (LH) produced by the pituitary gland. Large amounts of estrogens are synthesized by the developing follicle. At ovulation the mature follicle ruptures, breaking through the distended wall of the ovary and releasing the egg into the antrum of the fallopian tube. The remaining granulosa cells in the ruptured follicle proliferate and fill the residual cavity, forming the corpus luteum, which synthesises large amounts of progesterone.

## 1.2. Pathology and Aetiology of Ovarian Tumors

A large number of histological variants of tumors arise in the ovary because of the cellular diversity of this organ [Fox and Langley, 1976]. Tumors may develop from germ cells (e.g. teratoma, dysgerminoma), the sex-cord stromal cells (e.g. granulosa cell tumors, thecomas, fibromas) or the surface epithelium (serous, mucinous, clear cell, endometrioid). The latter group are known as the common epithelial tumors of the ovary and their incidence is much greater than those arising from the other cell types. These tumors may also be referred to as Müllerian tumors, the name indicating the relationship between the histology of the tumor and those parts of the female reproductive tract resulting from differentiation along the Müllerian pathway in the developing fetus (See Section 1.1). Thus the serous, mucinous and endometrioid tumors of the ovary resemble those tumors arising in the epithelium of the fallopian tubes, endocervix and endometrium respectively. The clear cell tumor and the Brenner tumors resemble tumors arising from mesonephroid tissue which is also derived from the celomic epithelium during embryonic development.

It is believed that the different histological sub-types of the common epithelial cancers of the ovary reflect the retained pluripotency of the surface epithelium of the ovary (derived from celomic epithelium) to differentiate along Müllerian pathways to give rise to tumors resembling the different epithelia of the female reproductive tract or the urinary system. There are benign counterparts for these tumors and it is not certain whether malignant tumors arise from foci of transformation within benign neoplasms, whether they arise *de novo* or whether from a combination of the two. The carcinogenic stimulus has not been determined either, but evidence has been cited indicating that the trauma of incessant ovulation to the surface of the ovary may provide a carcinogenic environment. Healing of the surface epithelium after ovulation gives rise to germinal inclusion cysts within the cortex of the ovary. The local environment within the cleft or cyst may contain transforming factors important in the carcinogenic process. Indirect evidence linking incessant ovulation with increased risk of ovarian cancer includes increased relative risk rates in nulliparous women, and decreased relative risk rates in women taking the contraceptive pill.

### 1.3. Clinical Features of Ovarian Cancer

The incidence of epithelial ovarian cancer varies in different countries from 1 per 100,000 to 17 per 100,000. Although it is not the most common gynaecological cancer it carries the highest death rate. Late presentation with advanced disease is usual because the disease is asymptomatic in its early stages. First-line treatment involves surgery to remove as much bulk of tumor as possible, followed by intensive chemotherapy with platinum-based regimes. The response rate to first-line chemotherapy is high but relapse usually occurs within one to two years. Development of resistance is common and although second-line chemotherapy can produce responses, the five year survival rate is very low. Newly developed drugs such as taxol have produced a small increase in survival rates but there is still a long way to go in understanding the biology of the disease and finding more effective ways of treating patients.

A particular problem associated with this is the heterogeneity of the disease. Within the main histological subtypes of mucinous, serous, endometrioid, and clear cell carcinomas there are three possible degrees of differentiation, (well, moderate, and poor), four main FIGO stages (Table 7.1) and a borderline category of tumor which carries a better prognosis than the frankly malignant tumors, but can

**Table 7.1. FIGO Staging system for ovarian cancer.**

| FIGO Stage | One ovary | Both ovaries | Ascites | Spread |
|---|---|---|---|---|
| Ia | **Yes**[1] | No | No | No |
| Ib | Yes | **Yes** | No | No |
| Ic | Yes | Yes/No | **Yes** | No |
| IIa | Yes | Yes/No | Yes/No | **To uterus and/or tubes** |
| IIb | Yes | Yes/No | Yes/No | **Extension to other pelvic tissues** |
| III | Yes | Yes/No | Yes/No | **Widespread intra-peritoneal metastases to the abdomen** |
| IV | Yes | Yes/No | Yes/No | **Distant metastases outside the peritoneal cavity** |

[1]The deciding factor for assigning FIGO Stage is shown in bold type.

still be fatal in a small percentage of cases. The potential number of variations is therefore in excess of 100 and it is very important when developing cell lines that as much information as possible is collected in order that use of the cell lines can be maximised and variations taken into consideration when comparing the behaviour of different cell lines. In this way, the cell lines can be better used to understand the biology of the disease and therefore develop more effective treatment strategies.

## 1.4. Overview of Ovarian Tumor Cell Culture and Cell Line Development

Early attempts to culture ovarian adenocarcinomas employed the plasma clot technique and obtained some success with outgrowth of epithelial cells using a balanced salt solution supplemented with 50% ascitic fluid and embryo extract [Parmley and Woodruff, 1974]. In the 1960's there was an increase in the number of attempts made to culture ovarian tumor cells which paralleled the introduction of anticancer drugs to clinical use. Reports were also appearing of correlations between in vitro drug sensitivity testing results on primary cultures from surgical specimens and clinical response to the drugs [Walker et al., 1965, Limburg and Heckmann, 1968]. The first reports of continuous cell lines derived from ovarian adenocarcinomas were published in the seventies together with detailed culture characteristics of ovarian tumor cells. [Ioachim et al., 1974, Ioachim et al., 1975, DiSaia et al., 1975, Kimoto et al., 1975]. In the last thirty years there has been a sharp increase in the number of ovarian tumor cell lines and the literature now contains reports of more than 200 [Wilson and Garner, 1999]. Some of these lines are used extensively as models for human ovarian cancer in ways which reflect the clinical problems presented by the disease. This includes mechanisms of drug resistance, screening for new drugs and drug development, analysis of oncogene/ tumor suppressor gene expression, development of models for adhesion and metastasis and studies on the control of proliferation via peptide growth factors, cytokines and hormones.

The available cell lines are broadly representative of the clinical spectrum of the disease, but there are gaps. These include underdocumentation of relevant clinical details (age, stage, specimen site, and treatment history) and either none or only one or two cell lines from Stage I disease, familial ovarian adenocarcinoma, benign neoplasms and mucinous ovarian carcinomas.

The success rate for establishing ovarian tumor cell lines from clinical specimens in different labs ranges from <1% to ~30%, though the percentage of tumors which give rise to primary cultures is much higher. Reasons for failure are usually due to absence of proliferation of epithelial cells, often accompanied by overgrowth of stromal cells. Stromal cell overgrowth can be minimized, but there are still many cultures which fail to show epithelial proliferation.

Media which have been most frequently used as a nutrient source for cell line development are Dulbecco's modification of Eagles medium (DMEM), RPMI 1640, Ham's F12 or a 1:1 mix of DMEM and Ham's F12. These are routinely supplemented with animal serum (usually fetal bovine serum, but sometimes horse or human umbilical cord serum) at concentrations ranging from 5–20%. Other supplements used include insulin, transferrin, selenium, [Wilson, 1998], cell-free-ascites [Uitendal et al., 1983, Broxterman et al., 1987, Mills et al., 1988] and various hormones and non-essential amino acids. Ascitic fluids and pleural effusions often contain large numbers of tumor cells and the success rate for establishing cell lines from these sources is greater than that for solid tumors.

## 2. PREPARATION OF MEDIA AND REAGENTS

### 2.1. Culture Medium PPIGSS

Cultures are routinely initiated using Dulbecco's modification of Eagles medium (DMEM), containing sodium pyruvate, penicillin, insulin, glutamine, fetal calf serum and streptomycin (PPIGSS) which are all added to a single-strength stock medium of DMEM immediately prior to use:

Dulbecco's modification of Eagle's medium (DMEM) containing—

| | |
|---|---|
| NaHCO$_3$ | 3.7 g/l |
| fetal calf serum | 10% |
| glutamine | 2 mM |
| sodium pyruvate | 1 mM |
| insulin | 20 IU/ml |
| streptomycin | 20 µg/ml |
| penicillin | 20 IU/ml |

is made up from single-strength DMEM. All other components are obtained as sterile stock solutions and added in appropriate amounts to give the final concentrations shown.

Experiments on the effects of insulin on growth of ovarian cancer lines have shown that, while not essential, it is mitogenic and significantly enhances proliferation. Hydrocortisone has been found to inhibit growth in some cell lines and it is certainly less effective than insulin in promoting growth [Wilson et al., 1991]. It is also a growth factor for mesothelial cells [Connell and Rheinwald, 1983] and its absence is therefore beneficial in reducing stromal cell contamination. Epidermal growth factor is not routinely used for the same reason [LaRocca and Rheinwald, 1985].

## 2.2. Double Strength Dulbecco's Medium 2× DMEM

Dilute fivefold in Ultra-pure water from 10× DMEM stock and buffer to pH 7.0–7.4 by the addition of an appropriate amount of 7.5% sodium bicarbonate such that the colour of the medium has changed from yellow to orange.

## 2.3. CMF

Hanks' balanced salt solution lacking $Ca^{2+}$ and $Mg^{2+}$.

## 2.4. Trypsin

Trypsin, 2.5% (10×) stored at $-20°C$ as 10 ml aliquots.

## 2.5. Trypsin/EDTA

Trypsin/EDTA (versene) is purchased as a combined stock solution from GIBCO and stored at $-20°C$ according to manufacturer's instructions. For use, the stock solution is thawed, diluted to single-strength in CMF, aliquotted to 5ml sterile bijoux and stored at $-20°C$ for use as required.

## 2.6. DNase

DNase I, 0.4%, made up in sterile HBSS and stored at $-20°C$ as 1 ml aliquots.

## 2.7. CFA

CFA is cell-free ascitic fluid derived from ascites after removal of the cells (see Protocol 7.7, Step (c)). Retain 100 ml (or more) of ascitic fluid after cells have been removed for culture. Dispense into 20 ml aliquots, label with patient laboratory number and freeze. Thaw out one aliquot when the stromal cell monolayer is ready for use (See 5.2 and Protocol 7.9).

## 3. COLLECTION OF TUMORS

### 3.1. Ethical Consent

The use of human tissue for research purposes requires informed patient consent and any study which involves the collection of clinical samples must obtain ethical approval. The United Kingdom Co-ordinating Committee on Cancer Research (UKCCR) has published set guidelines for the use of cell lines in cancer research which includes information on ethical and legal issues [UKCCCR Guidelines, 1999].

## 3.2. Collection System

The following system has worked well in practice. Critical factors in a successful collection procedure have been found to be:

(i)   Collaboration with an enthusiastic gynaecological surgeon who will provide tumor tissue.

(ii)   Collaboration with theatre staff for storage of specimen containers and contacting laboratory staff when tumor tissue is available for collection.

(iii)   Appropriate forms in theatre for documentation of patient details for follow-up and information on the specimens provided.

(iv)   Collaboration with a histopathologist. It is vital that tumor removal for laboratory use does not jeopardize the histological diagnosis.

(v)   Access to clinical follow-up (treatment, response, tumor markers, histopathology).

(vi)   Ethical approval from the local Clinical Ethics Committee, and informed consent from the patient or a close relative.

(vii)   A data manager who can liaise between consultant, laboratory staff, theatre staff and medical secretaries.

(viii) With the increased emphasis on the genetics of human cancer the collection of blood lymphocytes from patients who have donated tissue is a valuable asset which enhances the potential uses to which the cultures and cell lines can be put. It is also an invaluable source of patient's DNA which can be used subsequently to confirm the identity of any cell line derived from the specimen.

The system is shown in Table 7.2.

## 3.3. Collection of Solid Tumors and Ascites from Surgery

All reagents and equipment must be sterile. The addition of antibiotics is optional: some laboratories prefer to omit them so that low levels of contamination show up rapidly.

### Protocol 7.1.   Collection of Ovarian Tumor Samples

#### Reagents and Materials
*Sterile*
❑   Specimen containers: Universal containers (25 ml) or feces/ sputum pots. The latter are more convenient for larger specimens

**Table 7.2. Step by step guide to collecting tumor samples.**

| Location | Staff | Role |
|---|---|---|
| Gynaecology Clinic | Consultant Data manager | Identification of potential cases of ovarian neoplasia—benign/malignant |
| | Data manager with Medical secretaries | Checking theatre lists for dates of operation on identified cases |
| | Data manager | Obtaining informed patient consent |
| | Data manager Lab staff | Passing on identified cases and date of operation |
| | Lab staff Operating theatre staff | Informing theatre staff of interest in particular cases<br>Finding position of patient on theatre list<br>Checking supplies of specimen containers and forms; replenishing if necessary |
| Gynaecology Theatre | Consultant Theatre staff | Removal of tumor and transfer of sample to specimen pot<br>Storage in refrigerator<br>Contacting lab staff to inform that samples are ready for collection |
| | Lab staff | Collect specimens<br>Make sure patient form is completed before taking sample<br>Record details in laboratory log book<br>Process specimens |
| | Data manager Histopathology Lab staff | Obtain histopathology report for lab<br>Obtain patient follow-up details as and when relevant<br>Obtain tumor marker results<br>Obtain Copy of patient consent for filing |

because they have a wider diameter than a universal. Half-fill the storage containers with sterile HBSS, label with your name (or whoever the contact person will be) and contact phone number, and store at 4°C, preferably in the operating theatre refrigerator.

❏ Bottles for ascitic fluid, e.g. used media bottles which have been washed and re-sterilised.

❏ Patient forms for laboratory use; make sure they are clearly distinguishable from the routine histopathology forms e.g. by printing on brightly coloured paper, and have your name and contact number on them.

### Protocol

(a) Obtain a representative sample of unfixed tumor tissue from the operating surgeon. The site needs to be specified, e.g. primary left ovary or right ovary, secondary deposit from omentum, bowel. A minimum size of 2 cm$^3$ is recommended to ensure sufficient recovery of cells for setting up primary cultures; more is obviously desirable.

(b)   Either yourself, if in attendance, or a member of clinical staff should transfer the tumor sample into a sterile specimen pot containing HBSS. It is advisable to ensure that all staff involved with the collection are made aware that the tissue MUST NOT be formalin-fixed. Remember that the need of the Pathologist for adequate tumor tissue takes precedence.

(c)   Record the relevant information on the form. UKCCCR recommend that the patient's name is not recorded, other than by clinical staff and pathologist. For laboratory use patient identification should therefore be restricted to the hospital registration number. Also record the date of birth, the consultant's name, the nature of the specimen provided (e.g. solid tumor, metastatic deposit, peritoneal fluid, ascitic fluid etc.) and the site from which the tissue was taken (e.g. right ovary, left ovary, omentum, gut etc.)

(d)   Transfer the information on the form to a laboratory record book and assign the specimens individual ID numbers for laboratory use. Keep permanent records which ensure that the ID number can be retrospectively matched with the registration number, date of operation and site of origin. This also ensures that relevant clinical information can be matched with any tumour cell line which is developed.

**Ascites**

Ascitic fluid may be present in patients undergoing laparotomy. This is removed either by aspiration with a syringe or, if the volume is large, by use of a vacuum suction device which removes the fluid into large non-sterile glass bottles. Ascites can be a good source of tumor cells and it is worthwhile collecting large volumes if available. To ensure sterility it is preferable that a large syringe is used for the laboratory samples, transferring the fluid directly into the sterile containers provided. Fluid samples obtained from surgery may be labeled as ascitic fluid, peritoneal fluid or peritoneal washings. Normal volumes of peritoneal fluid range from 3–15 ml, whereas ascitic fluid volumes may range from 20 ml to four or more liters. The tumor cell concentration in ascites is very variable and a small volume of fluid may contain a large number of cells, whilst recovery from large volumes may produce virtually no cells. Peritoneal washings are obtained by washing out the peritoneal cavity with sterile saline and recovering the washings. It is carried out for detecting tumor cells free in the peritoneal cavity as part of the staging process. It is therefore a diagnostic procedure and washings are normally sent for cytological analysis.

Patients who relapse after chemotherapy commonly develop ascites and are admitted as in-patients for paracentesis. This involves insertion of a tube into the peritoneal cavity, using a local anesthetic, and drainage of the accumulated fluid into sterile plastic bags. These are fitted with a drainage device and the fluid can be drained from them directly into sterile bottles prior to centrifugation.

## 3.4. Understanding Pathology Reports

Generally patients undergo complete removal of the uterus and ovaries (bilateral salpingo-oophorectomy and hysterectomy (BLSOH) as well as removal of metastatic deposits. The omentum may also be biopsied for evidence of micrometastasis. Table 7.3 shows a hypothetical example of a histology report.

Information which can be gleaned from this report includes:

(i)  Presence of secretory endometrium indicates the patient is still menstruating.

(ii)  She has benign neoplasms in the uterus i.e. fibroids/leiomyata.

(iii)  She is at least FIGO Stage III since the omentum, located in the upper abdomen, contains metastatic tumor deposits (see Table 7.1).

**Table 7.3.  Example of information to be found on a hypothetical pathology report.**

| Source | Specimen | Microscopy | Summary |
|---|---|---|---|
| Uterus and cervix with both tubes and ovaries | The uterus is enlarged and distorted measuring 10 × 4 cms with attached fallopian tubes and right ovary. A separate cystic ovarian mass measures 15 cm on its maximum diameter. Numerous fibroids are visible in the cut surface of the myometrium. The cut surface of the left ovarian mass shows a multi-locular cyst. There are some areas of firm creamy white tissue in the wall. The external wall of the cyst is smooth, except for one area disrupted by solid tumor. | No abnormalities were seen in sections from the uterine cervix. Sections from the uterine corpus show secretory endometrium of normal appearance and confirm the presence of benign leiomyata in the myometrium. Follicular cysts are present in the R. ovary and the R tube is normal. Sections of the left ovarian mass show an adenocarcinoma which is moderately differentiated in some areas. Other areas are anaplastic with a high mitotic index. The ovarian capsule is penetrated by adenocarcinoma. | Uterus, leiomyata; left ovary, serous adenocarcinoma-anaplastic. |
| Omentum | A piece of fatty tissue 3 cm across, containing several areas of firm, creamy white tumor. | | Metastasis |

(iv) Some of the tumor is sufficiently differentiated to allow a diagnosis of serous cystadenocarcinoma.

(v) There are different levels of differentiation within the tumor.

The level of differentiation reported in the final summary is usually the lowest present in the tumor. i.e. that which carries the worst prognosis. It is therefore desirable to formalin-fix a representative sample of the tumor tissue used for culture so that a separate histology report can be obtained. This not only ensures that representative histology is available to complement the cell culture but will help to provide confirmation, if necessary, that the pathology report has not been compromised by removal of tissue for research purposes. Additionally, it may show reasons for culture failure, e.g. necrosis, low tumor cell content, or preponderance of normal tissue. In the early days of developing one's expertise in tissue culture such information is invaluable for making relationships between external appearance of tumor, the cell yield and culture outcome. The final tumor culture which arises may also have arisen from a particular level of differentiation.

## 4. PRIMARY CULTURE

### 4.1. Disaggregation of Solid Tumors

In a previous publication summarizing disaggregation methods it was reported that dispase appeared to produce the best success rate for establishing cell lines [Wilson, 1998]. More recently cold trypsinisation was found to give higher cell yields, higher cell density in primary culture, less stromal cell overgrowth and more rapid proliferation of epithelial cells. When the same tumor was disaggregated and culture outcome compared for monolayers derived from explant culture, washings, mechanical disaggregation, warm trypsinisation, cold trypsinisation, collagenase digestion and dispase digestion, the best results were obtained with cold trypsinisation (Table 7.4). This method is easy to perform and fits well into a normal working day since trypsinisation is carried out at 4°C overnight as is outlined in Protocol 7.3. The tumors must first be dissected clear of fat and necrotic tissue and chopped finely, as in the following protocol.

### Protocol 7.2.  Mechanical Disaggregation of Ovarian Tumors

**Reagents and Materials**

*Sterile*
- Culture medium: PPIGSS (see Section 2.1)
- CMF: see Section 2.3

**Table 7.4.   A comparison of different methods of obtaining free cells from a solid ovarian tumor (D729).**

| Method | Confluence at 7 days | Epithelial cells adherent at 7 days | Stromal cells adherent at 7 days | Outcome at ∼4 months |
|---|---|---|---|---|
| Explants | <1% | −/+ | −/+ | Discarded—no adherent cells |
| Washings | 10% | + | + | Discarded—very few adherent cells |
| Mechanical disaggregation | 10% | + | + | Discarded—few adherent cells |
| Warm trypsinisation | 20% | ++ | ++ | Discarded—low cell density and very slow proliferation |
| Dispase digestion | 50% | ++ | ++ | Discarded—stromal cells increased and epithelial cells slow to proliferate |
| Collagenase digestion | 50% | ++ | ++ | Discarded—as dispase |
| Cold trypsinisation | 80% | ++++ | + | Subcultured and 70 vials frozen between Pass 3 and Pass 12 |

❑ Glass Petri dishes, 90 cm diameter
❑ Dissecting instruments: pairs of large and small curved scissors, forceps, 2 scalpels
❑ Pastettes (plastic Pasteur pipettes)
❑ Teaspoon or large spatula
❑ Centrifuge tubes, 50-ml, screw-capped
❑ Tissue culture flasks: 25 cm$^2$ and 75 cm$^2$
*Non-sterile*
❑ Formol-saline, 10%

### Protocol

(a)   Transfer the tumor tissue from its transport container into a glass Petri dish.

(b)   Make a note of the size of the tissue and its appearance. Dissect away fat, capsule and necrotic tissue (typically friable and bloody, and can contain foci of pus). Tumors may be solid or have a cystic component, in which case solid areas protruding internally or externally from the cyst wall should be selected for disaggregation.

(c)   Remove a small piece of representative tissue and transfer to 10% formol-saline for fixation. This serves several purposes: (i) availability for confirmation of histology; (ii) storage as archival material for comparison with any cell lines which are developed; (iii) as a check for confirmation of suitability of material selected for disaggregation.

(d)   Wash exposed surface of the tissue using 3 ×10 ml CMF

(assuming a sample size of about 2 cm diameter) to remove blood. Discard the washings.

(e) Chop the tissue into fragments 3–5 mm across, using either crossed scalpels blades if the tumor is soft or a scalpel with forceps to anchor the tissue if it is harder. Fragments can be stored overnight at 4°C in culture medium (about 10:1, V/V, medium : tissue) if it is not convenient to complete the disaggregation procedure.

(f) Add 10 ml sterile CMF to the Petri dish and agitate fragments to remove any blood, which contains trypsin inhibitors.

(g) Tip the dish to an angle of 10–20° from the horizontal and allow the fragments to settle out into the angle of the dish.

(h) Gently aspirate the washings containing blood and mechanically loosened tumor cells using a Pastette, and put into a sterile universal container.

(i) Repeat this wash procedure twice more.

(j) Pool and centrifuge the washings at 175 g for 5 min.

(k) Resuspend the cell pellet in 10 ml culture medium, label as 'washings' and store at 4°C.

(l) Using curved scissors, gently mince the remaining tumor fragments into small pieces (2–4 mm) using only enough CMF to keep the tissue moist. (If too much liquid is added, the fragments float away from the scissors and mincing is harder to carry out).

(m) Transfer the tumor mince from the Petri dish to a sterile universal container using either the sterile teaspoon or spatula.


## Protocol 7.3.  Disaggregation of Solid Ovarian Tumor by Cold Trypsinisation

### Reagents and Materials

*Sterile*
- ❏ Trypsin, 2.5%: see Section 2.4
- ❏ DNase I, 0.4% see Section 2.6
- ❏ CMF: see section 2.3
- ❏ Wide-necked conical flask, 250 ml, with foil cap
- ❏ Sterile magnetic stirring bar
- ❏ Universal containers or 50 ml centrifuge tubes
- ❏ Tissue culture flasks: 25 cm$^2$ and 75 cm$^2$

*Non-sterile*
- ❏ Magnetic stirrer
  or
- ❏ Shaking water-bath at 37°C

❑ Viability stain: Trypan Blue, 0.4% or Nigrosin, 0.05%, in PBSA or HBSS

### Protocol

(a) Dilute 1 ml trypsin (2.5%) to 10 ml with CMF to give a working solution of 0.25% and add 100 μl of 0.4% DNase to give a final concentration of 0.004%. Aim for 2 vol of 0.25% trypsin/DNase solution to 1 vol of tissue fragments. (DNase is included in the disaggregation solution because denatured DNA released from damaged cells becomes viscous, sticks to the viable cells and causes reaggregation, which makes cell recovery very difficult).

(b) Add the trypsin solution to the tumor mince in the universal container from Protocol 7.2 and incubate at 4°C overnight.

(c) The following day, transfer the minced tumor tissue into a sterile wide-necked conical flask.

(d) Add 20–30 ml of warmed 0.25% trypsin/0.004% DNase solution, per 2 cm$^3$ of tissue, and gently agitate the flask to mix.

(e) Either add a sterile magnetic bar and stir the foil-capped flask for 20–30 min at 37°C, selecting a speed which keeps the tissue in continual suspension without producing frothing, or use a shaking water bath at 37°C which is again set at a suitable speed. If agitation is too vigorous, it will cause mechanical damage to the cells.

(f) After 15–20 min remove the flask and allow the fragments to settle out under gravity.

(g) Gently decant or aspirate the trypsin solution containing dissociated cells into a suitable sterile container, e.g., a universal container or 50 ml centrifuge tube.

(h) Add 30 ml of fresh trypsin/DNase solution to the remaining tumor in the flask and repeat Steps (d–g).

(i) Immediately add an equal volume of culture medium containing 10% FBS to the harvested cell-containing trypsin solution from Step (g) to inactivate the trypsin and protect the cells from continued proteolytic digestion.

(j) Centrifuge for 5 min at 175 g and re-suspend the cell pellet in 10 ml culture medium. Label as 'T-1' and store at 4°C.

(k) Repeat Steps (d–j) until the tumor fragments have been completely disaggregated, leaving only white floating connective tissue debris. Label the fractions consecutively as 'T-2', 'T-3', 'T-4' etc. Disaggregation is usually complete after two trypsinizations, if the tissue has been incubated overnight in cold trypsin. More may be needed if the tissue is being disaggregated by warm trypsinisation only. The time of each trypsinization may need to be modified to suit the consistency of the tissue, e.g. <20 min for very soft tissue, >30 min for hard tissue.

(l)  Remove all fractions from the refrigerator, re-suspend cells and check cell count and viability for each fraction using a hemocytometer and a viability stain such as trypan blue. Pool the fractions which have the highest viability and cell yield (typically T-1 and T-2; T-3 may show a low cell yield and is more likely to contain fibroblasts, and cells in the 'washings' are often of low viability). Centrifuge the selected pooled fractions at 175 g for 5 min.

## 4.2.  Removing Red Blood Cells

The resulting cell pellet will probably contain a large number of red blood cells (rbc). Cell adherence is much improved in the absence of rbc and there are several options available for their removal. These are i) osmotic shock (snap lysis) using sterile water followed by double strength medium to restore normal osmolality, ii) density gradient centrifugation using LymphoSep or a similar density medium. iii) lysis using a commercially available red cell lysing reagent. Density gradient centrifugation has the advantage of removing dead cells with the red blood cells, thereby improving viability of the cell suspension [Vries et al., 1973].

## Protocol 7.4.  Depletion of Red Blood Cells by Osmotic Shock (Snap Lysis)

### Reagents and Materials

*Sterile*
- ❑  Sterile water (UPW)
- ❑  Pastettes
- ❑  DMEM, 2x: see Section 2.2
- ❑  Culture medium: see Section 2.1

### Protocol

(a)  Remove medium without disturbing the cell pellet.

(b)  Add 10 ml sterile water and mix gently with a sterile Pastette.

(c)  After 15–20 sec add an equal volume of 2x DMEM at pH 7.0–7.4 to restore osmolality.

(d)  Mix well and centrifuge at 175 g for 5 min

(e)  Repeat if there are still large numbers of rbc present in the cell pellet.

(f)  Resuspend the cell pellet in 5–10 ml culture medium to give the final cell suspension.

## Protocol 7.5. Depletion of Red Blood Cells by Density Centrifugation

### Reagents and Materials

*Sterile*

- ❑ Density medium: Ficoll/Hypaque, LymphoSep, Histopaque, or equivalent medium of density 1.077 g/ml
- ❑ Culture medium: see Section 2.1
- ❑ Universal container, 30 ml
- ❑ Pastettes

### Protocol

(a) Resuspend the bloody cell pellet in culture medium, adding ~10 ml medium per 1 ml cell pellet.

(b) Dispense 15 ml of density medium into a sterile universal container.

(c) Carefully and slowly layer the resuspended cells over the density medium, taking care to avoid disturbing the interface by rapid or uncontrolled pipetting. The best separation is achieved when there is minimal disruption between the two layers at this stage.

(d) Centrifuge at 175 g for 20 min with the brake switched off on the centrifuge.

(e) Carefully remove the container from the centrifuge. The universal should now contain a pellet of rbc/dead cells at the bottom of the tube, a clear layer of density medium, and a clear layer of medium above a meniscus of viable cells at the interface between the two media. Use a Pastette to carefully remove the cells at the interface and transfer them to a fresh container.

(f) Add 20 ml of medium and mix the recovered cellular suspension to wash off the density medium.

(g) Centrifuge at 175 g for 5 min and repeat the wash step.

(h) Resuspend the resulting cell pellet in 5–10 ml culture medium to obtain the final cell suspension.

### 4.3. Initiation of Primary Cultures

Disaggregates from ovarian tumors will provide cultures containing both tumor and stromal cells. Selective adhesion is used in the following protocol to separate epithelial cells from mesothelial cells.

## Protocol 7.6. Primary Culture of Ovarian Tumor Cells

### Reagents and Materials

*Sterile*

- ❑ Culture medium (see Section 2.1)
- ❑ Tissue culture flasks, 25 cm$^2$ and 75 cm$^2$
- ❑ Universal containers or centrifuge tubes, 30–50 ml

*Non-sterile*

❏ Trypan Blue, 0.4% in PBSA or saline

**Protocol**

(a) Count the approximate number of viable cells in the final cell suspension using a hemocytometer and a viability stain such as trypan blue. Note the final volume, the viability and the cell yield.

(b) Adjust the final cell concentration to about $2 \times 10^5$ viable cells per ml of culture medium and add 20 ml of cell suspension per 75-cm$^2$ flask.

(c) Incubate the flask horizontally for 60 min at 37°C to allow stromal cells in the cell suspension to attach. These are usually mesothelial cells in cultures derived from ovarian tumors. Check the flask on an inverted phase-contrast microscope for flattened mesothelial cells and floating epithelial clumps. The incubation time may need to be adjusted to allow for variation between cell populations derived from different tumors; getting this right is a process of trial-and-error.

(d) Decant the medium containing non-adherent cells into a sterile universal or centrifuge tube, centrifuge at 175 g for 5 min and re-suspend cells in culture medium at ~5 $\times 10^5$ cells/ml. No attempt need be made to obtain single cells, because clumps are often a better source of cell growth and the process of disruption can significantly reduce the viability of the final cell population. Epithelial cells like to be in clumps.

(e) Set up primary cultures at 10 ml cell suspension/25-cm$^2$ growth area. It is preferable to have at least four flasks because this gives more scope for experimentation with conditions in the process of development of the cell line. Therefore, if the cell yield is low, use 25-cm$^2$ flasks rather than 75-cm$^2$ flasks.

## Protocol 7.7.  Culture of Ovarian Carcinoma Cells from Ascitic Fluids

**Reagents and Materials**

*Sterile*

❏ Ascitic fluid
❏ Heparin, 100 U/ml
❏ Culture medium: see Section 2.1
❏ Centrifuge tubes, 50 ml

**Protocol**

(a) If processing of the fluid is to be delayed for more than 8 h, heparin at 10 U/ml can be added to prevent clotting.

(b) Transfer ascitic fluid into sterile conical-bottomed 50-ml centrifuge tubes and centrifuge for 15 min at 390 g.

(c) Decant the supernate.

(d) Resuspend each pellet in 5 ml culture medium, pool the cell suspensions and store at 4°C.

(e) Follow one of the procedures described above to remove rbc (see Protocols 7.3 and 7.4).

(f) Count cells in the final cell suspension and re-suspend at $2 \times 10^5$ viable cells/ml in culture medium.

(g) Set up cultures as in Protocol 7.5, Steps (c) and (d) to reduce stromal cell contamination prior to setting up primary cultures.

## 5. FOLLOW-UP OF PRIMARY CULTURES

Following initiation of primary culture, the next step is dependent on the state of the cell monolayer after 4–7 days. Possibilities include:

(i) *Non-adherence of tumor cells.* This may occur with tumor cells from ascites which are heavily vacuolated and have a high buoyant density. Such cells do not proliferate in suspension, nor do they become adherent, even if left for a week or more. But non-vacuolated epithelial cells may also be very slow to adhere. Treat the flask as if it is a suspension culture when changing the medium, by centrifuging the non-adherent cells, discarding the old medium and resuspending the cells in fresh medium. If the cell density is low in the original flask, return the cells to the same flask. If the cell density is quite high seed a fresh flask with harvested floaters. This also has the advantage that the newly seeded flask will not be contaminated with stromal cells.

(ii) *Adherence of tumor cells and rapid growth to confluence.* This ideal event does happen occasionally, especially with cells from ascites. Subculture and expansion of cells from such cultures is straightforward.

(iii) *Development of a mixed adherent population consisting of epithelial islands interspersed with stromal cells (Figure 7.3).* This can still occur even if an initial separation was carried out prior to setting up the cultures and is probably the most frequent occurrence.

### 5.1. Cell Separation

If a mixed cell population is present, cell separation is required. Although some lines have been developed from mixed cell cultures, overgrowth by stromal cells goes hand in hand with either very slow proliferation of epithelial cells or epithelial cell death. When mixed adherent populations are present, differential enzyme treatment (DET) can be used to separate them. Mesothelial cells detach rapidly from the culture surface in response to treatment with enzymes,

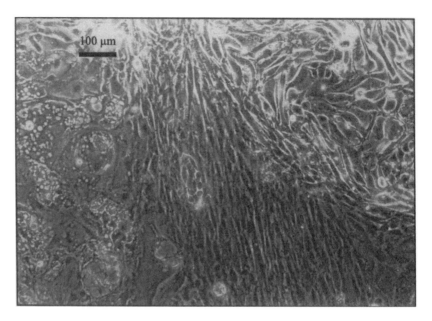

**Figure 7.3** A mixed monolayer of adherent cells. There are large vacuolated epithelial cells on the left, stromal cells in the centre and bipolar epithelial cells showing a swirling growth pattern on the right.

whereas epithelial cells usually take much longer. The difference is not always great enough to allow efficient separation particularly if the epithelial cell population is not fully adherent and well-spread on the culture surface. Separation may need to be repeated regularly until the stromal cell population has been reduced and is senescent.

### Protocol 7.8.  Separation of Cells from Adherent Cultures of Mixed Cell Type Using Differential Enzyme Treatment (DET)

***Reagents and Materials***
*Sterile*
- ❑ Trypsin /EDTA solution: see Section 2.5
- ❑ HBSS: Hanks' balanced salt solution
- ❑ CMF: HBSS without $Ca^{2+}$ and $Mg^{2+}$

***Protocol***
(a)  Rinse cell monolayer gently three times in sterile CMF to remove calcium and trypsin inhibitors present in the serum of the culture medium.

(b)  Add 1 ml trypsin/EDTA solution, warmed to 37°C, per 25 cm$^2$ growth area. Gently rock the flask so that the whole surface is bathed in the solution.

(c)  Place the flask onto the stage of an inverted phase contrast microscope and find an area of mesothelial cells adjacent to epithelial cells for observation. The mesothelial cells will retract and round up 1–3 min after adding the enzyme solution but may remain adherent to the plastic surface.

(d) Knock the side of the flask hard against the palm of the hand 3–6 times to achieve physical detachment; the shearing forces created are sufficient to detach the rounded-up cells.

(e) When detachment of the mesothelial cells has been achieved and before the epithelial cells have detached, wash the flask with 3 × 5 ml of HBSS to remove all detached cells.

(f) Add fresh medium to the residual epithelial cells.

(g) If a significant proportion of epithelial cells have detached in this process, they can be recovered by differential attachment (see Seeding Primary Cultures, Steps (c) and (d)).

(h) Resuspend the detached cells in culture medium, and incubate for 60 min at 37°C in an appropriately sized culture flask (e.g. if cells harvested from a confluent 25-cm$^2$ flask, use a 75-cm$^2$ flask) to allow adherence of mesothelial cells; remove the floating population and either initiate a new culture (if there are enough cells) or return to a flask containing the residual epithelial cells.

### 5.2. Use of Cell-Free-Ascitic Fluid (CFA) in Combination with DET

Some ascitic fluids will produce a loose fibrin mesh when added to a culture [Wilson 1987]. Mesothelial cells form a strong attachment to the fibrin, whereas the epithelial tumor cells do not. This difference can be effectively used to improve separation of the two cell populations. It is most effective when the mesothelial cell density is high. It also appears to have a mechanically damaging effect on the stromal cells so that those which are left behind after the separation procedure show impaired growth (Figure 7.4). It is very easy to grow mesothelial cells from ascitic fluids and if you are regularly attempting to culture ovarian tumor cells you will undoubtedly end up with some cultures which are pure mesothelial cells. These can be stored as frozen stock for the following procedure.

### Protocol 7.9. The Use of Monolayers of Mesothelial Cells for Selection of Ascitic Fluids Suitable for Epithelial-Stromal Separation

#### Reagents and Materials
*Sterile*
- ❑ CFA: see Section 2.7
- ❑ Mesothelial cells: confluent monolayer culture (see Protocol 7.6, Step (c))
- ❑ Trypsin/EDTA: see Section 2.5

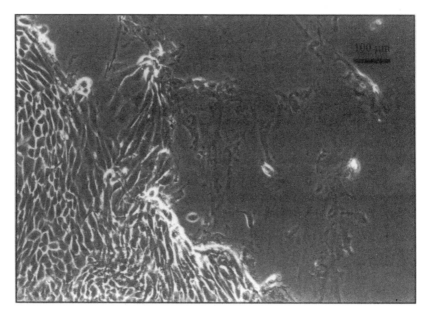

**Figure 7.4** A mixed cell population after DET and CFA treatment. The epithelial island on the left is still intact. The mesothelial cells on the right are sparse, well-spread and tatty in comparison with the monolayer of Figure 7.3.

### Protocol

(a) Remove the medium from a small flask containing a 70–100% confluent monolayer of mesothelial cells. Do not wash the monolayer because this removes tissue-derived clotting factors which are needed for formation of the fibrin mesh.

(b) Add 10 ml of CFA to the flask and incubate overnight at 37°C.

(c) The following day place the flask onto the stage of an inverted phase contrast microscope and check for fibrin mesh formation. The mesh may be seen by the naked eye as a gelatinous film overlying the cells. On the microscope a random mesh of fibers can be seen overlying the cells if the ascitic fluid is appropriate (Figure 7.5)

(d) Gently decant the remaining fluid CFA and rinse the monolayer carefully with three washes of CMF.

(e) Add 1 ml of warmed trypsin/EDTA solution per 25 cm$^2$ growth area, and replace the flask under the microscope.

(f) Check the flask for cell detachment. Within one to two minutes the stromal cells should round up and lift off as a sheet bound to the fibrin mesh, which retracts into a clot.

(g) Ascitic fluids which exhibit this property can be aliquotted and stored at −20°C for routine use in cell separation.

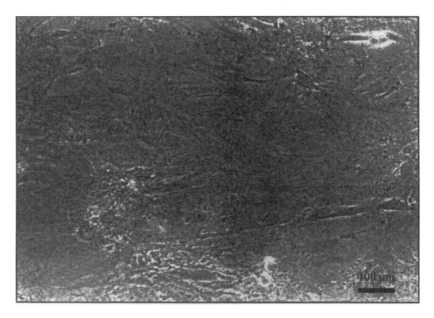

**Figure 7.5**  A reticular mesh overlies the monolayer of stromal cells.


## Protocol 7.10.   Separation of Ovarian Tumor Cells from a Mixed Monolayer with CFA and DET

### Reagents and Materials
*Sterile*
- ❑  CFA: cell-free ascitic fluid (see Section 2.7 and Protocol 7.7)
- ❑  CMF: Hanks' balanced salt solution without $Ca^{2+}$ and $Mg^{2+}$
- ❑  Trypsin/EDTA: see Section 2.5

### Protocol
(a)  Remove the culture medium and add 10 ml CFA per 25 $cm^2$ growth area. Do not wash the monolayer at this stage.

(b)  Incubate the flask overnight at 37°C.

(c)  The following day remove the flask and place on the stage of an inverted phase contrast microscope. Find an area of stromal cells surrounding an epithelial island of tumor cells and confirm that fibrin mesh formation overlying the stromal cells but not the epithelial island is present (Figure 7.6).

(d)  Gently decant the CFA and rinse the monolayer/mesh three times in CMF.

(e)  Add 1 ml of warmed Trypsin/EDTA solution per 25 $cm^2$ growth area, and replace the flask on the microscope.

(f)  Find an area of stromal cells surrounding an epithelial island and watch for rounding up of the stromal cells. There is an optimal

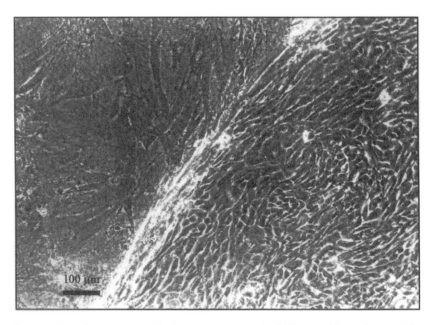

**Figure 7.6** A fibrin mesh overlies the stromal cells on the LHS of the Figure. The mesh is retracted around the edge of the epithelial island d on the RHS of the Figure and no fibres overlie the epithelial island.

window of time in which the stromal cells are rounded and loosely adherent, the tumor cells are just beginning to round and become less well-spread and the fibrin mesh is beginning to peel off at the edge of the flask (Figure 7.7). This is the time to use a Pastette to gently encourage the mesh to peel off the whole flask. On the microscope this can be seen as a retracting mesh containing rounded stromal cells, leaving completely clear areas of plastic interspersed with tumor cell islands.

(g)   Remove the fibrin clot and gently rinse the monolayer to remove any loose stromal cells. (Do not do this if the tumor cells are also beginning to detach ).

(h)   Add fresh culture medium to the flask and re-incubate.

(i)   This procedure can be repeated at regular intervals (e.g. weekly) if mesothelial cell overgrowth continues. Make sure that the epithelial cells have become fully adherent again, before repeating the process.


## 6.  SUBCULTURE

Once isolated from stromal cells, clumps of epithelial cells can survive in culture flasks for months, with only very slow proliferation; a period of 6 months may elapse before there are sufficient cells for

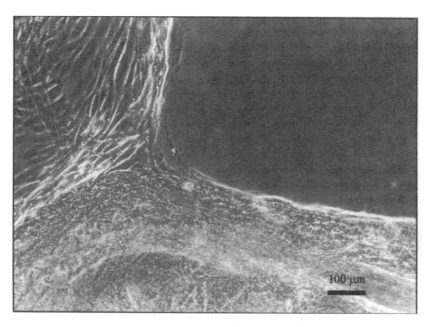

**Figure 7.7** A fibrin mesh attached to stromal cells is peeling away from an epithelial island, removing the stromal cells but leaving the epithelial cells.

subculture. Two different types of primary culture may be achieved once mesothelial cells have been eliminated.

(i)   Cells within epithelial islands may begin to enlarge without dividing. Changing the medium to PPIGSS/Ham's F-12 (1:1) with 5% FBS sometimes reverses this process and restores the culture to smaller proliferating cells. The addition of insulin, transferrin and selenium (ITS) to the mix is also beneficial in enhancing proliferation [Wilson, 1998].

(ii)   The culture may consist of numerous epithelial islands which do not spread by lateral growth but show extensive dense growth upwards, with loosely adherent clumps. More rapid expansion of cells can be achieved by pipetting the loosely adherent clumps of cells from areas of three-dimensional growth and either transferring them into new flasks or leaving them in the original flask to initiate new epithelial islands. When growth is three-dimensional in confluent or nearly confluent primary cultures, the cells may not subculture well using trypsin/EDTA. When use of trypsin/EDTA has failed to yield viable secondary cultures, enzyme-free dissociation medium (Sigma) allows successful subculture in the early stages of cell line development. Once two or three subcultures with enzyme-free dissociation medium have been successful, trypsin/EDTA can then be used routinely, though this

**Table 7.5. The culture history of D729.**

| Date | Appearance | Procedure | Pass Number and Culture Medium |
|---|---|---|---|
| 22/1/98–23/1/98 | Solid tumor for disaggregation | Trypsinisation and seeding of primary culture | P1 PPIGSS |
| 2/2/98 | 70–80% confluent with epithelial islands and debris | Medium change | P1 PPIGSS |
| 9/2/98 | Islands of swirly epithelial cells, some stromal cells. Confluency reduced | Medium change | P1 Medium changed to PPIGSS/F12 + ITS |
| 16/2/98 | Nice epithelial islands with cells of different morphology, some stromal cells | Medium change | P1 PPIGSS/F12 + ITS |
| 20/2/98 | Epithelial cells and stromal cells. Flask still not confluent | Medium change | P1 Medium changed back to PPIGSS |
| 24/2/98 | Epithelial islands and stromal cells. debris | Medium change | P1 PPIGSS |
| 2/3/98 | Still not confluent. Epithelial cells and stromal cells | Medium change | P1 PPIGSS |
| 9/3/98 | Swirly epithelial islands, misty patches, stromal cells, 3D growth | Medium change | P1 PPIGSS |
| 13/3/98 | Large epithelial islands, stromal cells, 3D growth | Differential enzyme treatment and medium change | P1 PPIGSS |
| 19/3/98 | Swirly epithelial cells, larger cells, debris, large epithelial island of small cuboidal cells, very few stromal cells | Medium change | P1 PPIGSS |
| 2/4/98 | Debris ++, swirling and cuboidal epithelial cells, 3D growth | No action | P1 |
| 8/4/98 | Large epithelial islands, swirling cells, cuboidal cells, 3D growth | Medium change | P1 PPIGSS |
| 16/4/98 | 3 cell types present, swirly, small cuboidal, small cuboidal and vacuolated | Medium change | P1 PPIGSS |
| 23/4/98 | Swirly cells, cuboidal cells, confluent in patches, very few stromal cells, 3D growth | Split 1:1 using trypsin-versene and check original flask to make sure that full cell recovery has been achieved | P2 PPIGSS |
| 29/4/98 | 80% confluent and looks good. 3D growth | Medium change | P2 PPIGSS |
| 5/5/98 | 90% confluent, 3D growth | Split 1:2 using trypsin-versene | P3 PPIGSS |
| 12/5/98 | Densely confluent with 3-D growth | Split 1:3 using trypsin-versene | P4 PPIGSS |
| 21/5/98 | Confluent | Freeze 5 flasks to 10 vials. Set up spot slides for identification and antibiotic-free flask for checking mycoplasma status. Split remaining flask 1:2 | P5 PPIGSS |
| 1/6/98 | Confluent | Split 1:3 | P6 PPIGSS |

**Table 7.5.** *(Continued)*

| Date | Appearance | Procedure | Pass Number and Culture Medium |
|------|-----------|-----------|-------------------------------|
| 8/6/98 | Confluent | Freeze 11 vials. Split remainder at 1:4 | P7<br>PPIGSS |
| 15/6/98 | 95% confluent | Split 1:4 | P8<br>PPIGSS |
| 22/6/98 | 95% confluent | Freeze 15 vials<br>Split 1:4<br>Freeze medium from Ab-free flask for checking mycoplasma status | P9<br>PPIGSS |
| 29/6/98 | 90% confluent | Split 1:4 | P10<br>PPIGSS |
| 6/7/98 | 90% confluent | Freeze 15 vials<br>Split 1:4 | P11<br>PPIGSS |
| 13/7/98 | 100% confluent | Split 1:4 | P12<br>PPIGSS |
| 20/7/98 | 100% confluent No 3D growth | Freeze 20 vials | Cell line held as frozen stock. No mycoplasma detected in frozen medium from P8 |

does result in loss of three-dimensional growth. The split ratio can be gradually increased as the cells adapt to culture conditions.

The culture history of the monolayers of tumor cells derived from cold trypsinisation of ovarian tumor sample D729 is shown in Table 7.5. The disaggregation and culture of this tumor is shown on Multimedia Methods in Cell Biology [Wilson et al., 1999] (see also Table 7.5).

## 7. IDENTIFICATION OF CELLS IN CULTURE

Identification of the cells growing in primary culture and subsequent sub-cultures must be carried out to confirm that they are i) epithelial, ii) from ovarian tumor, and iii) malignant. In addition, if the line becomes continuous further characterization is necessary. This includes confirmation that the cell line is of human origin, that it came from the patient source it was ascribed to, and that it is not cross-contaminated by other cell lines growing in the laboratory.

Some baseline characteristics need to be identified for any new cell line so that phenotypic and genotypic drift can be detected. The information provides a baseline for detection of phenotypic and geno-

**Table 7.6. A summary of methods used to characterize cultures developed from ovarian tumor cells.**

| Epithelial Markers | Ovarian Markers | Malignant Markers | Markers of Origin |
|---|---|---|---|
| Growth pattern in monolayer | Expression of CA125- a surface marker expressed by 82% of epithelial ovarian tumors [Bast et al., 1981]. | Modal Chromosome number <46 or >46 | Identification of human chromosomes |
| Presence of structured cell-cell junctions (tight junctions, desmosomes, hemi-desmosomes | Expression of placental alkaline phosphatase, as expressed by 94% of ovarian tumors [Nouven et al., 1985]. | Presence of marker chromosomes and malignancy-associated alterations | DNA fingerprint of tumor compares with DNA fingerprint of uncultured material from same patient |
| Expression of keratins | Expression of CEA, surface marker expressed by 70–80% of mucinous and clear cell tumors [Casper et al., 1984] | Clonogenic in soft agar (but not all malignant tumor cell lines are clonogenic in soft agar) | Marker chromosomes and modal chromosome number dissimilar from other cell lines cultured in same laboratory |
| Expression of HMFG2, surface marker expressed by 94% of epithelial ovarian tumors [Ward et al., 1987] | Expression of keratin 7, intermediate filament expressed by ovarian tumors but not tumors of gastrointestinal origin [Moll et al., 1982, Ping Wang et al., 1995] | Tumorigenic in animal models (but not all malignant tumor cell lines are tumorigenic) | |

typic drift during subsequent subcultures, and also for detection of possible cross-contamination at a later date. Some suggestions for ovarian epithelial cancers are outlined in Table 7.6.

The appearance of a range of cell types observed in culture from ascites and solid tumors is shown in Figure 7.8. Note that mesothelial cells can show both a fibroblast-like and an epithelial-like appearance in monolayer [Connell and Rheinwald, 1983, LaRocca and Rheinwald, 1985]. Senescent fibroblast-like cultures typically show enlarged cells with extensive stress fibre formation and decreased cell density. Senescent epithelial-like mesothelial cells typically show decreased cell density, large rounded cells and extensive multinucleation. Mesothelial cells and ovarian tumor cells can be differentiated using immunohistochemistry [Bradford and Wilson, 1996].

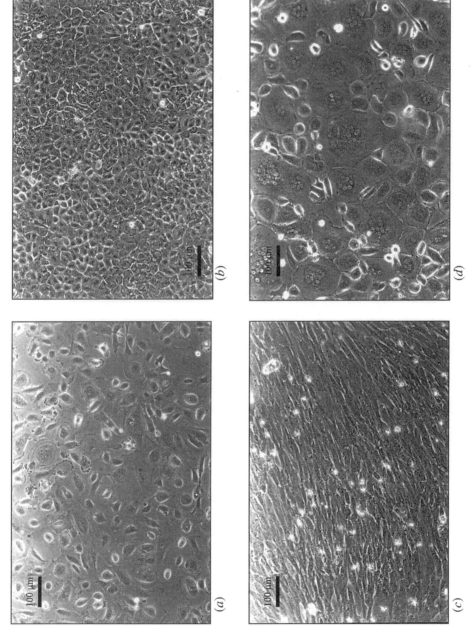

**Figure 7.8** Different patterns of growth shown by confluent monolayers of mesothelial cells derived from ascitic fluids and solid tumors. (a) Epithelial-like cells below saturation density. (b) Epithelial-like cells which have been at saturation density for several weeks. (c) Fibroblast-like cells. (d) Multinucleate cells.

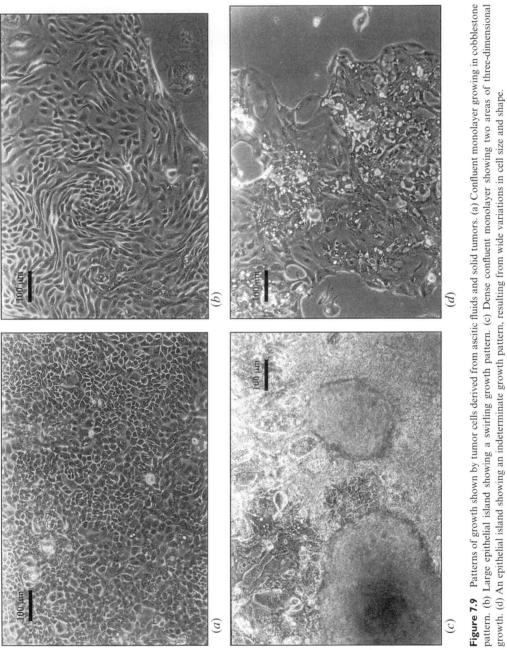

**Figure 7.9** Patterns of growth shown by tumor cells derived from ascitic fluids and solid tumors. (a) Confluent monolayer growing in cobblestone pattern. (b) Large epithelial island showing a swirling growth pattern. (c) Dense confluent monolayer showing two areas of three-dimensional growth. (d) An epithelial island showing an indeterminate growth pattern, resulting from wide variations in cell size and shape.

## SOURCES OF MATERIALS

| Item | Supplier |
| --- | --- |
| Centrifuge tubes | Nalge Nunc; Bibby; Corning |
| Collagenase | Worthing Biochemical Company |
| DMEM | Gibco |
| DNase | Sigma |
| F12 | Gibco |
| Fetal bovine serum (FBS) | Gibco |
| Ficoll/Hypaque | Amersham Pharmacia |
| Hanks' balanced salt solution (HBSS) | Gibco |
| HBSS without $Ca^{2+}$ and $Mg^{2+}$ (CMF) | Gibco |
| Heparin | From Hospital Pharmacy |
| Histopaque | Sigma |
| ITS | Sigma |
| LymphoSep | ICN |
| Nigrosin | Sigma |
| Pastette | Bibby |
| PBS | Gibco |
| Red cell lysing reagent | Sigma |
| RPMI 1640 | Gibco |
| Tissue culture plastic flasks and dishes | Nalge Nunc |
| Trypan Blue | Sigma; Gibco |
| Trypsin | Gibco |
| Universal containers | Bibby |

## REFERENCES

Bast RC, Feeney M, Lazarus H, Nadler LM, Colvin RB, Knapp RC (1981). Reactivity of a monoclonal antibody with human ovarian carcinoma. *J Clin Invest*, 68:1331–1337.

Broxterman HJ, Sprenkels-Schotte C, Engelen P, Leyva A, Pinedo HM (1987) Analysis of human ascites effect on clonogenic growth of human tumor cell lines and NRK-49F cells in soft agar. *Int J Cell Cloning* 5:158.

Casper S, van Nagell JR, Powell DF, Dubilier LD, Donaldson ES, Hanson MB, Pavlik EL (1984) Immunohistochemcal localization of tumour markers in epithelial ovarian cancer. *Amer J Obstets & Gynecol.* 149:154–158.

Connell ND and Rheinwald JG (1983) Regulation of the cytoskeleton in mesothelial cells: reversible loss of keratin and increase in vimentin during rapid growth in culture. *Cell* 34:245–253.

DiSaia PJ, Morrow M, Kanabus J, Piechal WT, Townsend DE. (1975). Two new tissue culture lines from ovarian cancer. *Gynaecol Oncol* 3:215–219.

Fox H and Langley FA (1976) Tumours of the Ovary. William Heinemann Medical Books Ltd.

Ioachim HL, Dorsett BH, Sabath M, Barber HR (1975) Electron microscopy, tissue culture, and immunology of ovarian carcinoma. *Nat Cancer Inst Monograph*, 42:45–62.

Ioachim HL, Sabbath M, Andersson B, Barner HRK (1974) Tissue cultures of ovarian carcinoma. *Lab Invest.* 31:381–390.

Kimoto, T. Ueki A, Nishitani K (1975) Phagocytosis of lymphoblastoid cells and cell destruction of human malignant tumor cells. *Acta Path Jap* 25:89–114.

LaRocca PJ and Rheinwald JG (1985). Anchorage-independent growth of normal human mesothelial cells: a sensitive bioassay for EGF which discloses the absence of this factor in fetal calf serum. *In vitro Cell and Developmental Biol.* 21:67–71.

Limburg HG. and Heckmann C (1968) Chemotherapy in the treatment of advanced pelvic malignant disease with special reference to ovarian cancer. *J Obstet Gynaecol* 75:1246–1255.

Mills GB, May C, McGill M et al. (1988) A putative new growth factor in ascitic fluid from ovarian cancer patients: identification, characterisation and mechanism of action. *Cancer Res.* 48:1066–1071.

Moll R Franke WW, Schiller DL, Geiger B, Krepler R (1982) The catalog of human cytokeratins: patterns of expression in normal epithelia, tumors and cultured cells. *Cell* 31:11–24.

Nouwen EJ, Pollett DE, Schelstraete JB, Eerdekens MW, Hansch C, Van de Voorde A, De Broe ME (1985). Human placental alkaline phosphatase in benign and malignant ovarian neoplasia. *Cancer Res.* 39:1185–1191.

Ovarian Cancer 2: Biology, Diagnosis & Management. Eds: F. Sharp, WP Mason & W. Creasman, Chapman & Hall Medical, London (1992).

Ovarian Cancer 3: Eds: F. Sharp, P. Mason, T. Blackett & J. Berek. Chapman & Hall Medical, London (1995).

Ovarian Cancer 4: Eds: F. Sharp, T. Blackett, R. Leake & J. Berek Chapman & Hall Medical, London (1996).

Ovarian Cancer: Biological and therapeutic challenges. Eds: F.Sharp, W.P. Mason, & R.E. Leake. Chapman & Hall, London (1990).

Parmley TH and Woodruff JD (1974). Ovarian mesothelioma. *Amer J Obstet Gynecol* 120:234–241.

Ping Wang N, Zee S, Zarbo RJ, Bacchi CE, Gown AM (1995). Coordinate expression of cytokeratins 7 and 20 defines unique subsets of carcinomas. *Appl. Immunohist.* 3:99–107.

Radford H and Wilson AP (1996). A comparison of immunohistochemical staining of human cultured mesothelial cells and ovarian tumour cells using epithelial and mesothelial cell markers. *Analytical Cellular Pathology* 11:173–182.

Uitendaal MP, Hubers HAJM, McVie JB, Pinedo HM. (1983). Human tumor clonogenicity is improved by cell-free ascites. *Br J Cancer* 48:55.

UKCCCR Guidelines for the use of Cell lines in Cancer Research. 1999. Published by UKCCCR, PO Box 123, Lincoln's Inn Fields London WC2A 3PX.

Vries JE, Benthem M, Runke P (1973). Separation of viable from non-viable tumour cells by flotation on a Ficoll-Triosil mixture. *Transplantation* 15:409–410.

Ward BG, Lowe DO, and Shepherd JH (1987). Patterns of expression of a tumour-associated antigen defined by the monoclonal antibody HMFG2, in human epithelial ovarian carcinomas. *Cancer* 60:787–793.

Wilson AP (1987) In vitro fibrin formation by ascitic and peritoneal fluids: a novel system for the study of fibrin-cell interactions. *Br J Cancer* 56:206.

Wilson AP (1998) Preparation of ovarian carcinoma cell lines. In: Cell and Tissue Culture Laboratory Procedures. Eds. JB Griffiths, A. Doyle, DG Newell. Jophn Wiley & Sons.

Wilson AP, Garner CM, Hubbold L (1999) Cell Culture in Multimedia Methods in Cell Biology. Chapman & Hall, CRCnetBASE, Eds. JR Harris, D Rickwood.

Wilson AP, Dent M, Lee H, Hubbold L (1991). The effects of insulin and hydrocortisone on the response of an ovarian tumour cell line to EGF and TGF-beta. *Br J Cancer*, 63, Suppl. XIII, 61.

Wilson AP, Garner CM (1999). Ovarian cancer. In Human Cell Culture Vol. II Cancer Cell Lines Part 2. Eds. John RW Masters & B Palsson. Kluwer Academic Publishers, Dordrecht.

Walker DG, Lyons MM, Wright JC (1965) Observations on primary short-term cultures of human tumors. A second 5-year study. *Eur J Cancer* 1:265–73.

# 8

# Culture of Cervical Carcinoma Tumor Cell Lines

Peter Stern[1], Catherine West[2] and Deborah Burt[1]

*CRC Department of Immunology[1] and Experimental Radiation Oncology[2], Paterson Institute for Cancer Research, Christie Hospital NHS Trust, Manchester, United Kingdom, pstern@picr.man.ac.uk*

*Culture of Human Tumor Cells*, Edited by Roswitha Pfragner and R. Ian Freshney.
ISBN 0-471-43853-7   Copyright © 2004 Wiley-Liss, Inc.

## I.  INTRODUCTION

Cervical carcinoma is the second most prevalent cancer that affects women worldwide, with more than 500,000 cases newly diagnosed each year. The introduction of national cervical screening programs, coupled with the availability of safe and effective treatment for premalignant disease of the cervix, has resulted in a dramatic reduction in the incidence of cervical cancer in Western countries over recent decades [Schoell et al., 1999]. In developing countries, where resources for screening and access to treatment for premalignant lesions are limited, cervical cancer remains a major cause of death.

DNA derived from oncogenic human papillomaviruses (HPV) can be detected in more than 99% of cervical cancers [Walboomers et al., 1999], which supports a causal role for these viruses in the etiology of the disease. This hypothesis is supported by a large number of experimental studies that have shown that high-risk HPV-type (most prevalent HPV 16 and 18) oncogene expression can induce the malignant transformation of cells in vitro. HPV infection is a necessary but insufficient component of the generation of malignant cervical cancers. The malignant transformation of cells infected with oncogenic HPV is a multistep process that occurs over more than a decade. Additional knowledge of the mechanisms by which HPV contributes in the process of carcinogenesis has been obtained from studies using cervical carcinoma cell lines.

The early gene products E6 and E7 are the oncogenic products of the HPV genome and their role in oncogenesis depends on their ability to bind to and inactivate the host tumor suppressor gene products p53 and retinoblastoma (Rb), respectively [Werness et al., 1990] and [Dyson et al., 1989]. These molecular events often coincide with integration of the HPV DNA into the host cell genome. In premalignant disease, HPV DNA is episomal and transcription of the E6 and E7 genes remains under the tight control of another HPV protein, E2. In carcinoma cells, HPV is often integrated into the cellular genome, a process that frequently results in the loss of functional E2 expression and the consequent overexpression of E6 and E7 [Howley, 1991].

Experimental work with cervical carcinoma cell lines has been critical in broadening our understanding of the molecular events that are essential in the malignant transformation of cervical cells. The cell line W12, for example, [Doorbar et al., 1990] contains episomal copies of HPV DNA, which can become integrated with increasing passage number in vitro and thereby provide a model for the process that occurs in vivo.

The design of effective chemo-, radio-, and novel immunotherapies for treating cervical cancer has also depended significantly on experimental studies with cervical carcinoma cell lines. Early-stage cervical cancer has traditionally been treated by surgical excision, either alone or in combination with radiotherapy, to achieve an overall cure rate of between 90 and 95% [Resbeut et al., 2000]. Cure rates are much lower for patients with late-stage or advanced cervical cancer, and treating these patients currently relies on the use of combined chemo- and radiotherapy regimens. Our understanding of the efficacy, mode of action, and sensitivity of cervical tumors to chemo- and radiotherapies has been derived from laboratory work by using cervical carcinoma cells grown in vitro.

More recently, research has focused on the mechanisms by which absent or inappropriate host immunity may be instrumental in the persistence of HPV infections and the subsequent development of tumors. Furthermore, the possible exploitation of the host immune response in the treatment of cervical cancer is the subject of great interest. HPV oncogenes E6 and E7 are expressed throughout the full spectrum of HPV-associated disease and thus may provide tumor antigens against which an immune response could be generated. Numerous different immunotherapeutic agents, including peptides, proteins, and vaccinia viruses genetically modified to encode HPV oncoproteins E6 and E7 have reached the clinic in early phase clinical trials with promising results [Borysiewicz et al., 1996]. The use of such vaccines in patients with high-grade cervical intraepithelial neoplasia (CIN 3) or invasive carcinomas, however, may be limited by the down-regulation or complete loss of HLA Class 1 molecules by these cells. Most therapeutic vaccines aim to generate oncogenic HPV E6 and/or E7-specific cytotoxic T lymphocytes (CTL) that can recognize and kill HPV-infected cells. T cell responses rely on the efficient processing of E6 and E7 proteins into peptides and their subsequent presentation on the surface of the tumor cell in conjunction with HLA Class 1 molecules. In many cervical tumors, these HLA molecules are down-regulated or absent [Garrido et al., 1997], and this may provide a mechanism by which cervical carcinoma cells can first evade the natural immune response and second to resist eradication by immune mechanisms generated by vaccination strategies. Cervical carcinoma cell lines have been used to gain an understanding of the mechanisms that cause such down-regulation of HLA expression [Brady et al., 2000].

This chapter will discuss the derivation and use of short-term cultures of cervical tumor cells and the establishment of passagable long-term lines. Short-term cultures are not exposed to the selective pressures that long-term cultures are subjected to and may reflect the heterogenous populations of cells present in the tumor in vivo. The number of cells obtained is limited, so the major advantage of long-term cultures is that a larger number of cells with identical properties to the original tumor are available for comprehensive study and analysis. Although the cell lines share properties of the original tumor they are subjected to selective pressures and may not fully represent the range of cell types that would have been present in the original tumor.

## 2. SAFETY CONSIDERATIONS

The major safety concern when establishing cervical tumor cell lines is that the starting material is from a primary human biopsy that may be a source of human pathogens. These pathogens may include Hepatitis B and HIV, which have the potential to cause ill health or death. Although samples should not be obtained from patients who are thought to belong in such high-risk groups, it is probable that information will not be provided on the status of patients with regard to such pathogens. Therefore, each biopsy sample should be treated as though potentially infectious.

Gloves should be worn at all times, and procedures should be performed in a Class II microbiological safety cabinet, which, in addition to providing a sterile environment essential to successful tissue culture, will provide a degree of user protection from the sample. One procedure during which extra care should be taken is the preliminary mincing of the biopsy material, as this involves the use of scalpels and the potential for sharps injuries to occur is unavoidably high.

One also needs to consider the hazardous nature of the growth factors that are included in the media. All of the growth factors used are considered to be harmful. Reagents should be weighed in a fume cupboard before the preparation of stock solutions, and gloves should be worn at all times.

## 3. PREPARATION OF MEDIA AND REAGENTS

### 3.1. Disaggregating Agents

#### 3.1.1. Enzyme Cocktail

Pronase, 0.25 mg/ml, collagenase Type I, 0.25 mg/ml, and deoxyribonuclease, 20 µg/ml, in transport medium. Pre-weighed aliquots of lyophilized enzymes are stored, and the cocktail is solubilized and filtered shortly before required. Warm to 37°C 5 min before use.

#### 3.1.2. Trypsin/EDTA

Trypsinize the tumor cell lines by using a solution of 0.05% trypsin, 0.02% EDTA. Dissolve trypsin at 0.5% in 0.3 mM EDTA in PBSA (see Section 3.1.5). Warm to 37°C 5 min before use.

#### 3.1.3. Purified Trypsin

0.02% solution of Worthington trypsin. This solution is prepared by dissolving purified trypsin powder in HBSS and then filter-sterilizing. Aliquots (20 ml) should be stored at 20°C and should be thawed fresh every two weeks.

#### 3.1.4. EDTA

Remove 3T3 from growing tumor cell cultures by using 0.3mM (0.02%) EDTA in PBSA (see Section 3.1.5).

#### 3.1.5. PBSA

Dulbecco's phosphate buffered saline without $Ca^{2+}$ and $Mg^{2+}$.

### 3.2. Culture Media

#### 3.2.1. Transport Medium

Eagle's basal medium supplemented with 200 µg/ml gentamycin, 25 µg/ml amphotericin, and 15 mM $N$-(2-hydroxyethyl)piperazine-$N'$-ethanesulfonic acid (HEPES).

### 3.2.2. Ham's F12 for Short Term Cultures

Ham's F12 (without glutamine) supplemented immediately before use with 2 μg/ml amphotericin, 25 μg/ml gentamycin, 10 ng/ml epidermal growth factor (EGF), 0.5 μg/ml hydrocortisone, 2.5 μg/ml transferrin, 10 μg/ml insulin, 1 ml glutamine (200 mM), and 15% fetal calf serum (FCS). Store all additives frozen in aliquots. Before use, thaw them rapidly and add to make a total of 100 ml medium. Warm to 37°C 5 min before required.

### 3.2.3. DMEM for 3T3 Cells

Dulbecco's modified Eagle's medium (DMEM) +10% donor calf serum (DCS), 5 mM l-Glutamine, 100 IU/ml penicillin, 100 μg/ml streptomycin.

### 3.2.4. Tumor Cell Line Culture Medium

*Full keratinocyte medium (FKM):* DMEM +5% FCS, 5 mM L-glutamine,100 IU/ml penicillin, 100 μg/ml streptomycin, 5 μg/ml insulin, 0.01 μg/ml cholera toxin, 0.4 μg/ml hydrocortisone, and 0.1 μg/ml EGF. Table 8.1 describes the preparation and storage conditions of the growth factors included in the tumor cell growth medium.

**Table 8.1.   Preparation of supplements for full keratinocyte medium.**

| Supplement + concentration | Preparation | Final concentration |
|---|---|---|
| Insulin 5 mg/ml | Dissolve at 5 mg/ml in 5 mM HCl. (The solution may need to be incubated at 37°C if the insulin is not fully dissolved.) Filter-sterilize through 0.2 μm filter. Store 0.5 ml aliquots at −20°C. | 5 μg/ml = 0.5 ml of 5 mg/ml into 500 ml medium. |
| Cholera toxin 1 μg/ml | Dissolve 1 mg cholera toxin in 1 ml sterile d.d H₂O, then dilute 1 in 1000 into HBSS +20% FCS = 1 μg/ml. Filter-sterilize through 0.2 μm filter. Store 5 ml aliquots at −20°C. | 0.01 μg/ml = 5 ml of 1 μg/ml into 500 ml medium. |
| Hydrocortisone 40 μg/ml | Make a 5 mg/ml stock in EtOH. Dilute 1 in 125 in HBSS + 5% FCS to give a 40 μg/ml stock. Filter-sterilize through 0.2 μm filter. Store 5 ml aliquots at −20°C. | 0.4 μg/ml = 5 ml of 40 μg/ml into 500 ml of medium. |
| Epidermal growth factor 10 μg/ml | Dissolve in HBSS + 5% FCS at 10 μg/ml. Filter-sterilize through 0.2 μm filter. Store 0.5 ml aliquots at −20°C. | 0.1 μg/ml = 0.5 ml of 10 μg/ml into 500 ml of medium. |

### 3.3.  Trypan Blue

Dissolve at 0.4% in PBSA.

### 3.4.  Agar Cloning Reagents

#### 3.4.1.  Rat Red Blood Cells

(i)  Take blood from August rats.

(ii)  When the blood arrives, centrifuge it for 10 min at 250 g and remove the clear supernatant.

(iii)  Add 5 ml phosphate buffered saline (PBSA) and centrifuge for 10 min at 250 g.

(iv)  After removing the clear supernatant, add 5 ml culture medium and centrifuge the cells for 10 min at 450 g.

(v)  Remove the clear supernatant and place the sample in a waterbath at 44°C for 1 h to remove any residual white blood cells.

(vi)  Add 5 ml culture medium to make up to the original volume.

(vii)  Dilute 1:8 in culture medium, and store in refrigeration for up to 2 weeks.

#### 3.4.2.  Soft Agar

Noble agar as 5% solution in PBSA made in small glass bottles and sterilized by autoclaving. During an experiment, place a bottle in a 60°C water bath to keep molten. Immediately before required, warm 18.4 ml culture medium to 37°C, and add 1.6 ml agar added to produce a 0.4% solution.

#### 3.4.3.  Iodonitrotetrazolium

Iodonitrotetrazolium (INT) Violet, 50 mg in 100 ml distilled water; filter-sterilize and store at 4°C for up to 2 months.

## 4.  ESTABLISHMENT OF SHORT-TERM CULTURES

The protocol for short-term culture of cervix carcinoma cells is based on a method first described by Courtenay and Mills [1978], which was subsequently modified for gynecological cancers [West and Sutherland, 1986].

A tumor biopsy is disaggregated by using an enzyme cocktail; cells are counted and plated in soft agar in test-tubes in media containing growth factors and rat red blood cells. Incubation in a 5% oxygen atmosphere and weekly feeding encourages tumor cell growth. The rat red blood cells lyse in agar at about 5–7 days releasing labile growth factors that improve colony-forming efficiency. Colonies are stained and counted under a microscope after 4 weeks.

Growth of human tumors can be difficult; finding a good batch of fetal calf serum (FCS) is crucial and extensive batch testing is often required. Good aseptic technique is essential, all materials must be sterile and it is important that all reagents are prepared/thawed shortly before use and that they are not stored in refrigeration for more than 1 week. The production of a single cell suspension is also critical.

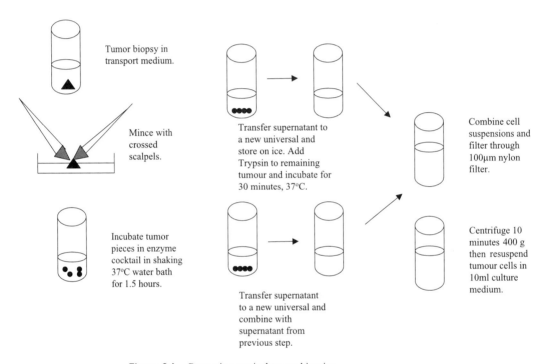

**Figure 8.1.** Processing cervical tumor biopsies.

## Protocol 8.1. Preparation of Cervical Tumor Cell Suspensions

### Reagents and Materials
*Sterile*
- ❏ Culture medium: Supplemented Ham's F12 (see Section 3.2.2)
- ❏ Enzyme cocktail
- ❏ Trypsin
- ❏ Plastic test-tubes
- ❏ Universal tubes containing sterile, high-antibiotic medium
- ❏ Petri dish, 10-cm diameter, preweighed
- ❏ Scalpels
- ❏ Duran bottle, 100 ml
- ❏ Nylon filter, 100-μm

*Non-sterile*
❏ 4% formaldehyde: dilute 40% saturated solution of formaldehyde I in 10 inPBSA
   Note: This must be done in a fume cupboard.
❏ Trypan Blue, 0.4%

**Protocol** (Fig. 8.1.)

(a) Store universal tubes containing sterile, high-antibiotic medium in a refrigerator in the surgical theater.

(b) Place tumor biopsies immediately in a universal, and take to the laboratory. Either process immediately, disaggregate, and cryo-preserve as a single cell suspension or mince and cryopreserve as minced tumor tissue.

(c) In a Class II safety cabinet, transfer the sample into a pre-weighed 10-cm diameter Petri dish that can be re-weighed if information on cell yield is required.

(d) Mince the tissue by using crossed scalpels.

(e) Place the sample in enzyme cocktail (approximately 50 ml per 0.5 g tissue) in a 100 ml Duran bottle, and place in a shaking water bath for 1.5 h.

(f) Remove the cell suspension and place on ice.

(g) Disaggregate the tumor further in trypsin for 0.5 h.

(h) Combine the two cell suspensions and filter by using a 100-$\mu$m nylon filter.

(i) Centrifuge for 10 min at 400 g.

(j) If there is blood in the pellet, lyse by adding 9 ml double-distilled water, pipette up and down twice, and mix rapidly. Immediately add 1 ml of 10 PBSA followed by 10 ml culture medium.

(k) Centrifuge for 10 min at 120 g.

(l) Resuspend the pellet in culture medium (the amount depends on the size of the pellet and is gained with experience). View the cells under a hemocytometer, and dilute further if required.

(m) Filter through a 37-$\mu$m pore filter to remove any remaining cell clumps.

(n) Remove 100 $\mu$l of cell suspension, and add 100 $\mu$l 0.4% Trypan Blue. Leave for 10 min, and count the number of single cells, doublets, triplets, clumps, and dead cells. If there are <95% single cells, filter by using a 20-$\mu$m filter.

(o) Calculate the viable cell number, viable cell yield, percentage of dead cells, and cell multiplicity.

## Protocol 8.2. Soft Agar Assay of Cervical Tumor Cells

*Reagents and Materials*

*Sterile*
- ❑ Culture medium (see Section 3.2.2)
- ❑ Agar 0.4% (see Section 3.4.2)
- ❑ Rat red blood cells (see Section 3.4.1)
- ❑ Falcon #2057 tubes
- ❑ Metal test-tube rack

*Non-sterile*
- ❑ Humidified incubator with gas mixture 5% $O_2$:5% $CO_2$:90% $N_2$

*Protocol*

(a)  Place 0.5 ml cell suspension at 10 times that required into Falcon tubes.

(b)  Add 0.5 ml of rat red blood cells.

(c)  Add 4 ml of 0.4% agar, pipette up and down twice, and aliquot 1 ml into each of 4 pre-labeled Falcon tubes (in pre-sterilized metal test-tube rack).

(d)  Place the tip of the pipette in the bottom of the tube, taking care that agar is not pipetted along the tube sides and that air bubbles are not produced.

(e)  Repeat for other dilutions as required. As it is important to ensure linearity of colony formation (i.e., a linear relationship between the number of cells plated and colonies formed) three cell densities are usually plated for control cultures and two or three per treated culture. The number depends on the viable cell yield. Examples for controls would be to plate $2 \times 10^4$, $5 \times 10^4$ and $10^5$ cells per tube with eight replicates per cell density and $5 \times 10^4$ and $1 \times 10^5$ per tube for treated culture.

(f)  Once the agar has set, place the rack containing the tubes into a 5% $O_2$:5% $CO_2$:90% $N_2$ humidified incubator. It is important that the lids of the tubes are loose in order to maintain the correct pH.

(g)  Add 1 ml freshly prepared, prewarmed culture medium after 1 week.

(h)  On weeks 2 and 3, remove 1 ml spent medium and add 1 ml fresh culture medium.

(i)  After 4 weeks incubation, add 0.2 ml pre-warmed INT solution and incubate for another 24 h.

(j)  Remove from the incubator and, in a safety cabinet, pipette off the medium above the agar.

(k)  Add 1 ml 4% formaldehyde to each tube in a fume cupboard, and leave the tubes with caps removed overnight.

(l)　In the fume cupboard, tip the agar from each tube onto a microscope slide (labeled), and gently squash the agar plug with a coverslip.

(m)　Count colonies under a low-power light microscope, and use a graticule to ensure that only colonies greater than 60 μm in diameter are counted.

### 4.1.　Applications of Short-Term Cultures

The short-term culture of cervical tumor cells allows assessment of growth potential, clonogenicity, and analysis of sensitivity to cytotoxic or cytostatic agents. For example, the assays can be used to assess radiosensitivity [West et al., 1997], anti-estrogen effects [DeFriend et al., 1994], and the effect of growth factors on tumor growth [Lang et al., 1997]. Although a serum-free medium formulation was developed [Wilks and West, 1991] that worked well with human breast tumors [DeFriend et al., 1994], it was not found to be as effective as fetal calf serum for cervix tumors. The assay is the gold standard method for assessing radiosensitivity, mainly because assessment by clonogenic assay is essential due to the mechanism of radiation-induced cell death (i.e, not always immediate but sometimes after several cell divisions). Alternative short-term assays have been studied for chemosensitivity testing, and theses include measuring cell growth (e.g., MTT assay) and cell viability e.g., tritiated thymidine uptake, adenosine triphosphate (ATP) tumor chemosensitivity assay. The disadvantage of these non-clonogenic assays is that they do not discriminate between the tumor and normal cells, which can make up a high proportion of the tumor cell suspensions.

## 5.　ESTABLISHMENT OF LONG-TERM CULTURES

The protocol for long-term culture of cervix carcinoma tumor cells is based on those described by Freshney [2000] and Rheinwald and Green [1975] for the culture of normal human keratinocytes.

The major difficulty in culturing epithelial cells is that fibroblasts present in the starting material have a growth advantage and will grow in preference to the epithelial cells. The method therefore involves the use of an irradiated feeder layer of a clone of Swiss Murine 3T3 fibroblasts [Parkinson, 1985], which in addition to suppressing fibroblast growth also provides growth factors for the cervical tumor cells. The medium also contains additional growth factors, which will support the proliferation of cervix carcinoma tumor cells. The method is represented diagrammatically in Figure 8.2.

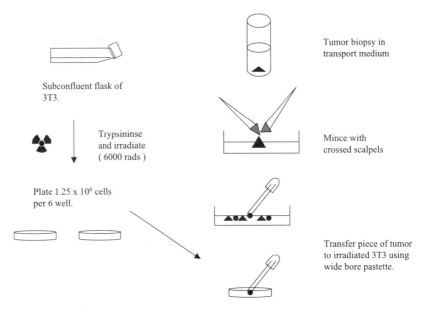

**Figure 8.2.** Utilization of 3T3 feeder layer in primary culture of tumor biopsies.

## Protocol 8.3. Production of 3T3 Feeder Cell Layer

### Reagents and Materials

*Sterile*

❑ Culture medium for 3T3 cells: Dulbecco's modified Eagle's medium (DMEM) + 10% donor calf serum (DCS), 5 mM L-Glutamine, 100 IU/ml penicillin, 100 μg/ml streptomycin.

❑ Purified trypsin: 0.02% solution of Worthington trypsin. (see Section 3.1.2)

❑ Trypsin/EDTA (see Section 3.1.2)

❑ Tumor cell line culture medium, full keratinocyte medium (FKM) (see Section 3.2.4)

### Protocol

(a) Remove medium from a subconfluent flask of 3T3 cells.

(b) Wash the monolayer with PBSA.

(c) Remove the PBSA, and add 2 ml of 0.02% Worthington trypsin per 75 cm$^2$.

(d) Incubate at 37°C for 2–3 min, or until the cells detach.

(e) Tap the flask to detach the cells completely.

(f) Add 8 ml DMEM + 10% DCS to neutralize the trypsin and disperse the cells.

(g) Transfer the contents of the flask to a universal, and spin it at 600 $g$ for 5 min.

(h) Remove the medium, then resuspend the cell pellet in 4 ml DMEM + 10% FBS.

(i) Plate I ml into a fresh 75 cm$^2$ flask for continued culture, and irradiate the remaining cells.

(j) The cells should be irradiated with 6000 rads (60 gy) from a sealed $^{135}$Cs source.

(k) After irradiation, wash the cells in PBSA, then plate into sixwell plates.

(l) Each sixwell plate should contain 1.25 × 10$^6$ cells in DMEM + 10% DCS, and the cells should be allowed to attach for a minimum of 3 h before use.

Irradiated 3T3 cells can be stored at 4°C for up to one week before addition to tumor cell cultures. It has also been reported that irradiated feeder cells can be stored in liquid nitrogen and thawed immediately before use [Hunyadi et al., 1989].

Tumor biopsy samples should be collected in a universal containing transport medium. To increase the chances of successfully establishing a tumor cell line, the samples should be processed as soon as possible after collection.

## Protocol 8.4.  Establishment of Primary Cervical Tumor Cultures

### Reagents and Materials
*Sterile*
- ❑ Irradiated feeder cells
- ❑ FKM (see Section 3.2.4)
- ❑ PBSA
- ❑ 0.02% EDTA
- ❑ Six-well plate
- ❑ Petri dish, 10-cm
- ❑ Scalpels, forceps
- ❑ Fine-tip and wide bore Pastette

### Protocol
(a) Pour the biopsy, and transport medium into a 10-cm Petri dish containing a small amount of FKM.

(b) Using a pair of scalpels, remove and discard any fat and necrotic tissue, and then finely mince the tumor into pieces less than I mm$^2$.

(c) Remove the DMEM + 10% DCS from the irradiated feeder cells, and carefully transfer a piece of tumor by using a wide bore Pastette or sterile forceps onto the feeder layer.

(d) Add 3 ml of FKM per 6-well plate (0.5 ml/well), then incubate at 37°C, 5% $CO_2$.

(e) The medium should be replaced every 2–3 days (taking care not to disturb the explanted tumor).

(f) The 3T3 feeder cells should be replaced at least weekly (or when gaps in the feeder layer appear). The existing 3T3 should be removed as follows:

    (i) Remove the medium from the well and carefully wash with 1 ml PBSA.

    (ii) Add 1 ml per well of 0.02% EDTA, which will detach the feeder cells but not the tumor cells.

    (iii) Using a fine-tip Pastette, carefully wash the 3T3 from the well and leave any developing tumor cell colonies intact).

    (iv) EDTA can also be used to remove any normal fibroblasts, which appear in the tumor cell cultures. The normal fibroblasts can be cultured in DMEM + 10% FCS and can provide a useful counterpart when studying molecular events that occur in the cervical tumor cell lines.

It is important that the primary cultures are examined frequently and that the medium is changed and feeder cells are replaced whenever necessary. One should establish several different cultures with different explants to increase the chances of establishing a long-term line. Once the tumor cells begin to migrate out from the explants and distinct colonies can be seen and are confluent, they should be subcultured on to freshly prepared feeder cell layers.

### Protocol 8.5.  Subculture of Cervical Tumor Cells

***Reagents and Materials***
*Sterile*
- ❏ PBSA
- ❏ 0.02% EDTA
- ❏ 0.05% Trypsin, 0.02% EDTA
- ❏ Full keratinocyte medium (FKM)

***Protocol***
(a) Remove the medium from the well, and wash with PBSA.

(b) Add 1 ml of 0.02% EDTA, and remove the feeder layer by using a Pasteur pipette (see Protocol 8.4).

(c) Add 1 ml of trypsin/EDTA, and incubate at 37°C until the tumor cells detach. (The time taken for the tumor cells to detach varies and depends on the morphology of the tumor cell line. Cells that grow in tightly packed colonies generally take longer to trypsinize).

(d)  Add 4 ml of FKM, and transfer to a universal.

(e)  Spin at 600 g for 5 min, then resuspend the cells in fresh FKM.

(f)  Replate half into a new 6-well plate containing an irradiated feeder layer.


## 5.1.  Cryoconservation

Following trypsinization and neutralization of trypsin, the tumor cells should be resuspended in a freezing mixture consisting of 90% FCS + 10% DMSO. A maximum number of $1 \times 10^7$ cells/vial should not be exceeded. The cells should be frozen at a controlled rate of $-1°C$/minute before being transferred to liquid nitrogen. The controlled cooling can be achieved by using a Nalgene cryo container placed in a $-80°C$ freezer overnight.

The cells should be thawed rapidly at 37°C, transferred drop-wise to warm FKM, and washed before plating.


## 5.2.  Alternative Methods for Establishing Long-Term Cell Lines

Variations exist in the use of feeder layers, the method used to establish primary cultures and the media in which cervical tumor cell lines are grown.

Tumor cell lines may be established without the use of an irradiated feeder layer [Koopman et al., 1999], and tumor cell lines, which are established on feeder layers, may become feeder-independent after several passages in culture. The ability to establish cell lines in the absence of feeder cells may be due to differences in the original tumor specimens. Thus, tumors that grow more rapidly in vivo may be propagated more easily in vitro, or the starting material may contain fewer fibroblasts, thus reducing the need for feeder cells to suppress them. Similarly the ability of some cell lines to become feeder-independent may reflect the different growth potential of tumors.

Although we found the explant method useful, many groups have established long-term lines from disaggregated tumor tissue [Rheinwald and Green, 1975]. In order to establish cell lines from disaggregated tumor material, one should follow a disaggregation procedure as described for the short-term culture of cervical tumor cells, but should plate them onto irradiated 3T3 rather than into agar.

Many variations on the media used to culture the tumor cells can be found, and there is a commercially available medium designed to promote the growth of keratinocytes (Invitrogen) that can also be used for tumor cell lines. The media have been used for the culture of cervical epithelial cells by [Pirisi et al., 1988]. Many commercially available cervical tumor cell lines grow in relatively simple media, which does not contain additional growth factors; but this may reflect

**Table 8.2.  General properties of cervical tumor cell lines.**

| Cell line | Feeders required | Morphology | Time to trypsinize |
|---|---|---|---|
| 808 | No | Regular pavement | 3 min |
| 778 | No | Regular pavement | 3 min |
| 879 | No | Irregular pavement | 5 min |
| 866 | Sometimes | Irregular pavement | 5 min |
| 877 | Yes | Irregular colonies within feeder layer | 3 min |
| 878 | Yes | Irregular colonies within feeder layer | 5 min |
| 873 | Yes | Regular colonies within feeder layer | 3 min |
| 915 | Yes | Tightly packed colonies within feeder layer | 8 min |

the fact that these cell lines have adapted to grow without additional growth factors during long periods in culture.

## 5.3.  Characterization of Tumor Cell Lines

The main aim is to confirm that the cell line obtained is derived from cervical tumor material and not normal tissue, which may also be present in the starting material. There is no single marker that can identify the cells as tumor cells, but this can be determined by examining a variety of properties, some of which are listed below.

### 5.3.1.  Growth Characteristics

For long-term cervical tumor cell cultures, the fact that the cells can be cultured long-term suggests that they are derived from tumor cells because any normal, non-transformed cells would be expected to senesce [Hayflick, 1961].

The individual growth properties of our bank of cervical cell lines vary. Some lines initially consisted of heterogenous populations of cells; in some cases the cell lines were cloned by limiting dilution on feeder cells to establish a more homogenous population. The heterogeneity of early cultures probably reflects the heterogenous nature of the original tumor material or mixtures of normal and malignant cells. The growth rate of individual lines varies, as does the split ratio used for subculture.

Table 8.2 shows some of the basic properties of our bank of cell lines.

### 5.3.2.  Morphology

Morphologically, the cervical tumor cell lines usually resemble their normal epithelial counterparts. The morphology varies depending on whether the cell line is still dependent on feeder cells; with feeder layers the cells grow as colonies within the feeder layer while without feeders they form the characteristic cobblestone appearance of epithelial cells.

Examples of feeder independent cell lines are shown in Figure 8.3.

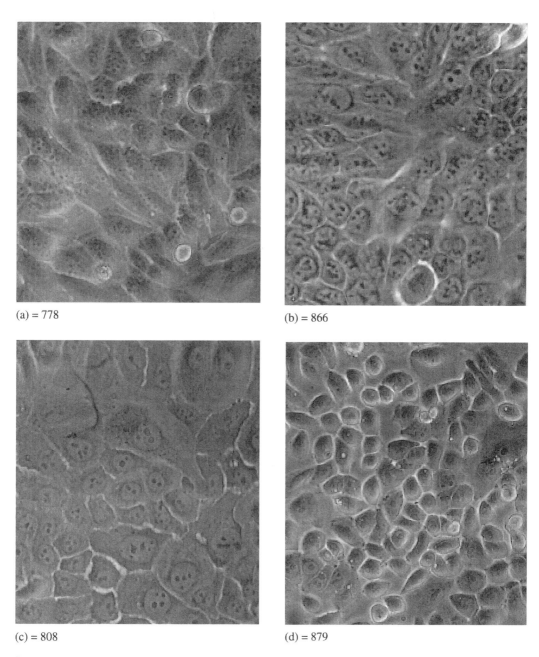

(a) = 778

(b) = 866

(c) = 808

(d) = 879

**Figure 8.3.** Feeder independent cell lines from cervical tumors.

### 5.3.3. Karyotypes

The karyotype of the tumor cell lines would not be expected to be normal, and in our bank of cell lines it varied between hyperdiploid to near pentaploid. The karyotypes are complex with numerous numerical and structural abnormalities. It is interesting that several of our tumor cell lines had extra copies of isochromosome 5p, and this finding has been reported to be characteristic of tumor cell lines derived from squamous cell carcinoma of the cervix but is rare in other tumor types [Mitra et al., 1994].

### 5.3.4. Presence of HPV DNA

As previously mentioned, HPV is found in more than 99% of all cervical carcinomas [Walboomers et al., 1999]. One would therefore expect HPV DNA to be present in a high percentage of cervical tumor cell lines. The presence of HPV DNA can be determined by using the general primer (Gp5+/Gp6+)-based PCR method [de Roda Husman et al., 1994]. The oncogenic HPV types, which are most prevalent, are HPV 16 and 18; it is these types that are found in the majority of our and other cervical tumor cell lines.

The number of copies of HPV 16 per cell was determined for each of the HPV16-positive cell lines by using quantitative real-time PCR [Josefsson et al., 2000]. With this technique, a 180 bp fragment of the HPV16 E1 open reading frame was amplified in the presence of a specific hybridization probe linked to a fluorescent dye and the real-time accumulation of fluorescence detected by an ABI Prism 7700 Sequence Detection System. The simultaneous amplification of the nuclear -actin gene allowed the HPV 16 copy number to be compared directly with the number of cells in the sample. Using this technique, we have conformed that the widely used Caski and SiHa cervix cancer lines contain HPV 16 copy numbers of approximately 750.6 and 1.7, respectively. Table 8.3 shows the number of copies of HPV 16 per cell in each of the HPV 16-positive cell lines isolated by this laboratory.

## 6. APPLICATIONS OF LONG-TERM CULTURES

It is known that HPV infection alone is insufficient to lead to malignant conversion, but the other events that may occur are complex and

Table 8.3. HPV copy number.

| Cervical carcinoma cell line | HPV 16 copy number |
| --- | --- |
| 866 | 27.5 |
| 879 | 1.5 |

poorly understood. Recent studies utilizing cervical tumor cell lines have provided some useful insights to these molecular events.

The classic model of carcinogenesis is that first a cell is immortalized, and then genetic changes accumulate, which may cause the activation of specific oncogenes or the inactivation of tumor suppressor genes, and eventually lead to the development of a malignant cell type. Much work involving cervical tumor cell lines has studied the way in which HPV contributes to the immortalization of cells. In particular, studies have concentrated on the role of telomerase. In normal somatic cells, telomeres, which are the ends of linear chromosomes, gradually shorten during the life span of the cell and eventually lead to senescence, followed by crisis and ultimately death of the cell.

Immortalization of human cells has recently been linked to activation of the telomere-lengthening enzyme telomerase [Harley et al., 1994], and telomerase is known to be activated by transcription of the human telomerase transcription gene (hTERT; Bodnar et al., 1998]. In 1998, Snijders et al. noted that telomerase activity could be detected in the majority of cervical carcinomas and in a subset of high-grade pre-neoplastic lesions. Other studies have documented that the loss of alleles for chromosome 6 and 3p is common in cervical carcinomas [Kersemaekers et al., 1999]. The Snijders group has recently suggested that chromosome 6 is the site of a putative telomerase repressor locus and has tested this theory by molecular analysis of both HPV-transfected normal human keratinocytes and the HPV 16-positive cervical carcinoma cell line SiHa. These studies have revealed that chromosome 6 does indeed harbor a telomerase repressor locus, as reintroduction of chromosome 6 into cells by microcell-mediated chromosome transfer, represses expression of hTERT, thus reducing telomerase activity and inducing growth arrest in the cells. Consequently, loss of chromosome 6 in cervical carcinoma tumor cells may explain why the cells become immortalized [Steenbergen et al., 2001]. Other studies using normal human keratinocytes have suggested that telomerase is activated by the binding of HPV E6 through myc and GC-rich Sp1 binding sites [Oh et al., 2001]. Once the cervical cells are immortalized, genetic changes occur and may eventually lead to malignant conversion.

Koopman et al. [1999] have used a bank of cervical tumor cell lines to perform detailed analysis of the chromosomal rearrangements and sites of HPV integration into the host genome. It has been suggested that the integration site may reveal how HPV contributes to carcinogenesis, as it may lead to the activation or mutation of specific oncogenes or to the inactivation of tumor suppressor genes. These studies utilized a combined binary ratio-fluorescent analysis and showed that HPV is frequently integrated at translocation breakpoints.

Another area of interest is the role of the host immune response in the development of cervical carcinoma. It has been known for several years that down-regulation or loss of expression of HLA Class 1 molecules is a frequent event in cervical neoplasia [Keating et al., 1995], and this event has important consequences for the immune response. What has not been understood until recently is how these events occur. Cervical tumor cell lines have presented the opportunity to systematically study mechanisms of such HLA down-regulation [Koopman et al., 1999; Brady et al., 2000].

The HLA Class 1 expression on the cell lines was determined by using immunohistochemistry and flow cytometry, where specific antibodies were available, or by isoelectric focusing, where they were not, and was compared with the original tumor phenotype and genotype. The genotypes of the cell lines were determined by molecular tissue typing. There was 83% congruency for HLA expression between cell lines and the primary biopsy [Brady et al., 2000]. This supports the relevance of the cell lines for these types of studies. However the derived cell lines are not necessarily a complete picture of the range of different cell types in the primary tumor. Table 8.4 shows the HLA Class 1 expression of our tumor cell lines and their corresponding biopsies.

Molecular studies revealed that five basic mechanisms were responsible for the HLA loss or down-regulation:

(i) HLA allelic transcription but no protein production.
(ii) Abnormal HLA Class 1 allelic transcription
(iii) Loss of heterozygosity
(iv) No HLA-B locus transcription
(v) No IFN-$\gamma$ up-regulation of HLA expression
(vi) Different combinations of defect were found in 5/8 of our cell lines.

HLA down-regulation is thought to decrease the ability of the tumor cell lines to be recognized by T cells. We used the 778 cell line, which lacks expression of HLA-A2, to study whether re-introduction of the latter could allow identification of any anti-tumor cytotoxic T cells (CTL) from the patient. In other words, the effectors that might have immunoselected the HLA-A2 negative phenotype in vivo. HLA-A2 expression was re-introduced into cells by transfection with HLA-A2 genomic DNA. The resulting cell line 778.A2 was then used as a stimulator cell to generate cytotoxic T cell responses in vitro with autologous T peripheral blood mononuclear cells. No anti-778 CTL were generated when untransfected tumor cells were used as stim-

**Table 8.4. HLA phenotypes of cervical tumor cell lines.**

| Tumor/allele | HLA-A | HLA-A | HLA-B | HLA-B |
|---|---|---|---|---|
| 778 | 02 | 32 | 07 | 18 |
| Biopsy | − | + | + | − |
| Line | − | + | + | − |
| 808 | 02 | 24 | 35 | 63 (1517) |
| Biopsy | − | − | − | − |
| Line | − | + | − | − |
| 866 | 02 | 68 (28) | 27 | 62 |
| Biopsy (NA) | | | | |
| Line | + | No mAb | + | + |
| 873 | 01 | 24 | 07 | 08 |
| Biopsy | No mAb | + | − | + |
| Line | No mAb | + | + | + |
| 877 | 02 | 29 | 35* | 44 |
| Biopsy | − | + | + | + |
| Line | − | + | − | − |
| 878 | 01 | 03 | 08 | 44 |
| Biopsy | No mAb | +/− | − | − |
| Line | No mAb | + | − | − |
| 879 | 02 | 31 | 39 | 60 |
| Biopsy | + | + | Bw6 | No mAb |
| Line | + | + | No mAb | No mAb |
| 915 | 03 | ** | 65 (1042) | 47* |
| Biopsy | + | | − | +/− |
| Line | + | | − | − |

*Allele not detectable in tumor cell line by molecular tissue typing (SSP).
** Only one HLA-A allele detected, presumed HLA-A3 homozygous.
NA = not available. Shaded area = not applicable for testing.

ulators. However, specific T cells were generated with the HLA-A2 transfected stimulator cells. The T cells were cloned by limiting dilution and their ability to kill autologous tumor cells tested by replica plating against 778 or 778.A2 targets in $^{51}$Cr release assays. The frequency of responding cells generated with 778.A2 cells was higher than those stimulated by the original cell lines. It was possible to expand three of the positive clones and show that they were killing the tumor cells in an HLA Class 1 restricted manor, as the level of killing decreased when the tumor cells were pre-incubated with an anti HLA Class 1 antibody. This finding suggests that the loss of HLA-A2 does influence the ability of the host to deliver an effective immune response. The results of these studies are shown in Figure 8.4.

More recently Evans et al. [2001] have shown that Class 1 antigen processing defects limit the presentation of a CTL epitope from human papillomavirus E6. They generated HPV-specific CTL by using

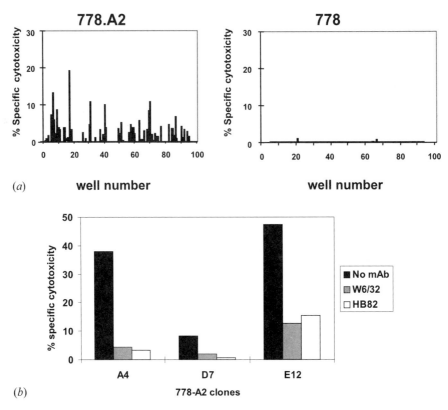

**Figure 8.4.** **a)** Replica plating of T cell clones tested against 778 or 778.A2. **b)** Antibody blocking of cytotoxic T cell clones A4, D7, and E12 killing of 778-A2 clones.

dendritic cells pulsed with HPV-expressing adenoviruses as stimulator cells and showed that the T cells that were produced could kill HPV E6, HLA-A2 expressing B-lymphoblastoid cells (B-LCL), but the cells could not kill HPV 16-expressing tumor cell lines, even when the level of HPV 16 E6 was increased by transfection. They found that the cervical tumor cell lines had low HLA Class 1 expression and also had defects in antigen processing in particular the proteosomes and TAP complex.

Other groups have looked at the effect of increasing the expression of co-stimulatory molecules on cervical carcinoma tumor cells. For instance Gilligan et al. [1998], showed that the immunogenicity of cervical carcinoma tumor cells could be increased by adenoviral delivery of the co-stimulatory molecule B7-1 (CD80). Co-stimulatory molecules provide an important second interaction between cytotoxic T cells and their targets.

## 7. CONCLUSIONS

It is clear that, by using the methods described in this chapter, it is possible to establish short-term explant and clonogenic cultures, and long-term cell lines, from cervical carcinoma biopsy material. A major use of short-term cultures has been in studying the radiation response of human cervical carcinoma cells. Studies have revealed [Kelland and Steel, 1989] that large differences in the intrinsic radiation recovery capacity exist within cervical carcinomas. It was only possible to reach this conclusion by examining cells grown in short-term clonogenic assays. These cells are a true representation of the range of cells present in the original tumor and not clones of cells with a growth advantage, which may preferentially grow in long-term cultures. Such short-term cultures should allow rapid predictive testing of radiosensitivity, which may allow radiotherapy regimens to be tailored to individual patients and will also prevent patients undergoing unnecessary chemotherapy if they have a radiosensitive tumor [Achary et al., 2000].

Combined radiochemotherapy is proving to be a useful alternative for patients whose tumors have been resistant to other treatments, and much of the knowledge of the chemosensitivity of cervical carcinoma cells has been obtained in vitro by using cell lines. Recent study of the cervical carcinoma cell line Caski, as well as tumor material, has led to the proposal that elevated topoisomerase I activity may be a potential target for chemoradiation therapy [Chen et al., 2000].

Long-term cell lines have been useful in studies to determine the role that HPV plays in the etiology of the disease. One exciting discovery has been the role of HPV and telomerase in the immortalization of cervical carcinoma cells because immortalization is an important early event in the carcinogenic process. Further studies are required to determine the other key factors that contribute to carcinogenesis. Some of the risk factors that have been implicated in cervical carcinoma are the use of steroid hormones, the influence of external carcinogens, such as cigarette smoking, and the influence of genetic changes.

Long-term cervical tumor cell lines will be particularly useful for genetic studies because large amounts of material are required, and many studies cannot be performed with the small amounts of primary tumor material that are available. The increased knowledge of the natural history of HPV infections and the mechanisms responsible for malignant transformation mean that several potential viral antigens for immunotherapeutic interventions have been identified [Stern et al, 2001]. However, studies of HLA modulation undertaken with cervical tumor cell lines have also identified potential problems with immunotherapeutic approaches.

## SOURCES OF MATERIALS

| Item | Supplier |
|------|----------|
| Amphotericin | Sigma |
| August rats | Harlan Sera-Lab |
| Basal medium | Invitrogen |
| Cholera toxin | Sigma |
| Collagenase Type I | Sigma |
| Deoxyribonuclease | Sigma |
| DMEM | Sigma |
| Donor calf serum (DCS) | Invitrogen |
| EGF | Sigma |
| Fetal calf serum (FCS) | PAA |
| Gentamycin | Sigma |
| Glutamine | Invitrogen |
| Ham's F12 medium | Invitrogen |
| Hanks' balanced salt solution (HBSS) | Sigma |
| Hydrocortisone | Sigma |
| Insulin | Sigma |
| Iodonitrotetrazolium (INT) violet | Sigma |
| Nylon filter, 100-µm | Tekmar-Dohrmann |
| PBSA | Sigma |
| Penicillin). | Invitrogen |
| Plastic test-tubes | BD Biosciences |
| Pronase | Sigma |
| Streptomycin | Invitrogen |
| Transferrin | Sigma |
| Trypsin, purified | Worthington |

## REFERENCES

Achary, M.P., Jaggernauth, W., Gross, E., Alfieri, A., Klinger, H.P., Vikram, B. (2000) Cell lines from the same cervical carcinoma but with different radiosensitivities exhibit different cDNA microarray patterns of gene expression. *Cytogenet. Cell Genet.* 91:39–43.

Bodnar, A.G., Oulette, M., Frolkis, M., Holt, S.E., Chiu, C.P., Morin, G.B., Harley, C.B., Shay, J.W., Lichtsteiner, S., Wright, W.E. (1998) Extension of life-span by introduction of telomerase into normal human cells. *Science* 279:349–352.

Borysiewicz, L.K., Fiander, A., Nimakom M., Wilkinson, G.WE., Westmoreland, D., Evans, A.S., Adams, M., Stacey, S.N., Boursnell, M.E., Rutherford, E., Hickling, J.K., Inglis, S.C. (1996) A recombinant vaccinia virus encoding human papilloma types 16 and 18, E6 and E7 proteins as immunotherapy for cervical cancer. *Lancet* 347:1523–1527.

Brady, C.S., Bartholomew, J.S., Burt, D.J., Duggan-Keen, M.F., Glenville, S., Telford, N., Little, A.M., Davidson, J.A., Jimenez, P., Ruiz-Cabello, F., Garrido, F., Stern, P.L. (2000) Multiple mechanisms underlie HLA dysregulation in cervical cancer. *Tissue Antigens* 55:401–411.

Chen, B.M., Chen, J.Y., Kao, M., Lin, J.B., Yu, M.H., Roffler, S.R. (2000) Elevated topoisomerase I activity in cervical cancer as a targert for chemoradiation therapy. *Gynecol. Oncol.* 79:272–280.

Courtenay, V.D., and Mills, J. (1978) An in vitro colony assay for human tumours grown in immune-suppressed mice and treated in vivo with cytotoxic agents. *Br. J. Cancer* 37:261–268.

De Roda Husman, A.M., Walboomers, J.M., Meijer, C.J., Risse, E.K., Schipper, M.E., Helmerhorst, T.M., Bleker, O.P., Delius, H., van den Brule, A.J. (1994) Analysis of cytomorphologically abnormal cervical scrapes for the presence of 27 mucosotropic human papuillomavirus genotypes, using poymerase chain reaction. *Int. J. Cancer* 56:802–806.

DeFreind, D.J., Anderson, E., Bell, J., Wilks, D.P., West, C.M., Mansel, R.E., Howell, A., (1994) Effects of 4–hydrotamoxifen and novel pure anti-oestrogen (ICI 182780) on the clonogenic growth of human breast cancer cells in vitro. *Br. J. Cancer* 70:204–211.

Doorbar, J., Parton, A., Hartley, K., Banks, L., Crook., T., Srtanley, M., Crawford, L. (1990) Detection of novel splicing patterns in a HPV16–containing keratinocyte cell line. *Virology* 178:254–262.

Dyson, N., Howley, P.M., Munger, K., Harlow, E. (1989) The human papilloma virus-16 E7 oncoprotein is able to bind to the retinoblastoma gene product. *Science* 243:934–937.

Evans, M., Borsiewicz, L.K., Evans, A.S., Rowe, M., Jones, M., Gileadi, U., Cerundolo, V., Man, S. (2001) Antigen processing defects in cervical carcinomas limit the presentation of a CTL epitope from human papillomavirus 16 E6. *J. Immunol.* 167:5420–5428.

Freshney, R.I. (2000) Culture of Animal Cells: A Manual of Basic Techniques, 4th Edition. Wiley Liss Inc., New York, NY.

Garrido, F., Ruiz-Cabello, F., Cabrera, T., Perez-Villar, J.J., Lopez-Botet, M., Duggan-Keen, M., Stern, P.L. (1997) Implications for immunosurveillance of altered HLA class I phenotypes in human tumours. *Immunol. Today* 18:89–95.

Gilligan, M.G., Knox, P., Weedon, S., Barton, R., Kerr, D.J., Searle, P, Young, L.S. (1998) Adenoviral delivery of B7–1 (CD80) increases the immunogenicity of human ovarian and cervical carcinoma cells. *Gene Ther.* 5:965–974.

Harley, C.B., Kim, N.W., Prowse, K.R., Weinrich, S.L., Hirsch, K.S., West, M.D., Bacchetti, S., Hirte, H.W., Counter, C.M., Greider, C.W. et. al. (1994) Telomerase, cell immortality, and cancer. *Cold Spring Harb. Symp. Quant. Biol.* 59:307–315.

Hayflick. L.M.P.S. (1961) The serial cultivation of human diploid cell strains. *Exp. Cell Res.* 25:585–621.

Howley, P.M. (1991) Role of the human papillomaviruses in human cancer. *Cancer Res.* 51:5019s–5022s.

Hunyadi, J., Simon, M., Jr., Dobozy, A. (1989) Cryopreserved 3T3 fibroblasts retain their capacity to enhance the growth of human keratinocyte cultures. *Acta Derm. Venereol.* 69:509–512.

Josefsson, A.M., Magnusson, P.K., Ylitalo. N., Sorensen, P., Qwarforth-Tubbin, P., Andersen, P.K., Melbye, M., Adami, H.O., Gyllensten, U.B. (2000) Viral load of human papilloma virus 16 as a determinant for development of cervical carcinoma in situ: a nested case-control study. *Lancet* 355:2189–2193.

Keating, P.J., Cromme, F.V., Duggan-Keen, M., Snijders, P.J., Walboomers, J.M., Hunter, R.D., Dyer, P.A., Stern, P.L. (1995) Frequency of down-regulation of individual HLA-A and B alleles in cervical carcinomas in relation to TAP-1 expression. *Br. J. Cancer* 72:405–411.

Kelland, L.R., and Steel, G.G. (1989) Recovery from radiation damage in human squamous carcinoma of the cervix. *Int. J. Radiat. Biol.* 55:119–127.

Kersemaekers, A.M., van de Vijver, M.J., Kenter, G.G., Fleuren, G.J. (1999) Genetic alterations during the progression of squamous cell carcinomas of the uterine cervix. *Genes Chrom. Cancer* 26:346–354.

Koopman, L.A., Szuhai, K., van Eendenburg, J.D., Bezrookove, V., Kenter, G.G., Schuuring, E., Tanke, H., Fleuren, G.J. (1999) Recurrent integration of human papillomaviruses 16, 45, and 67 near translocation breakpoints in new cervical cancer cell lines. *Cancer Res.* 59:5615–5624.

Lang, S.H., West, C.M., Jones, L., Brooks, B., Kasper, C., deWynter, E., Testa, N.G. (1997) In vitro effects of recombinant human megakaryocyte growth and development factor on primary human tumour colony growth. *Oncology* 54:141–145.

Mitra, A.B., Rao, P.H., Pratap, M. (1994) i(5p) and del(6q) are nonrandom abnormalities in carcinoma cervix uteri. Cytogenetics of two newly developed cell lines. *Cancer Genet. Cytogenet.* 76:56–58.

Oh, S.T., Kyo, S., Laimins, L.A. (2001) Telomerase activation by human papillomavirus type 16 E6 protein: induction of human telomerase reverse transcriptase expression through Myc and GC-rich Sp1 binding sites. *J. Virol.* 75:5559–5566.

Parkinson, E.K. (1985) Defective responses of transformed keratinocytes to terminal differentiation stimuli. Their role in epidermal tumour promotion by phorbol esters and by deep skin wounding. *Br. J. Cancer* 52:479–493.

Pirisi, L., Creek, K.E., Doniger, J., DiPaolo, J.A. (1988) Continuous cell lines with altered growth and differentiation properties originate after transfection of human keratinocytes with human papillomavirus type 16 DNA. *Carcinogenesis* 9:1573–1579.

Resbeut, M., Haie-Meder, C., Alzieu, C., Gonzague-Casabianca, L. (2000) Radiochemotherapy of uterine cervix cancers. Recent data. *Cancer Radiother.* 4:140–146.

Rheinwald, J.G., and Green, H. (1975) Serial cultivation of strains of human epidermal keratinocytes: the formation of keratinizing colonies from single cells. *Cell* 6:331–343.

Schoell, W.M., Janicek, M.F., Mirhashemi, R. (1999) Epidemiology and biology of cervical cancer. *Semin. Surg. Oncol.* 16:203–211.

Snijders, P.J., van Duin, M., Walboomers, J.M., Steenbergen, R.D., Risse, E.K., Helmerhorst, T.J., Verheijen, R.H., Meijer, C.J. (1998) Telomerase activity exclusively in cervical carcinomas and a subset of cervical intraepithelial neoplasia grade III lesions: strong association with elevated messenger RNA levels of its catalytic subunit and high-risk human papillomavirus DNA. *Cancer Res.* 58:3812–3818.

Steenbergen, R.D., Kramer, D., Meijer, C.J., Walboomers, J.M., Trott, D.A., Cuthbert, A.P., Newbold, R.F., Overkamp, W.J., Zdzienicka, M.Z., Snijders, P.J. (2001) Telomerase suppression by chromosome 6 in a human papillomavirus type 16–immortalized keratinocyte cell line and in a cervical cancer cell line. *J. Natl. Cancer Inst.* 93:865–872.

Stern, P.L., Faulkner, R., Veranes, E.C., Davidson, E.J. (2001) The role of human papillomavirus vaccines in cervical neoplasia. *Best Pract. Res. Clin. Obstet. Gynaecol.* 15:783–801.

Walboomers, J.M., Jacobs, M.V., Manos, M.M. Bosch, F.X., Kummer, J.A., Sha, K.V., Snijders, P.J., Peto, J., Meijer, C.J., Munoz, N. (1999) Human papillomavirus is a necessary cause of invasive cervical cancer worldwide. *J. Pathol.* 180:12–19.

Werness, B.A., Levine, A.J., Howley, P.M. (1990) Association of human papillomavirus types 16 and 18 E6 proteins with p53. *Science* 248:76–79.

West, C.M., Davidson, S.E., Roberts, S.A., Hunter, R.D. (1997) The independence of intrinsic radiosensitivity as a prognostic factor for patient response to radiotherapy of carcinoma of the cervix. *Br. J. Cancer* 76:1184–1190.

West, C.M., and Sutherland, R.M. (1986) A radiobiological comparison of human tumor soft-agar clonogenic assays. *Int. J. Cancer* 37:897–903.

Wilks, D.P., and West, C.M. (1991) A serum-free medium for the Courtenay-Mills soft agar assay. *Int. J. Cell Cloning* 9:559–569.

# 9

# Primary Culture of Human Mammary Tumor Cells

Valerie Speirs

*Molecular Medicine Unit, University of Leeds, Clinical Sciences Building, St. James's University Hospital, Leeds, LS9 7TF, U.K. Phone: 0113 2065261. Fax: 0113 2444475, v.speirs@leeds.ac.uk*

*Culture of Human Tumor Cells*, Edited by Roswitha Pfragner and R. Ian Freshney.
ISBN 0-471-43853-7  Copyright © 2004 Wiley-Liss, Inc.

## I. BACKGROUND

### 1.1. Breast Cancer Models: Cell Lines

The majority of breast cancer researchers use immortalized human breast cancer cell lines as experimental models. The advantages of using cell lines include their relative homogeneity and ease of use. Although they have proven to be an important research tool over the years, there are disadvantages: Cell lines are prone to genetic drift with time in culture and many are derived from metastatic deposits rather than the primary tumor itself. A classic example, used in laboratories worldwide, is the MCF-7 cell line, which was actually established from a metastatic deposit 30 years ago [Soule et al., 1973], as is the case with the majority of breast cancer cell lines currently in use today. Furthermore, Osborne et al. [1987] have highlighted discrepancies between MCF-7 cells obtained from four different laboratories. Although morphologically identical, striking differences were observed in growth rate, karyotype, hormone receptor content, and clonogenicity. Despite this, a review of PubMed [http://www.ncbi.nlm.nih.gov/PubMed/] from 1990–2000 revealed 4136 citations for studies using MCF-7, making it the most commonly used breast cancer cell line in the world.

### 1.2. Breast Cancer Models: Primary Cultures

A more clinically relevant model is the use of epithelial cells derived directly from human breast tumors, as drug therapies are predominantly targeted against the primary tumor. Thus, routine primary culture of tumor epithelial cells has been a goal of many laboratories. Such culture models can be established either in the form of explant cultures, where mixed cell populations grow out from small fragments of tissue [Telang et al., 1991; Hood and Parham, 1998] or (more desirable) enriched populations of defined cell types [Emerson and Wilkinson, 1990; Hofland et al., 1995; Dairkee et al., 1997]. However, although the former method is relatively easy, the latter has met with variable success over the years. This inconsistency is due to difficulties in establishing the cultures, which often have a relatively slow doubling time. Rapid overgrowth with fibroblasts is also a significant problem.

These difficulties may be explained partly when the architecture of the mammary gland is considered [Millis et al., 1999]. The principal component of the gland is the stroma, which comprises about 80% of its volume. The stroma is made up of adipose and connective tissue, blood vessels, and extracellular matrix. The remaining 20% of the breast mass comprises epithelial cells, which branch to form the structural unit, the terminal duct lobular unit (TDLU). The epithelium is arranged in a bilayer, consisting of an inner apical layer of luminal epithelial cells, which abut an outer layer of myoepithelial cells. The epithelial bilayer is polarized with the luminal cells facing the lumen of the ducts and alveoli, whereas the basal myoepithelial layer is in contact with the basement membrane. Most breast lesions arise from epithelial cells, and as these are clearly in the minority, it is perhaps not surprising that attempts by many investigators at culturing these cells have been abortive.

Despite these problems, we have worked to establish a robust technique for the culture of primary breast cancer epithelial cells from tumor biopsies. This chapter describes our protocol, based on a modification of an original method [Emerson and Wilkinson, 1990], which utilizes differential centrifugation followed by culture in selective media to establish short-term epithelial-enriched preparations from primary breast tumors [Speirs et al., 1996a; 1996b; 1998a].

## 2. PREPARATION OF MEDIA AND REAGENTS

### 2.1. Antibiotic Solution (ABC-PBSA)

PBSA + 200 U/ml penicillin, 200 µg/ml streptomycin, 5 µg/ml fungizone.

### 2.2. Organoid Medium (OM)

DMEM supplemented with 100 U/ml penicillin, 100 µg/ml streptomycin, 2 mM glutamine, 10 mM $N$-(2-hydroxyethyl)piperazine-$N'$-ethanesulfonic acid (HEPES), 0.075% bovine serum albumin (BSA), 10 ng/ml cholera toxin, 0.5 µg/ml hydrocortisone, 5 µg/ml insulin, and 5 ng/ml EGF.

### 2.3. Collection Medium

DMEM with 200 U/ml penicillin, 200 µg/ml streptomycin, 5 µg/ml Fungizone.

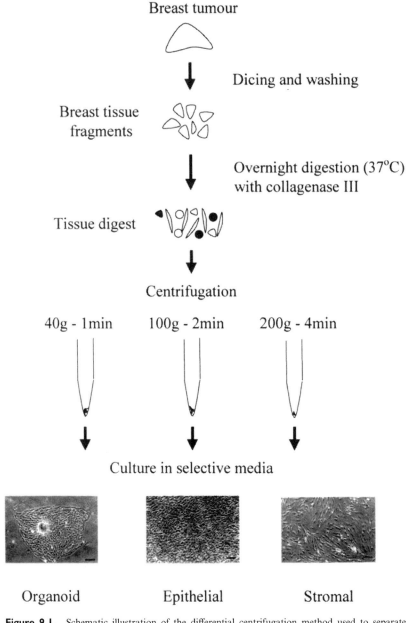

Breast tumour

Dicing and washing

Breast tissue
fragments

Overnight digestion (37°C)
with collagenase III

Tissue digest

Centrifugation

40g - 1min          100g - 2min          200g - 4min

Culture in selective media

Organoid            Epithelial            Stromal

**Figure 9.1.** Schematic illustration of the differential centrifugation method used to separate collagenase-digested breast tumor into individual organoid, epithelial, and stromal fractions. Typical morphological appearances of each of the fractions viewed under phase contrast microscopy are shown. Scale bar = 15 μm.

### 2.4. Primary Culture Medium

OM and DMEM, 3:1.

### 2.5. Complete Culture Medium

DMEM/10FB: DMEM containing 10% heat-inactivated fetal bovine serum (HIFBS).

### 2.6. Tris-buffered Saline (TBS)

Tris-HCl, 200 mM, NaCl, 7.3 mM, BSA, 1%, Triton X-100, 1%, pH 6.

## 3. CULTIVATION OF MAMMARY TUMOR CELLS

### Protocol 9.1. Dispersal of Breast Tumors in Collagenase

A schematic diagram of the breast tissue dispersal method is shown in Figure 9.1.

***Reagents and Materials***
*Sterile*
❏  ABC-PBSA (see Section 2.1)
❏  Collection medium (see Section 2.3)
❏  0.1% Collagenase Type III in complete culture medium
❏  Scalpels, number 11 blades
❏  Universal tubes

***Protocol***
(a)  With approval from your local ethics committee, obtain breast tumor biopsies from pathology and transport to the laboratory in Collection Medium. If the tissue cannot be processed immediately (recommended), it may be stored refrigerated for up to 24 h without significant loss of viability.
(b)  In a Class II laminar flow cabinet, wash tissue extensively in phosphate buffered saline lacking $Ca^{2+}$ and $Mg^{2+}$ (PBSA) supplemented with 200 U penicillin, 200 μg/ml streptomycin, and 5 μg/ml fungizone (ABC-PBSA).
(c)  Mince finely by using crossed scalpels (number 11 blades are recommended).
(d)  Wash twice more in ABC-PBSA.
(e)  Disaggregate for 18–20 h in 0.1% collagenase Type III (made up in complete culture medium) in a 37°C incubator. We have evaluated numerous collagenase enzymes for this purpose and find

that Collagenase III gives a superior cell yield without compromising viability [Speirs et al., 1996a].

(f)　Remove tissue from incubator, and shake vigorously by hand. This will break up any remaining large clumps of partially digested tissue.

(g)　Centrifuge at 40 g for 1 min by using a swing-out rotor.

(h)　Carefully remove supernate and place in a fresh universal. The pellet contains single cells and small fragments of partially digested tissue, which we term the *organoid fraction*.

(i)　Retain and resuspend the pellet in 5–10 ml ABC-PBSA.

(j)　Re-centrifuge the supernate at 100 g for 2 min.

(k)　Remove supernate and transfer to a fresh universal. The pellet contains the *epithelial fraction*. Resuspend in 5–10 ml ABC-PBSA and retain.

(l)　Re-centrifuge the supernate at 200 g for 4 min. Remove supernate (do not retain). The pellet contains the *fibroblast fraction*, which should be retained and resuspended in 10 ml ABC-PBSA.

### Protocol 9.2. Cell Culture from Disaggregated Breast Tumor

***Reagents and Materials***
*Sterile*
❑　ABC-PBSA (see Section 2.1)
❑　Primary Culture Medium: (see Section 2.4)
❑　Organoid medium (see Section 2.2)
❑　Complete culture medium (see Section 2.5)

***Protocol***

(a)　Pellet each fraction by centrifuging at 200 g for 4 min.

(b)　Aspirate supernates, and wash pellets twice more with ABC-PBSA.

(c)　For the first 24 h, plate cells from the organoid and epithelial fractions in primary culture medium.

(d)　After 24 h remove media and replace with OM. Maintain in this medium for the duration of the culture. The components of OM selectively inhibit fibroblast overgrowth and encourage the growth of epithelial cells.

(e)　Seed stromal cells in complete culture medium and maintain in this medium throughout.

Because of the heterogeneity of the organoid fraction, we do not culture this routinely, as it has been reported that tumor organoids give rise to rapidly proliferating epithelial cells, most of which are genetically normal [Wolman et al., 1985]. However, it can represent a good source of normal epithelial cells if these are desired. Typically, fragments of tissue adhere to the surface of the culture vessel and, within about 3–5 days, epithelial cells migrate radially and undergo proliferation. Cells from the epithelial fraction adopt a classic cobblestone morphology, whereas a characteristic bipolar or spindle-shape morphology is seen in the fibroblast fraction. The choice of culture vessel depends on the particular application. For routine preparation and subsequent passage, we use 25-cm$^2$ flasks, whereas for experimental purposes 4-, 6- or 24-well plates are more convenient.

## 4. CHARACTERIZATION

### 4.1. Confirmation of Culture Phenotype

Two important considerations in this type of work are a) confirming the phenotype of the epithelial cells and b) knowing whether the isolated cells are of neoplastic origin, as breast tumors are not homogeneous and contain cells of normal as well as malignant origin [Petersen et al., 1990]. Thus, phenotypic characterization is essential. When possible, we recommend complementing immunostaining with flow cytometry of specific surface or intracellular antigens studies to further confirm the phenotype.

### Protocol 9.3. Phenotyping Breast Cell Cultures by Immunostaining

**Reagents and Materials**

*Sterile*

☐ Four-well plates containing a sterile glass coverslip in each well

*Non-sterile*

☐ Antibodies: vimentin (1:100), cytokeratin (clone MNF 116, 1:100), keratin 19 (K19, 1:100), epithelial membrane antigen (EMA, 1:40)
☐ Vectastain Quick Kit
☐ 3,3′-Diaminobenzidine
☐ Hematoxylin
☐ Absolute methanol
☐ Tris-buffered saline (TBS)

**Protocol**

(a)  Grow a representative amount of each fraction in four-well plates containing a sterile glass coverslip in each well (a few drops of the cell suspension is adequate plus an appropriate volume of culture medium). In our experience, the cells do not grow well on chamber slides, for example, Nunc LabTek chambers or similar; therefore, although more time-consuming, the coverslip method is recommended.

(b) Allow cultures to reach 50% confluency, then fix with absolute methanol for 10 min at room temperature.

(c) Block endogenous peroxidase by incubating with 3% hydrogen peroxide in Tris-buffered saline for 10 min at room temperature. PBSA is also suitable if preferred.

(d) Block nonspecific binding sites by incubating with 1:5 normal serum (the species derivation will depend on the primary antibody) in TBS for 20 min.

(e) Incubate with appropriate primary antibodies for 1 h at room temperature. These should include the fibroblast marker vimentin (1:100) and the epithelial markers cytokeratin (clone MNF116, 1:100) keratin 19 (K19, 1:100) and epithelial membrane antigen (EMA, 1:40). Optimum concentrations of each primary antibody should be determined empirically: the concentrations provided are for guidance.

(f) Wash with TBS and incubate with biotinylated secondary antibody followed by streptavidin/peroxidase complex. We find the Vectastain Quick Kit, which comes with a universal blocking serum, convenient; however, the same result will be achieved if the correct individual reagents are used.

(g) Visualize with 3,3′-diaminobenzidine.

(h) Counterstain with hematoxylin.

(i) Dehydrate and mount coverslip.

*Note:* K19 is a luminal epithelial marker, which is expressed in breast tumors [Bartek et al., 1985]. Although this antigen is expressed in tissue sections by luminal epithelial cells within the terminal ductal lobular units of both normal and malignant origin, under in vitro conditions it is generally associated with tumor epithelial cells only [Taylor-Papadimitrou et al., 1989]. In accordance with other studies, isolated epithelial cells are not universally positive for K19 but focal positivity is observed with areas of K19 positivity in single cells or islands of cells adjacent to colonies, which are K19-negative [Shearer et al., 1992; Bergstraesser and Weizman, 1993; Ethier et al., 1993; Speirs et al., 1998b]. Some examples of epithelial and stromal culture stained with cytokeratin and vimentin are shown in Figure 9.2.

## 4.2. Flow Cytometry

This analysis can be performed by using freshly prepared aliquots of isolated cell fractions before plating in culture or after the cells are

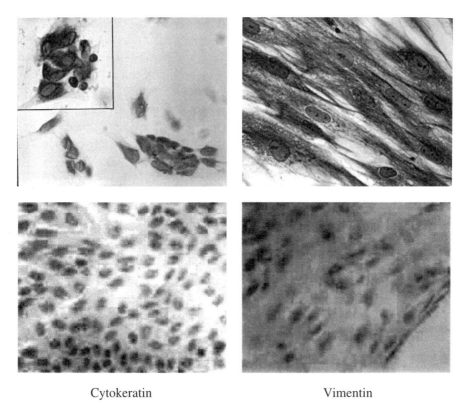

|  Cytokeratin  |  Vimentin  |

**Figure 9.2.** Immunohistochemical analysis of the epithelial and stromal fractions (left and right panels respectively) stained with relevant primary antibody (top) or a negative control antibody (bottom).

established in culture. We have used cells from both sources to compare their phenotype over time.

## Protocol 9.4. Flow Cytometric Analysis of Breast Cell Cultures

### Reagents and Materials

❏ PBSA-BSA-azide (PBSA containing 0.25% (w/v) BSA, 10 mM sodium azide

❏ LEUCOperm Kit

❏ Primary antibody (as described in Protocol 9.3.)

❏ FITC-conjugated secondary antibody, 5 μg/ml

❏ FACSCalibur, BD (Oxford, U.K.); Cell Quest acquisition software

### Protocol

(a) Detach cells from the culture vessel by treatment with PBSA containing 3 mM EDTA at 37°C or by scraping (this step is not necessary for freshly isolated cells).

(b)   Wash in PBSA-BSA-azide.

(c)   Fix and permeabilize (LEUCOperm Kit) according to the manu-
      facturer's instructions. If surface antigens are being investigated
      (e.g., EMA), this step is not necessary, and the cells can be incu-
      bated directly with primary antibody.

(d)   Incubate cells for 1 h at room temperature with 25 μl of appro-
      priate primary antibody (10 μg/ml; suitable antibodies are de-
      scribed in Section 3.1). Include an irrelevant Ig-subtype matched
      negative control antibody.

(e)   Wash in PBSA-BSA-azide, and incubate for 1 h with 50 μl FITC-
      conjugated secondary antibody (5 μg/ml).

(f)   Wash and resuspend in PBSA-BSA-azide.

(g)   Count and analyze a total of 5000 cells by using a flow cytometer
      (e.g., FACSCalibur equipped with Cell Quest acquisition soft-
      ware). Set gates such that 5% of cells from the irrelevant negative
      control antibody are positive. Using the test antibodies, the per-
      centage of positively stained cells above this marker can be
      determined.


Representative flow cytometry profiles for epithelial and fibroblast
populations from passage 1 cultures stained with cytokeratin and
vimentin are shown in Figure 9.3, illustrating that each fraction represents a distinct population of cells. However, it is not uncommon to
have a mixture of fibroblasts and epithelial cells immediately after
dispersal and differential centrifugation, but with increasing time in
defined media the relevant cell populations are expanded preferentially, illustrating the selective effects of the media (Fig. 9.4).


### 4.3.   Growth Control

Growth rates of epithelial cells derived from different tumors are
variable and probably reflect biological variability between samples.
Typical population doubling times range from 3–6 days, and in most
cases cells began to reach plateau after 21 days. Cell viability, assessed
by Trypan Blue exclusion, ranges from 94–97% when the cultures are
first established to 80–85% after 21 days in vitro. We do not routinely
continue the cultures beyond 21 days, although we have shown similar
cultures to be both phenotypically and genotypically stable in vitro for
up to 6–8 weeks [Speirs et al., 1996b], so longer-term culture is possible, if required.

   If further confirmation of the epithelial nature of the cultures is
required, reverse transcriptase polymerase chain reaction (RT-PCR)
analysis of the NB-1 gene, which encodes a calmodulin-like protein of

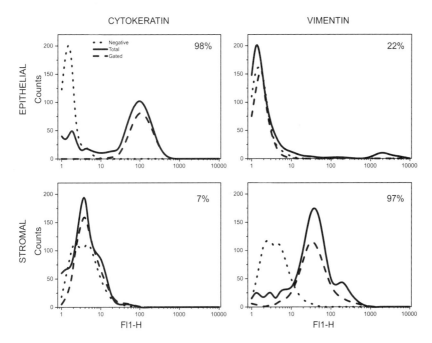

**Figure 9.3.** Intracellular flow cytometric analysis of keratin 19 and vimentin in representative epithelial and stromal fractions at passage one. The percentage refers to the number of positively stained cells.

unknown function [Yaswen et al., 1990; 1992] is suitable. NB-1 is expressed in normal breast epithelial cultures but is not found in those derived from tumors [Yaswen et al., 1990; 1992; Stamper et al., 1993]. Clearly it would be more desirable to have a suitable gene that is preferentially expressed in tumor rather than normal epithelial cells, but to date we have been unable to identify such a gene. If this method is used, it is essential that relevant controls (both positive and negative) are included. As a positive control, the organoid fraction is suitable because, as outlined above, this tends to contain cells of normal rather than tumor origin. To avoid false positive results, all samples should be tested in parallel for the expression of housekeeping genes (e.g., $\beta$-actin, glyceraldehyde phosphate dehydrogenase, etc.). This testing will ensure that the samples are truly negative and that the negative result, which should be observed when tumor epithelial cDNA is amplified with NB-1 oligonucleotide primers, is not due to degradation of the cDNA template used in the PCR reaction. This, coupled to the sensitivity of the PCR, in theory capable of amplifying messages from a single cell, should unequivocally demonstrate that the cultures contain tumor and not normal epithelial cells.

We recommend using the following criteria used to ascertain culture success:

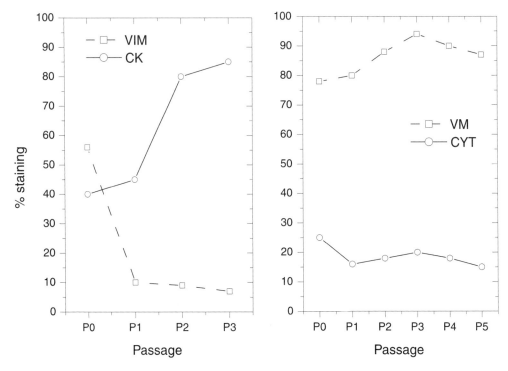

**Figure 9.4.** Expression of epithelial (cytokeratin) and stromal (vimentin) markers over time in epithelial and stromal cultures obtained by differential centrifugation followed by culture in selective media. Note that the percentage of positively stained cells in each cell fraction, but particularly the epithelial fraction increases over time, illustrating the selective effects of the media.

(i)   Attachment and growth of phenotypically epithelial-like cells to the culture vessel after 3–5 days in vitro in defined medium.

(ii)   Continued growth under these conditions for up to 6–8 weeks, with a doubling time of approximately 3–6 days, with no evidence of fibroblast infiltration.

(iii)  Immunopositivity for broad-spectrum cytokeratin, clonal reactivity for K19, and EMA, with no expression of vimentin.

## 5.  APPLICATIONS OF BREAST CANCER MODELS

It is important to note that the methods described herein are not intended to support the long-term culture of primary breast cancer epithelial cells (i.e., months rather than weeks) or to establish a new cell line(s), but rather to provide sufficient quantities of cells of defined phenotype suitable for cell and molecular biology studies. The method offers a rapid, simple, and reproducible way of culturing epithelial cell enriched populations derived from human breast tumors, which retain many of the characteristics associated with breast tumors in vivo [Speirs et al., 1996b; 1998a]. The technique additionally separates fibroblasts from epithelial cells, and the ability to culture these cells independent of epithelial cells offers the possibility of recombination experiments, allowing the re-creation, under controlled conditions, of cell interactions, which clearly exist in vivo.

We have found no correlation between culture success and clinical details, including tumor histology, grade, stage, or lymph node involvement. Notably, however, we have observed a gradual increase over time in the number of successful cultures established in our laboratory, when this procedure is performed by one or two specific individuals, which suggests that culture success is attributable to the acquisition of technical expertise rather than a particular clinical feature.

This method has been used successfully in our laboratory to isolate tumor epithelial cells suitable for use in a range of cell and molecular biology applications, including studies on angiogenesis [Speirs and Atkin, 1999], steroid biosynthesis [Speirs et al., 1998b], and gene expression [Speirs et al., 1996b; Speirs et al., 1998b]. The isolated cells are also suitable for transient transfections [Speirs et al., 2000]. Since establishing the technique in 1994, we have developed cultures from well over 300 primary breast tumors, demonstrating the robustness of the technique and its ease of use. For those with some previous cell culture experience, with a little patience and careful attention to detail this technique should be straightforward.

## ACKNOWLEDGMENTS

I thank A.R. Green, D.S. Walton, R.L. Loveday, and E.C. Norman for contributing to the development of this technique and to Yorkshire Cancer Research and the Liz Dawn Breast Cancer Research Fund for financial support.

## SOURCES OF MATERIALS

| Item | Supplier |
| --- | --- |
| Antibodies | Sigma, Dako |
| Bovine serum albumin (BSA) | Sigma-Aldrich |
| BSA | Sigma |
| Cell Quest acquisition software | Becton Dickinson |
| Cholera toxin | Sigma-Aldrich |
| Collagenase Type III | Worthington Biochemical |
| Culture dishes, flasks and plates | Nalge Nunc |
| Culture media, salts, etc. | Invitrogen |
| Cytokeratin (clone MNF 116) | Dako |
| 3,3'-Diaminobenzidine | Vector or Sigma |
| DMEM | Invitrogen |

## SOURCES OF MATERIALS *(Continued)*

| Item | Supplier |
|------|----------|
| EGF | Sigma-Aldrich |
| Epithelial membrane antigen (EMA) | Dako |
| FACSCalibur, | Becton Dickinson |
| FITC-conjugated secondary antibody | Serotec |
| Four-well plates | Nalge Nunc |
| Fungizone | Invitrogen |
| Glutamine | Sigma-Aldrich |
| Heat-inactivated fetal bovine serum (HIFBS) | Invitrogen |
| Hematoxylin | Merck-BDH |
| HEPES | Sigma-Aldrich |
| Hydrocortisone | Sigma-Aldrich |
| Insulin | Sigma-Aldrich |
| Keratin 19 (K19) | Dako |
| LEUCOperm Kit | Serotec |
| NaCl | Sigma |
| PBSA | Invitrogen |
| PBSA | Invitrogen |
| Penicillin | Invitrogen |
| Sodium azide | Sigma |
| Streptomycin | Invitrogen |
| Tris-HCl | Sigma |
| Triton X-100 | Sigma |
| Universal tubes | Bibby Sterilin |
| Vectastain Quick Kit | Vector |
| Vimentin antibody | Sigma |

## REFERENCES

Bartek, J., Taylor-Papadimitrou, J., Miller, N., Millis, R. (1985) Patterns of expression of keratin 19 as detected with monoclonal antibodies in human breast tissues and tumours. *Int. J. Cancer* 35:359–364.

Bergstraesser, L.M., and Weitzman, S.A. (1993) Culture of normal and malignant human mammary epithelial cells in a physiological manner simulates in vivo growth patterns and allows discrimination of cell type. *Cancer Res.* 53:2644–2654.

Dairkee, S.H., Deng, G., Stampfer, M.R., Waldman, F.M., Smith, H.S. (1997) Partial enzymatic degradation of stroma allows enrichment and expansion of primary breast tumour cells. *Cancer Res.* 57:1590–1596.

Emerson, J.T., and Wilkinson, D.A. (1990) Routine culturing of normal, dysplastic and malignant human mammary epithelial cells from small tissue samples. *In Vitro Cell Dev. Biol.* 26:1186–1194.

Ethier, S.P., Mahacek, M.L., Gullick, W.J., Frank, T.S., Weber, B.L. (1993) Differential isolation of normal luminal mammary epithelial cells and breast cancer cells from primary and metastatic sites. *Cancer Res.* 53:627–635.

Hofland, L.J., van der Burg, B., van Eijk, C.H.J., Sprij, D.M., van Koetsveld, P.M., Lamberts, S.W.J. (1995) Role of tumour-derived fibroblasts in the growth of primary culture of human breast cancer cells: Effects of epidermal growth factor and the somatostatin analogue octreotide. *Int. J. Cancer* 60:93–99.

Hood, C.J., and Parham, D.M. (1998) A simple method of tumour culture. *Pathol. Res. Pract.* 194:177–181.

Millis, R.R., Hanby, A.M., Oberman, H.A. (1999) The Breast. In Sternberg, S.S.

(ed) Diagnostic Surgical Pathology, 3rd Edition. Lippincott Williams & Wilkins: Philadelphia, PA, pp. 319–385.

Osborne, C.K., Hobbs, K., Trent, J.M. (1987) Biological differences among MCF-7 human breast cancer cells lines from different laboratories. *Breast Cancer Res. Treat.* 9:111–121.

Petersen, O.W., van Deurs, B., Nielsen, K., Madsen, M.W., Laursen, I., Balslev, I., Briand, P. (1990) Differential tumorigenicity of two autologous human breast carcinoma cell lines, HMT-390951 and HMT-390958, esatblished in serum-free medium. *Cancer Res.* 50:1257–1270.

Shearer, M., Bartkova, J., Bartek, J., Berdichevsky, F., Barnes, D., Millis, R., Taylor-Papadimitriu, J. (1992) Studies of clonal cell lines developed from primary breast cancers indicate that the ability to undergo morphogenesis *in vitro* is lost early in malignancy. *Int. J. Cancer* 51:602–612.

Soule, H.D., Vasquez, J., Long, A., Albert, S., Brennan, M. (1973) A human cell line from a pleural effusion derived from a breast carcinoma. *J. Natl. Cancer Inst.* 51:1409–1413.

Speirs, V., Green, A.R., White, M.C. (1996a) Collagenase III: A superior enzyme for complete disaggregation and improved viability of normal and malignant human breast tissue. *In Vitro Cell Dev. Biol.* 32:72–74.

Speirs, V., Green, A.R., White, M.C. (1996b) A comparative study of cytokine gene transcripts in normal and malignant breast tissue and primary cell culture derived from the same tissue samples. *Int. J. Cancer* 66:551–556.

Speirs, V., Green, A.R., Walton, D.S., Kerin, M.J., Fox, J.N., Carleton, P.J., Desai, S.B., Atkin, S.L. (1998a) Short-term primary culture of epithelial cells derived from breast tumours. *Br. J. Cancer* 78:1421–1429.

Speirs, V., Green, A.R., Atkin, S.L. (1998b) Activity and expression of 17$\beta$-hydroxysteroid dehydrogenase type I in primary cultures of epithelial and stromal cells derived from normal and tumourous human breast tissue: The role of IL-8. *J. Steroid Biochem. Mol. Biol.* 67:267–274.

Speirs, V., and Atkin, S.L. (1999) Production of VEGF and expression of the VEGF receptors Flt-1 and KDR in primary cultures of epithelial and stromal cells derived from breast tumours. *Br. J. Cancer* 80:898–903.

Speirs, V., Green, A.R., Walton, D.S., Kerin, M.J., Fox, J.N., Carleton, P.J., Desai, S.B., Atkin, S.L. (1998a) Short-term primary culture of epithelial cells derived from breast tumours. *Br. J. Cancer* 78:1421–1429.

Stampfer, M., and Yaswen, P. (1993) Culture systems for study of human mammary epithelial cell proliferation, differentiation and transformation. *Cancer Surveys* 18:7–34.

Taylor-Papadimitriou, J., Stampfer, M., Bartek, J., Lewis, A., Boshell, M., Lane, E.B., Leigh, I.M. (1989) Keratin expression in human mammary epithelial cells cultures from normal and malignant tissue: relation to *in vivo* phenotypes an influence of medium. *J. Cell Sci.* 94:403–413.

Telang, N.T., Axelrod, D.M., Wong, G.Y., Bradlow, H.L., Osborne, M.P. (1991) Biotransformation of estradiol by explant culture of human mammary tissue. *Steroids* 56:37–43

Wolman, S.R., Smith, H.S., Stampfer, M., Hackett, A.J. (1985) Growth of diploid cells from breast cancers. *Cancer Genet. Cytogenet.* 16:49–64.

Yaswen, P., et al. (1990) Down-regulation of a calmodulin-related gene during transformation of human mammary epithelial cells. *Proc. Natl. Acad. Sci. USA* 87:7360–7364.

Yaswen, P., Smoll, A., Hosada, J., Parry, G., Stampfer, M.R. (1992) Protein product of a human intronless calmodulin-like gene shows tissue-specific expression and reduced abundance in transformed cells. *Cell Growth Diff.* 3:335–346.

# 10

# Myoepithelium: Methods of Culture and Study

Sanford H. Barsky and Mary L. Alpaugh

Department of Pathology and Revlon/UCLA Breast Center, University of California, School of Medicine, Los Angeles, California.
Email: sbarsky@ucla.edu

*Culture of Human Tumor Cells*, Edited by Roswitha Pfragner and R. Ian Freshney.
ISBN 0-471-43853-7   Copyright © 2004 Wiley-Liss, Inc.

## 1.  INTRODUCTORY REVIEW

Paracrine regulation of tumor progression by host cells is an important determinant of tumor growth, invasion, and metastasis. However, one cell that has largely been ignored in this regulation is the myoepithelial cell. In any organ where there is significant branching morphogenesis, such as the breast, myoepithelial cells ubiquitously accompany and surround epithelial cells and are thought to keep in check (negatively regulate) the process of branching. Myoepithelial cells surround both normal ducts and precancerous lesions, especially of the breast (so-called DCIS, ductal carcinoma in situ), and form a natural border separating proliferating epithelial cells from proliferating endothelial cells (angiogenesis). Myoepithelial cells, by forming this natural border, are thought to negatively regulate tumor invasion and metastasis. Whereas epithelial cells are susceptible targets for transforming events leading to cancer, myoepithelial cells are resistant. Indeed tumors of myoepithelial cells are uncommon and, when they do occur, are almost always benign. Therefore, it can be said that myoepithelial cells function as both autocrine as well as paracrine tumor suppressors.

Our laboratory has found that myoepithelial cells secrete numerous suppressor molecules: high amounts of diverse proteinase inhibitors, which include maspin, TIMP-1, protease nexin-II, and $\alpha$-1 antitrypsin; low amounts of proteinases and high amounts of diverse angiogenic inhibitors, which include maspin, thrombospondin-1, and soluble bFGF receptors, but low amounts of angiogenic factors compared with common malignant cell lines. Whereas carcinoma cells secrete more proteinases than proteinase inhibitors and more angiogenic factors than angiogenic inhibitors, myoepithelial cells then do just the opposite. This observation holds in vitro, in mice and in humans, and suggests that myoepithelial cells exert pleiotropic suppressive effects on tumor progression.

This constitutive gene expression profile of myoepithelial cells may largely explain the pronounced anti-invasive and anti-angiogenic effects of myoepithelial cells on carcinoma and pre-carcinoma cells and may also account for the low-grade biology of myoepithelial tumors, which are devoid of appreciable angiogenesis and invasive behavior. Many of the secretory gene products of myoepithelial cells are present in body fluids, such as in breast ductal fluid and in saliva, reflecting the structural and functional integrity of the ductal-lobular units of the mammary and salivary glands, respectively. Some of the myoepithelial gene products, for example, maspin, in ductal fluid may serve as a surrogate (intermediate) end-point marker (SEM) to estimate the risk of DCIS progression to invasive cancer in the breast and, alternatively, in saliva, may serve as a tumor marker to detect the presence of incipient myoepithelial tumors occurring within the salivary glands of the head and neck.

In order to study the cell and molecular biology of myoepithelial cells further, methods to isolate and characterize human myoepithelial cells must be devised and perfected. Two sources of human myoepithelial cells are 1) the normal ductal-lobular units of organs rich in myoepithelial cells, such as the breast and salivary glands; and 2) benign tumors of myoepithelial cells where large numbers of myoepithelial cells, albeit transformed, can be obtained and used as normal myoepithelial cell surrogates.

## 2. STUDIES OF MYOEPITHELIAL CELLS

It has become clear that cancer cells come under the influence of important paracrine regulation from the host microenvironment [Cavenee, 1993]. Such host regulation may be as great a determinant of tumor cell behavior in vivo as the specific oncogenic or tumor suppressor alterations occurring within the malignant cells themselves, and may be mediated by specific extracellular matrix molecules, matrix-associated growth factors, or host cells themselves [Liotta et al., 1991; Safarians et al., 1996]. Both positive (fibroblast, myofibroblast, and endothelial cell) and negative (tumor-infiltrating lymphocyte and cytotoxic macrophage) cellular regulators exist, which profoundly affect tumor cell behavior in vivo [Folkman and Klagsbrun, 1987; Cornil et al., 1991]. However, one host cell, the myoepithelial cell, has escaped the paracrine onlooker's attention. The myoepithelial cell, which lies on the epithelial side of the basement membrane, is thought to contribute largely to both the synthesis and remodeling of this structure. This cell lies in juxtaposition to normally proliferating and differentiating epithelial cells in health and to abnormally proliferating and differentiating epithelial cells in precancerous disease states, such as DCIS of the breast. This anatomical relationship suggests that myoepithelial cells may exert important paracrine effects on normal glandular epithelium and may regulate the progression of DCIS to invasive breast cancer. Circumstantial evidence suggests that the myoepithelial cell naturally exhibits a tumor suppressive phenotype. Myoepithelial cells rarely transform, and, when they do, they generally give rise to benign neoplasms that accumulate rather than degrade extracellular matrix [Guelstein et al., 1993]. Myoepithelial cells directly or indirectly through their production of extracellular matrix and proteinase inhibitors, including maspin, are thought to regulate branching morphogenesis that occurs in the developing breast and salivary gland during embryological development [Cutler, 1990]. There has been a paucity of studies on myoepithelial cells because they has been relatively difficult to culture and because tumors that arise from these cells are rare.

In previous studies we have been extremely fortunate to have successfully established immortalized cell lines and transplantable xenografts from benign or low-grade human myoepitheliomas of the salivary gland and breast [Sternlicht et al., 1996; Sternlicht et al., 1997]. These cell lines have been designated HMS-1 to HMS-6, and the xenografts HMS-X to HMS-6X. These initials stand for human matrix secreting line and xenograft, respectively and refer to the chronological order of establishment. These lines and xenografts are available to investigators on request. They displayed an essentially normal diploid karyotype and expressed identical myoepithelial markers as their in situ counterparts, including high constitutive expression of maspin. Unlike the vast majority of human tumor cell lines and xenografts, which exhibited matrix-degrading properties, these myoepithelial cell lines and xenografts, like their myoepithelial counterparts in situ, retained the ability to secrete and accumulate an abundant extracellular matrix composed of both basement membrane and non-basement membrane components. When grown as a monolayer, one prototype myoepithelial cell line, HMS-1, exerted profound and specific effects on normal epithelial and primary carcinoma morphogenesis [Sternlicht et al., 1996]. These studies support our position that our established myoepithelial lines/xenografts recapitulate a normal differentiated myoepithelial phenotype and can therefore be used experimentally as a primary myoepithelial cell surrogate. Prompted by these studies and by the conspicuous absence of studies examining the role of the myoepithelial cell in tumor progression, we decided to examine the myoepithelial cell from this perspective. Experiments with these cell lines/xenografts together with relevant in situ observations form the cornerstone of our studies, which observe that the human myoepithelial cell is a natural tumor suppressor.

## 3. INHIBITION OF TUMOR INVASION

Breast ducts and acini are surrounded by a circumferential layer of myoepithelial cells exhibiting strong immunoreactivity for S100, smooth-muscle actin, the common acute lymphocytic leukemia antigen (CALLA), calponin, and diverse proteinase inhibitors, including maspin, $\alpha$1-AT, PNII/APP, and TIMP-1 (Fig. 10.1A). In DCIS, the myoepithelial layer appeared either intact or focally disrupted, but the myoepithelial cells themselves exhibited the same pattern of immunoreactivity (Fig. 10.1B). In DCIS, although proliferations of vWf immunoreactive blood vessel capillaries were observed focally within the supporting stroma, such blood vessels were not observed within the proliferating DCIS cells on the epithelial side of the myoepithelial layer (Fig. 10.1C). The human tumoral-nude mouse xenografts derived from the human myoepitheliomas of the salivary gland, HMS-X and HMS-3X, and breast, HMS-4X demonstrated immunocytochemical profiles identical to each other and to that exhibited by the myoepithelial cells surrounding normal ducts and DCIS with especially intense maspin immunoreactivity (Fig. 10.1D).

Not only was strong proteinase inhibitor immunoreactivity present within the myoepithelial cells of these xenografts, but strong proteinase inhibitor immunoreactivity could be demonstrated within their extracellular matrix as well. Due to this matrix, the myoepithelial xenografts appeared white and cartilaginous in nature (Fig. 10.1E). Within this abundant extracellular matrix deposited by the different human myoepithelial xenografts, murine blood vessels were not observed (Fig. 10.1F). Through the use of a mouse-specific Cot-1 DNA probe (Fig. 10.1G), human myoepithelial xenografts HMS-X, HMS-3X, and HMS-4X demonstrated absent or near-absent angiogenesis. Human non-myoepithelial xenografts of breast cancer cell lines MDA-MB-231 and MDA-MB-468, in contrast, showed a comparatively large murine component of angiogenesis. In a two-dimensional matrix, myoepithelial cell lines grew as a confluent monolayer with self-forming spheroids at superconfluency (Fig. 10.1H). In a three-dimensional matrix, myoepithelial cell lines branched and budded (Fig. 10.1I). In monolayer culture, ultrastructural studies confirmed the cells' myoepithelial identity (Fig. 10.1J).

Detailed studies [Sternlicht et al., 1997] conducted with HMS-1, a

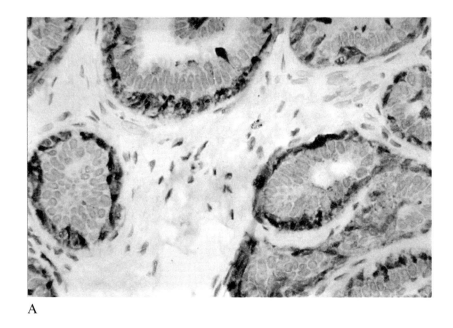

A

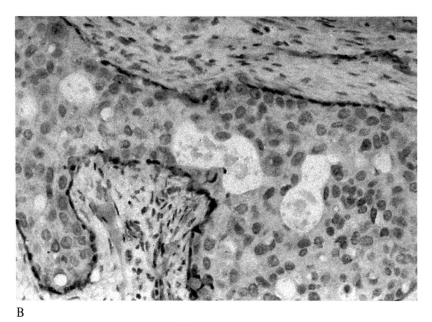

B

**Figure 10.1.** In situ immunocytochemistry profile of myoepithelial cells and their derived cell lines/xenografts. A) Differential maspin immunoreactivity of myoepithelial cells surrounding breast ducts and acini. B) Differential maspin immunoreactivity of myoepithelial cells in DCIS. C) Angiogenesis demonstrated by vWf immunoreactivity limited to stromal side of DCIS. D) Cytoplasmic maspin immunoreactivity of myoepithelial xenograft, HMS-X. E) Gross appearance of one myoepithelial xenograft, HMS-X. F) Microscopic appearance of one typical myoepithelial xenograft, HMS-4X, devoid of apparent angiogenesis. G) Murine Cot-1 dot blot. With a mouse-specific Cot-1 DNA probe, human myoepithelial xenografts, HMS-X, HMS-3X,

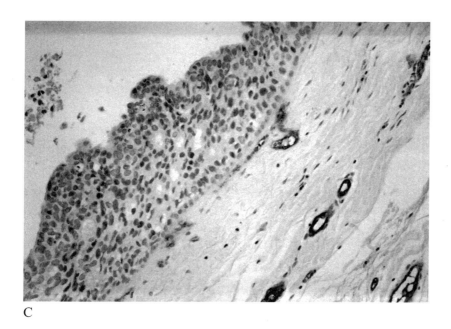

C

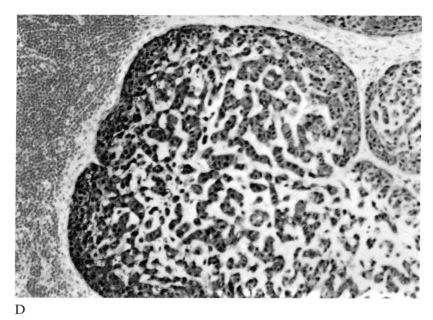

D

**Figure 10.1.** *(Continued)*
and HMS-4X are devoid of a murine DNA angiogenic component in contrast to the angio-
genic-rich MDA-MB-231-X and MDA-MB-468-X breast carcinoma xenografts (right column);
control dot blots of varying murine DNA percentages are also depicted (left column). H) The
prototype myoepithelial cell line, HMS-1, in culture, grows as self-inducing spheroids on top of
its own monolayer. I) HMS-1 undergoes branching morphogenesis and budding when grown
on either Matrigel or a myoepithelial-derived matrix, *Humatrix.* J) Myoepithelial cells in culture, ultrastructurally, give the impression of "smiling," which is symbolic of the cells' natural
tumor-suppressive function.

Myoepithelium: Methods of Culture and Study     227

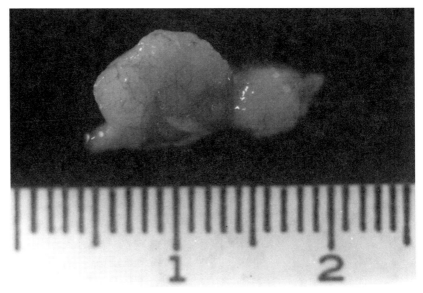

E

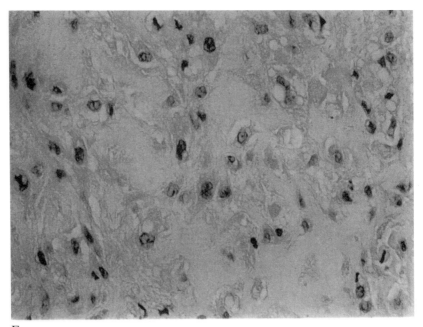

F

**Figure 10.1.** *(Continued)*

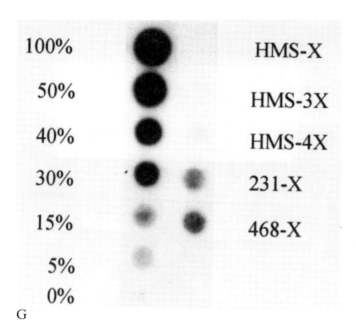

| | | |
|---|---|---|
| 100% | | HMS-X |
| 50% | | HMS-3X |
| 40% | | HMS-4X |
| 30% | | 231-X |
| 15% | | 468-X |
| 5% | | |
| 0% | | |

G

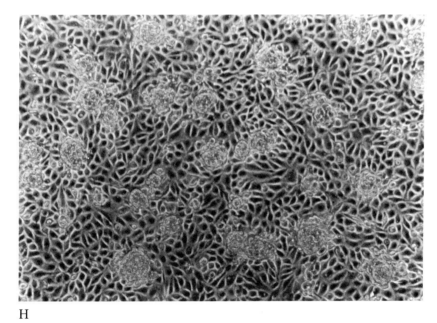

H

**Figure 10.1.** *(Continued)*

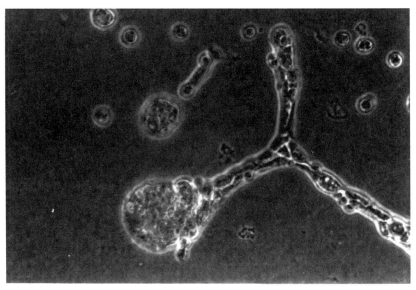

I

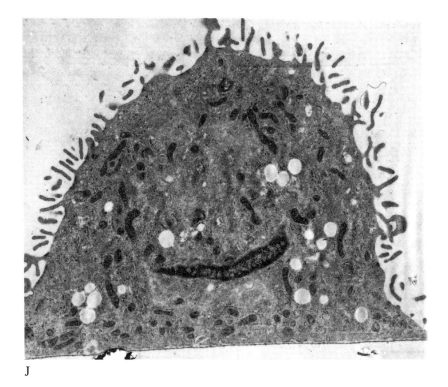

J

**Figure 10.1.** *(Continued)*

prototype myoepithelial cell line, revealed a constitutively high proteinase inhibitor to proteinase ratio in strong contrast to the high proteinase to proteinase inhibitor ratio observed in a number of malignant human cell lines (Fig. 10.2A). Marker studies with this cell line and corresponding xenograft (HMS-X) reflected the constitutive gene expression profile of myoepithelial cells in situ (Fig. 10.2B). This finding was especially true with respect to maspin. Direct gelatin zymography of conditioned media (CM) revealed only low levels of the 92 and 72 kDa type IV collagenases (MMP-9 and MMP-2, respectively) in HMS-1; the 72 kDa collagenase was reduced sixfold in HMS-1 compared with the levels in the majority of the malignant lines; direct fibrin zymography revealed visibly lower levels of the 54 kDa urokinase plasminogen activator (uPA) in HMS-1. This condition was also observed in casein/plasminogen gels.

Tissue-type plasminogen activator was not detected in any cell line, nor was plasmin detected in control gels lacking plasminogen. Stromelysin-1 (MMP-3) was also not detected in HMS-1. The proteinase inhibitor expression profile of HMS-1, in contrast, was characterized by high constitutive expression in CM of several proteinase inhibitors, including TIMP-1; PAI-1; three trypsin inhibitors: α1-AT, PNII/APP, and an unidentified 31 kDa inhibitor detected initially on reverse zymography; and the tumor suppressor maspin. With respect to the trypsin serine proteinase inhibitors, the conspicuous doublet at 116 kDa consistently greater in HMS-1 than in any of the other lines examined was confirmed on Western blot as PNII/APP. These bands represented the 770 and 751 amino acid isoforms of PNII/APP, which possessed a Kunitz-type serine proteinase inhibitor domain.

Interestingly, in 2 M urea extracts of HMS-X, HMS-3X, and HMS-4X, a novel 95 kDa band of trypsin inhibition was detected by reverse-zymography and was confirmed by Western blot to represent an active breakdown product of PNII. This 95 kDa PNII breakdown product was completely absent from HMS-1 CM and urea extracts of HMS-1 cells, suggesting that it was produced in situ within the myoepithelial extracellular matrix to which it bound. The retention of proteinase inhibitor activity by this breakdown product indicated that it retained the Kunitz-type serine proteinase inhibitor domain responsible for its ability to inhibit trypsin. In contrast to PNII/APP, protease nexin I was not detected. The second trypsin serine proteinase inhibitor was present at 54 kDa and was α1-AT. This inhibitor appeared nearly equivalent in HMS-1 compared with the malignant lines examined on reverse-zymography, but by Western blot its signal was markedly stronger and slightly more mobile in HMS-1 than in the malignant lines. These data were reconciled with the fact that α1-AT was probably less glycosylated in HMS-1. This relative underglycosylation

## A

| Enzymes/Inhibitors | HMS-1 | C8161 | MCF-7 | T47D | BT-549 | MDA-MB-157 | MDA-MB-231 | Hs578T | A253 | A431 | Hs578Bst | HMEC | Methods |
|---|---|---|---|---|---|---|---|---|---|---|---|---|---|
| **PROTEINASES** | | | | | | | | | | | | | |
| 72-kDa Gelatinase A | | | | | | | | | | | | | Z |
| 92-kDa Gelatinase B | | | | | | | | | | | | | Z |
| Stromelysin-1 | | | | | | | | | | | | | Z |
| u-PA | | | | | | | | | | | | | Z |
| t-PA | | | | | | | | | | | | | Z |
| Plasminogen | | | | | | | | | | | | | Z |
| **INHIBITORS** | | | | | | | | | | | | | |
| Maspin | | | | | | | | | | | | | W,N |
| TIMP-1 | | | | | | | | | | | | | Z,N |
| Protease Nexin II | | | | | | | | | | | | | Z,W |
| α1-Antitrypsin | | | | | | | | | | | | | Z,W |
| 31-kDa Inhibitor | | | | | | | | | | | | | Z |
| PAI-1 | | | | | | | | | | | | | Z,W |
| PAI-2 | | | | | | | | | | | | | W |
| PAI-3 | | | | | | | | | | | | | W |
| Protease Nexin I | | | | | | | | | | | | | W |
| α2-Antiplasmin | | | | | | | | | | | | | W,C |

## B

| | HMS-X,3X,4X | | Normal Breast | | DCIS | |
|---|---|---|---|---|---|---|
| | Cells | Matrix | ME† | Epi‡ | ME | Epi |
| S-100 | ++++* | - | ++++ | - | ++++ | - |
| Maspin | ++++ | - | ++++ | + | ++++ | ± |
| α1-AT | ++ | ++ | ++ | - | ++ | - |
| PNII | ++ | +++ | ++ | - | ++ | ± |
| TIMP-1 | ++ | + | ++ | - | ++ | - |
| PAI-1 | + | ± | + | + | + | + |
| vWf | - | - | -§ | -§ | -§ | -§ |

**Figure 10.2.** A) Relative constitutive expression of diverse proteinase inhibitors and proteinases in myoepithelial cells (HMS-1) compared with various malignant cell lines. Z, direct or reverse zymography; W, Western blot; N, Northern blot; C, chromogenic substrate assay.

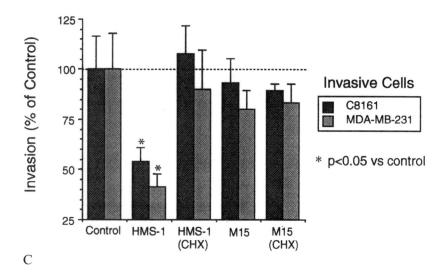

C

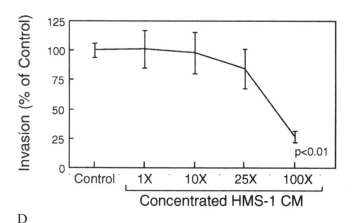

D

**Figure 10.2.** *(Continued)*
B) Myoepithelial-related immunoreactivity In Situ. [†]myoepithelial cells; [‡]epithelial cells; [*]++++, intensely positive; +++, strongly positive; ++, positive; +, weakly positive; ±, equivocally positive; −, negative; [§]on epithelial side of basement membrane. C) Effects of HMS-1 cells on C8161 (melanoma) and MDA-MB-231 (breast carcinoma) invasion. D) Effects of HMS-1 CM on invasion of C8161 cells. Results with MDA-MB-231 cells were similar. Assays in both (C, D) were performed in quadruplicate and show mean percentage of control invasion ± standard deviation. E) Effects of pharmacological treatment of HMS-1 cells with various agents inducing invasion-permissive and non-permissive phenotypes: CHX, cyclo-heximide; DEX, dexamethasone; dB-cAMP, $N^6$,2'-O-dibutyryladenosine 3':5'-cyclic mono-phosphate; Na-But, sodium butyrate; RA, all *trans* retinoic acid; 5-azaC, 5-azacytidine; PMA, phorbol 12-myristate 13-acetate. Dexamethasone induced invasive-permissive phenotype, whereas PMA induced a non-permissive phenotype. Invasion (percentage of control) of C8161 melanoma cells is depicted.

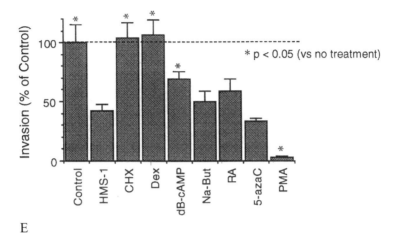

**Figure 10.2.** *(Continued)*

caused α1-AT from HMS-1 to migrate slightly further into the gel and accounted for its poorer reactivation following sodium dodecyl sulfate (SDS)-denaturation on reverse-zymography compared with the more highly glycosylated isoforms present in the malignant lines.

The third trypsin serine proteinase inhibitor detected at 31 kDa was clearly not a degradation product of either PNII or α1-AT, as demonstrated by negative Western blot. The 31 kDa inhibitor was strongly expressed in HMS-1 and was either absent or nearly absent in all of the malignant lines examined. It is being determined whether this unidentified inhibitor is a novel inhibitor. In contrast to the above inhibitors, PAI-1 was expressed only slightly greater in HMS-1 compared with the majority of the malignant lines by both reverse-zymographic and Western blot analysis. Neither PAI-2, PAI-3 or α2-antiplasmin were detected by Western blot analysis in any of the cell lines. Antiplasmin activity, as determined by photometric assay, was completely absent as well.

The most striking difference, however, between the strong proteinase inhibitor profile of HMS-1 and the profile of the malignant cell lines examined was in the expression of maspin. Intense maspin transcripts (3.0 and 1.6 kb) and protein (42 kDa) were identified in HMS-1 and HMS-1 CM, respectively, but were completely absent in all of the malignant lines examined (Fig. 10.2A). With its proteinase inhibitor profile of increased maspin, TIMP-1, PNII, α1-AT, and the 31 kDa inhibitor, HMS-1 bore strong resemblance to normal human mammary epithelial cells (HMEC, Clonetics) (Fig. 10.2A) except that the expression of all of these proteinase inhibitors, including maspin, was even more enhanced in HMS-1. Being derived from normal ducts and acini of the human breast, HMEC cultures likely contain myoepithelial as well as epithelial cells. Thus, the resemblance of HMS-1 to HMEC further supported our contention that HMS-1, though immortal, expressed a well-differentiated myoepithelial phenotype. In addition, because HMS-1 was a clonal line expressing a pure myoepithelial phenotype, it would be predicted to express certain myoepithelial-associated proteins, such as maspin, $\alpha$1-AT, PNII/APP, and TIMP-1, to a greater degree than HMEC. Predictably, the myofibroblast line Hs578Bst was strongly expressive of TIMP-1 but did not express maspin, PNII, or the 31 kDa inhibitor (Fig. 10.2A). The strong proteinase inhibitor profile exhibited by HMS-1 was shared by all of the myoepithelial xenografts, including HMS-X, HMS-3X, and HMS-4X.

In the modified Matrigel invasion chamber used in this study, HMS-1 cells and their CM dramatically inhibited invasion of two invasive melanoma and breast carcinoma cell lines (Fig. 10.2C, D). The HMS-1 line was itself non-invasive in this chamber. Predictably, the anti-invasive effects of HMS-1 could be abolished by CHX (40 $\mu$g/ml) 24-h pretreatment. HMS-1 CM inhibited invasion in a dose-response fashion up to 30% $\pm$ 8% of control ($P < 0.01$) (Fig. 10.2D). Pretreatment of HMS-1 with dexamethasone (0.25 $\mu$M) produced a complete invasion-permissive phenotype (100% of control), whereas pretreatment with phorbol 12-myristate 13-acetate (PMA; 5 $\mu$M) produced an essentially nonpermissive phenotype (2% of control) ($P < 0.05$) (Fig. 10.2E). The effects of dexamethasone and PMA were quite dramatic. The effects of other agents, including RA, dB-cAMP, Na-But, and 5-azaC, showed either permissive or non-permissive trends but were less dramatic. PMA's induction of the nonpermissive phenotype began after 20 min pretreatment, was almost complete after 2 h, and was maximized after 24 h ($P < 0.05$). The induction of this nonpermissive phenotype correlated with the induction of a dramatic fivefold increase in maspin secretion measured in HMS-1 CM. As a result of PMA treatment, both an immediate release (within 2 min) of maspin from HMS-1 cells occurred, as well as a more sustained secretion for at least 24 h following PMA pretreatment. The increased maspin secretion was not on the basis of an increase in steady-state maspin transcripts. PMA also resulted in a less dramatic twofold increase in both MMP-9 and TIMP-1 secretion. Dexamethasone's induction of an invasion-permissive phenotype in HMS-1 was not associated with a change in either maspin transcription or secretion. Immunoprecipitation with anti-maspin antibody at 1/100 dilution successfully removed all detectable maspin from myoepithelial cell CM. This CM lost its ability to inhibit invasion. Similar results were observed with the CM from the other myoepithelial lines (HMS-3 and HMS-4) studied. None of the non-myoepithelial cell CM inhibited invasion.

## 4. INHIBITION OF TUMOR ANGIOGENESIS

Human myoepithelial cells, which surround ducts and acini of certain organs such as the breast, form a natural border separating epithelial cells from stromal angiogenesis. Myoepithelial cell lines (HMS-1–6), derived from diverse benign myoepithelial tumors, all constitutively express high levels of active angiogenic inhibitors, which include maspin, TIMP-1, thrombospondin-1, and soluble bFGF receptors but very low levels of angiogenic factors [Nguyen et al., 2000] (Fig. 10.3A). Recently, maspin has been shown conclusively to be an angiogenesis inhibitor [Zhang et al., 2000]. As expected, our myoepithelial cell lines inhibited endothelial cell chemotaxis (Fig. 10.3B) and proliferation. These myoepithelial cell lines sense hypoxia and respond to low $O_2$ tension by increased hypoxia-inducible factor-1$\alpha$ (HIF-1$\alpha$) but with only a minimal increase in VEGF and nitric oxide synthetase (iNOS) steady-state mRNA levels. Their corresponding xenografts (HMS-X–6X) grow very slowly (Fig. 10.3C) compared with their non-myoepithelial carcinomatous counterparts and accumulate an abundant extracellular matrix devoid of angiogenesis but containing bound angiogenic inhibitors. These myoepithelial xenografts exhibit only minimal hypoxia but extensive necrosis compared with their non-myoepithelial xenograft counterparts. These former xenografts inhibit local and systemic tumor-induced angiogenesis and metastasis, presumably from their matrix-bound and released circulating angiogenic inhibitors. These observations collectively support the hypothesis that the human myoepithelial cell (even when transformed) is a natural suppressor of angiogenesis.

Myoepithelial cells in situ separate epithelial cells from stromal angiogenesis, and this seemingly banal observation serves to illustrate the fact that stromal angiogenesis never penetrates this myoepithelial barrier and raises the hypothesis that myoepithelial cells are natural suppressors of angiogenesis. This observation was reinforced by a microscopic, immunohistochemical, and DNA analysis of our myoepithelial xenografts. Our diverse myoepithelial xenografts secrete and accumulate an abundant extracellular matrix, which is devoid of blood vessels in routine hematoxylin and eosin staining and vWf immunocytochemical staining in contrast to non-myoepithelial xenografts, which show bursts of blood vessels. Microscopic quantitation of vessel density in 10 high power fields reveals absent to low vessel density in the myoepithelial xenografts compared to the non-myoepithelial xenografts ($p < 0.01$) (Fig. 3D). As mentioned previously, murine DNA Cot-1 analysis further reveals the absence of a murine component in the myoepithelial xenografts. Because, in the xenografts, angiogenesis would be murine in origin, the absence of a mur-

A) Table — *Relative constitutive gene expression profiles* (cell values are graphical pie-symbols indicating relative expression):

| Angiogenic Factors/Inhibitors | HMS-1 | C8161 | MCF-7 | T47D | BT-549 | MDA-MB-157 | MDA-MB-231 | Hs578T | A253 | A431 | Hs578Bst | HMEC | HT-29 |
|---|---|---|---|---|---|---|---|---|---|---|---|---|---|
| **FACTORS** | | | | | | | | | | | | | |
| bFGF | | | | | | | | | | | | | |
| aFGF | | | | | | | | | | | | | |
| TGF α | | | | | | | | | | | | | |
| TGF β | | | | | | | | | | | | | |
| TNF α | | | | | | | | | | | | | |
| VEGF | | | | | | | | | | | | | |
| Angiogenin | | | | | | | | | | | | | |
| HGF | | | | | | | | | | | | | |
| Placental GF | | | | | | | | | | | | | |
| Platelet-derived ECGF | | | | | | | | | | | | | |
| Heparin-binding ECGF | | | | | | | | | | | | | |
| **INHIBITORS** | | | | | | | | | | | | | |
| Thrombospondin-1 | | | | | | | | | | | | | |
| Soluble bFGF Receptor | | | | | | | | | | | | | |
| Plasminogen fragments* | | | | | | | | | | | | | |
| Prolactin fragments* | | | | | | | | | | | | | |
| Interferon- α | | | | | | | | | | | | | |
| Platelet Factor 4 | | | | | | | | | | | | | |
| TIMP-1 | | | | | | | | | | | | | |

A

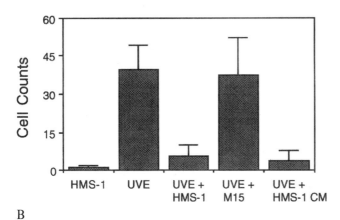

B

**Figure 10.3.** A) Relative constitutive gene expression profiles of diverse angiogenic inhibitors and angiogenic factors in HMS-1 compared with numerous other non-myoepithelial cell lines. All measurements were made by Western blot on either CM or HMS-X matrix extracts* and

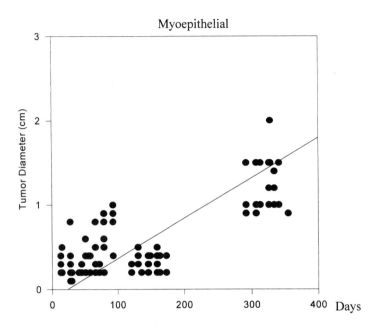

Myoepithelial

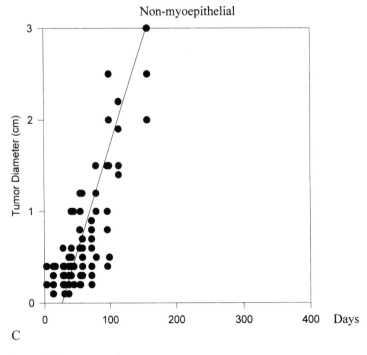

Non-myoepithelial

C

**Figure 10.3.** *(Continued)*
depicted as relative levels of expression. HMS-1 (HMS-X*) uniquely expressed a balance of
angiogenic inhibitors over angiogenic factors. B) Inhibition of HUVEC chemotaxis to bFGF
by HMS-1 cells and HMS-1 CM is depicted as cell counts collected on the undersurface of a
dividing filter. HMS-1 cells themselves were non-migratory. A control non-myoepithelial
human melanoma cell line, M15, did not inhibit UVE chemotaxis. C) Growth rates of myo-

## vWf DENSITY

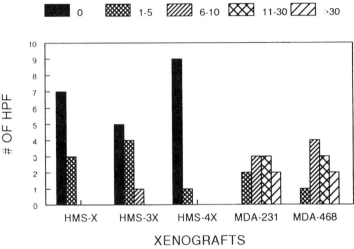

D

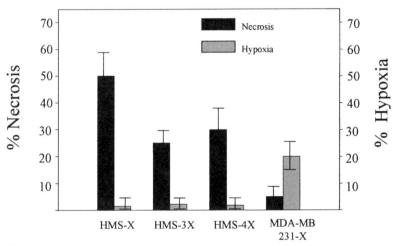

E

**Figure 10.3.** *(Continued)*
epithelial xenografts, for example, HMS-X, compared with growth rates of non-myoepithelial xenografts, for example, MDA-MB-231, revealed comparatively slow myoepithelial growth. This finding suggests a link to endogenously low levels of angiogenesis. D) Density of vWf-positive vessels in 10 H.P.F.'s of myoepithelial versus non-myoepithelial xenografts reveals absent-to-significantly fewer blood vessels in the former xenografts. E) Percentages of hypoxia and percentages necrosis in the myoepithelial versus non-myoepithelial xenografts are contrasted. In the myoepithelial xenografts, necrosis is prominent; whereas, hypoxia is inconspicuous where the reverse is true in the non-myoepithelial xenografts. Under low-$O_2$ tension, myoepithelial cells; for example, HMS-1 (HMS), like non-myoepithelial carcinoma cells; for example, MDA-MB-231 (CA), show an increase in HIF-$1\alpha$ (F) but, unlike carcinoma cells (CA), show less of an increase in VEGF (G) and iNOS (H) steady-state mRNA levels. Other myoepithelial and carcinoma lines tested exhibited a similar pattern of findings.

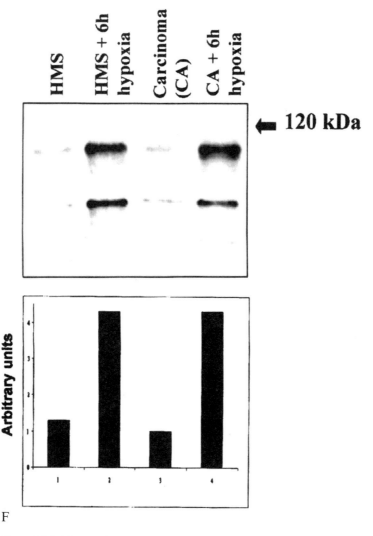

**Figure 10.3.** *(Continued)*

ine DNA component is another indication that angiogenesis is minimal. As also mentioned, the myoepithelial xenografts grew slowly compared with the non-myoepithelial xenografts, a feature that was not found in comparison between the myoepithelial versus the non-myoepithelial cell lines in vitro.

To explain these in vivo observations, we analyzed the gene expression profiles of our myoepithelial cell lines versus non-myoepithelial cell lines with respect to known angiogenic inhibitors and angiogenic factors. HMS-1, as a prototype myoepithelial cell line, constitutively expressed none of the known angiogenic factors, including bFGF, aFGF, angiogenin, TGF-$\alpha$, TGF-$\beta$, TNF-$\alpha$, VEGF, PD-ECGF, pla-

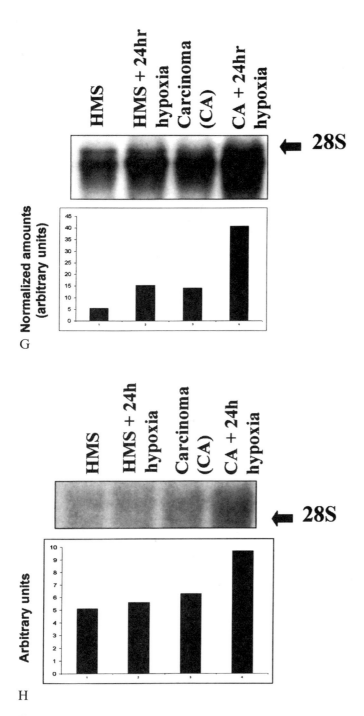

Figure 10.3. *(Continued)*

cenal growth factor (PlGF), IFα, HGF, and HB-EGF, but rather expressed maspin, thrombospondin-1, TIMP-1, and soluble bFGF receptors at high levels; this was in contrast to a high angiogenic factor (which included bFGF, VEGF, TGF-α, TGF-β, HB-EGF, and PD-ECGF) to angiogenic inhibitor gene expression profile, which was observed in non-myoepithelial cell lines.

Other myoepithelial cell lines (HMS-2–6) exhibited an angiogenic inhibitor/angiogenic factor profile similar to that of HMS-1. Interestingly, in 2 M urea extracts of the myoepithelial xenografts, but not in any of the non-myoepithelial xenografts, strong thrombospondin-1, TIMP-1, as well as plasminogen and prolactin fragments could be detected by Western blot. HMS-1 and HMS-1 CM (concentrated 10- to 100-fold) exerted a marked inhibition of endothelial migration (Fig. 10.3B) and proliferation, both of which were abolished by pretreatment of the myoepithelial cells with cycloheximide or dexamethasone. HMS-1 cells themselves did not migrate in response to either K-SFM, FBS, or bFGF.

When mixed with HUVEC, HMS-1 cells reduced endothelial migration to $12\% \pm 6\%$ of control ($P < 0.01$) (Fig. 10.3B). HMS-1 concentrated CM reduced migration to $8\% \pm 7\%$ of control ($P < 0.01$). All of the non-myoepithelial malignant human cell lines studied stimulated both endothelial migration and proliferation. Concentrated CM from HMS-1, when fractionated on a heparin-Sepharose column, inhibited endothelial proliferation to $47\% \pm 10\%$ of control ($P < 0.01$). This inhibitory activity was present only in the 1.5–2.0 M gradient fraction. Pretreatment of HMS-1 cells with PMA resulted in a two- to fivefold increase in endothelial antiproliferative inhibitory activity in both unfractionated CM, as well as in the heparin-Sepharose fraction. Western blot of the heparin-Sepharose column fractions revealed that the 1.5–2.0 M NaCl fraction contained thrombospondin-1. Immunoprecipitation of this fraction with anti-thrombospondin was effective at removing all thrombospondin-1, but it decreased endothelial antiproliferative activity by only 50% and raised the possibility that other angiogenic inhibitors, including maspin, were present in this fraction. The other myoepithelial cell lines (HMS-2–6) exhibited similar antiangiogenic inhibitory activity in their fractionated and unfractionated CM. Therefore, it is likely that both maspin and thrombospondin-1 are anti-angiogenesis effector molecules of myoepithelial cells.

To further explain our in vivo observations of minimal angiogenesis in our myoepithelial xenografts, we performed in vitro and in vivo hypoxia studies. Non-myoepithelial xenografts; for example, MDA-MB-231, exhibited florid hypoxia but only minimal necrosis when they reached a size of 2.0 cm (Fig. 10.3E). In contrast, the myoepithelial xenografts exhibited only minimal hypoxia but prominent necrosis ($P < 0.001$) at the same size of 2.0 cm (Fig. 10.3E). Quantitation of the areas of hypoxia (pimonidazole immunoreactivity) and areas of necrosis in the myoepithelial versus non-myoepithelial xenografts suggested that, in the myoepithelial tumors where angiogenesis is minimal, hypoxic areas progress to necrosis rapidly. In the non-myoepithelial tumors, however, hypoxic areas accumulate but do not progress to necrosis, presumably as a result of the angiogenesis which the hypoxia elicits (Fig. 10.3E).

Comparative analysis of myoepithelial versus non-myoepithelial cell lines to low $O_2$ tension revealed that, although both cell lines sense hypoxia in that they responded by increasing HIF-1$\alpha$ (Fig. 10.3F), the myoepithelial lines up-regulated their steady-state mRNA levels of the downstream genes, VEGF (Fig. 10.3G) and iNOS (Fig. 10.3H) to a lesser extent than the carcinoma lines, suggesting the possibility of decreased transactivation of hypoxia response element (HRE). Specifically, we observed an approximate 1.7-fold increase in VEGF (1.1-fold increase in iNOS) in myoepithelial cells in response to hypoxia compared to an approximate 2.5-fold increase in VEGF (1.5-fold increase in iNOS) in carcinoma cell lines in response to hypoxia. Although these fold differences by themselves were not impressive, the absolute levels of VEGF (and iNOS) expressed in carcinoma cells in response to hypoxia were 2.5-fold greater for VEGF (and 1.7-fold greater for iNOS) than the levels of VEGF (and iNOS) expressed in myoepithelial cells in response to hypoxia. Therefore, it can be concluded that myoepithelial cells did not express VEGF or iNOS in response to hypoxia to nearly the same extent as carcinoma cells.

To study both local and systemic effects of myoepithelial cells on metastasis, we injected spontaneously metastasizing tumor cells into our myoepithelial xenografts. The highly metastatic neoC8161 cells injected into the myoepithelial xenografts could be recovered in significant numbers, although the numbers of clones recovered were less than those recovered from the non-myoepithelial xenografts. Histological analysis of the extirpated xenografts revealed neoC8161 cells actively invading through all of the non-myoepithelial xenografts in contrast to the appearance in the myoepithelial xenografts, where the neoC8161 cells were confined to the immediate areas around the injection site. Pulmonary metastases of neoC8161 were completely absent in the myoepithelial xenograft-injected group, whereas they were quite numerous in the non-myoepithelial group ($P < 0.001$). Analysis of extirpated myoepithelial xenografts containing injected neoC8161 cells showed no evidence of murine angiogenesis by either vWf immunocytochemical studies or murine DNA Cot-1 analysis, whereas a similar analysis of extirpated neoC8161 injected-non-myoepithelial xenografts showed an increase in murine angiogenesis by both methods. This suggested that either the matrices of our myoepithelial xenografts or gene product(s) of the myoepithelial cells or both were inhibiting *neo*C8161-induced angiogenesis in vivo. We, in fact, found evidence of maspin, thrombospondin-1, TIMP-1, soluble bFGF receptors, prolactin, and plasminogen fragments within 2 M urea extracts of our myoepithelial xenografts.

In tail-vein injection studies of *neo*C8161 in mice harboring the myoepithelial xenografts, *neo*C8161 formed smaller pulmonary colonies than in mice harboring non-myoepithelial xenografts or in control mice (no xenografts) ($P < 0.01$). In a vWf factor immunocytochemical analysis of these smaller colonies in the mice harboring the myoepithelial xenografts, angiogenesis was minimal. These latter studies suggest the presence of circulating angiogenesis inhibitors released by the myoepithelial xenografts. Just recently we have demonstrated circulating maspin in mice harboring myoepithelial xenografts.

## 5. PHYSIOLOGICAL AND PHARMACOLOGICAL MANIPULATIONS

Since PMA and dexamethasone were effective at pharmacologically altering maspin levels and the myoepithelial phenotype, we wondered whether physiological agents could do so as well. Because previous basic and clinical studies had examined the role of estrogen agonists and antagonists on human breast cancer cells and because issues of hormone replacement therapy (HRT) and tamoxifen chemoprevention are such timely issues in breast cancer, we wondered whether hormonal manipulations might affect myoepithelial cells in vitro as far as their paracrine suppressive activities on breast cancer were concerned. We recently demonstrated [Shao et al., 2000] that treatment of myoepithelial cells with tamoxifen, but not 17$\beta$-estradiol, increases both maspin secretion and invasion-blocking ability. 17$\beta$-Estradiol, however, competes with these suppressive effects of tamoxifen, which suggests that the mechanism of tamoxifen action is estrogen-receptor mediated. Myoepithelial cells lack ER-$\alpha$ but express ER-$\beta$. Tamoxifen, but not 17$\beta$-estradiol, increases AP-1 CAT but not ERE-CAT activity. Again, 17$\beta$-estradiol competes with the transcription-activating effects of tamoxifen. These experiments collectively suggest that the actions of tamoxifen on the increased secretion of maspin by myoepithelial cells may be mediated through ER-$\beta$ and the *trans*-activation of an ER-dependent AP-1 response element.

As mentioned previously, immunoprecipitation of maspin from HMS-1 CM reversed the anti-invasive effects of myoepithelial CM on breast carcinoma cell invasion in vitro. Tamoxifen treatment of HMS-1 resulted in a two- to threefold increase in maspin secretion with increasing doses of tamoxifen and increasing times of exposure. $17\beta$-Estradiol, in contrast, exerted no effects on maspin secretion and completely abolished the maspin stimulatory effects of tamoxifen in competition experiments. Tamoxifen's increase in maspin secretion was not due to an increase in steady-state maspin mRNA levels, which were essentially unchanged by this treatment. Myoepithelial cell lines lacked ER-$\alpha$ expression but uniformly expressed ER-$\beta$. Because estrogen agonists/antagonists bound to estrogen receptors (either ER-$\alpha$ or ER-$\beta$) activate downstream genes containing either a classical ERE or an ER-dependent AP-1 response element, myoepithelial cell lines were transfected with CAT-reporter constructs fused to heterologous promoters containing the human estrogen response element (ERE-tk-CAT) or AP-1-tk-CAT. Tamoxifen ($10^{-7}$ M) increased AP-1-CAT activity threefold. This effect was not observed with $17\beta$-estradiol. Furthermore $17\beta$-estradiol ($10^{-5}$ M) competed with and blocked the effects of tamoxifen ($10^{-7}$ M). $17\beta$-estradiol ($10^{-7}$ M) did increase ERE-CAT activity but tamoxifen ($10^{-7}$ M) did not.

## 6. MYOEPITHELIAL GENE PRODUCTS AS SURROGATE END-POINT MARKERS

Because myoepithelial cells are ubiquitous components of the ductal-lobular units of the breast and other organs, which exhibit branching morphogenesis, we hypothesized that gene products of myoepithelial cells might be detectable in fluid secreted by these ductal-lobular units. As there has been a lot of recent interest in breast ductal fluid and breast nipple aspirates, especially, we measured one myoepithelial gene product, maspin, by Western blot, and found it to be present in both nipple aspirates and ductal fluid but not blood or urine [unpublished observations]. These observations indicate that ductal fluid is not a mere transudate of blood or serum and that it is not a product only of epithelial cells (although epithelial protein products, such as casein, lactalbumin, and carcinoembryonic antigen (CEA), are certainly present). Ductal fluid also reflects a significant contribution from myoepithelial cells. From this observation, we are currently studying groups of patients to see whether their maspin levels serve to stratify them. We are currently analyzing ductal fluid collected following cannulation and washing of selected ducts in patients with microcalcifications on screening mammography who are about to undergo either excisional or core biopsy. Paired comparisons of maspin levels in ductal fluid obtained from ducts harboring microcalcifications or DCIS and normal ducts from the same patients are also being made. Maspin levels can be correlated with the histopathology surrounding the microcalcifications. It is anticipated that some of these patients will exhibit normal ductal histopathology surrounding their microcalcifications, some will harbor proliferations, such as hyperplasia, adenosis, ADH, and DCIS, and still others invasive carcinoma. The screening value of maspin levels in all of these patients can be determined. Measurements of myoepithelial maspin in ductal fluid will be compared with levels of a breast epithelial cell marker, such as CEA. Increased CEA has been observed in nipple secretions and in ductal fluid in patients with ductal hyperplasia. Hence, the maspin/CEA ratio might be predictive of risk with increased maspin/CEA correlating with normalcy and decreased maspin/CEA correlating with high risk, microcalcifications, and/or precancerous histopathology. In this sense, maspin can be used as a surrogate end-point marker to predict either the risk of DCIS or the likelihood that DCIS will progress to invasive breast cancer.

Another interesting observation with respect to the use of myoepithelial maspin as a marker—this time, a tumor marker—is the observation that maspin can be detected in normal saliva but that it is markedly elevated in saliva secreted from a salivary gland neoplasm and that it is also elevated in murine serum in mice harboring human myoepithelial xenografts [unpublished observations]. Most salivary gland neoplasms are thought to be myoepithelial in origin. These include pleomorphic adenomas, basal cell adenomas, basal cell adenocarcinomas, and adenoid cystic carcinomas. It was human tumors of these types that originally gave rise to our myoepithelial cell lines/xenografts that led to a dissection of the myoepithelial phenotype and to our observations concerning myoepithelial maspin. If screening saliva for maspin shows promise for detecting small incipient salivary gland neoplasms, then myoepithelial maspin will show its utility as a tumor marker. So, in summary, our findings indicate that the gene products of myoepithelial cells; for example, maspin reflect the structural and functional integrity of the ductal-lobular units of different organs, and alterations in the levels of these myoepithelial gene products in fluid from these units may reflect disease states.

## 7. TRANSFORMED MYOEPITHELIAL CELLS

As has been shown, transformed myoepithelial cells can be derived from benign myoepithelial tumors. The most common site of myoepithelial neoplasia is the salivary gland. Myoepithelial tumors include the pleomorphic adenoma, the basal cell adenoma, the basal cell adenocarcinoma, and the adenoid cystic carcinoma. Although the cells of these tumors are transformed, karyotype analysis is very close to normalcy. Analysis of the gene products of these cells reveals minimal differences from normal myoepithelial cells; therefore, we believe that these transformed myoepithelial cells can serve as normal myoepithelial cell surrogates. Our laboratory has been successful in establishing six different human myoepithelial cell lines and xenografts using the following methods.

## 8. PREPARATION OF MEDIA AND REAGENTS

### 8.1. Supplemented K-SFM

K-SFM keratinocyte serum-free medium supplemented with recombinant epidermal growth factor (EGF, 5 ng/ml) and bovine pituitary extract (BPE, 50 µg/ml)

### 8.2. Attachment Medium

K-SFM with 0.5% FBS.

### 8.3. CMF-HBSS

$Ca^{2+}/Mg^{2+}$-free Hanks' balanced salt solution.

### 8.4. Disaggregation Medium

F12:DMEM (1:1 mixture of Ham's F12: Dulbecco's modified Eagle's medium) with 10mM HEPES buffer, 2% bovine serum albumin, fraction V, 5 µg/ml Insulin, 300 U/ml Collagenase, and 100 U/ml Hyaluronidase.

### 8.5. Trypsin-EDTA

Trypsin, 0.05%, 0.7 mM EDTA (disodium ethylene diamine tetraacetate) in CMF-HBSS.

### 8.6. F12/DMEM/H

F12/DMEM with 15 mM Hepes.

### 8.7. Dispase Medium

F12/DMEM/H with a reduced $Ca^{2+}$ concentration (0.06 mM), and containing 5 U/ml dispase.

### 8.8. Serum-free F12/DMEM/H

F12:DMEM:H medium supplemented with 1 mg/ml BSA, 1 g/ml insulin, 0.5 g/ml hydrocortisone, 10 ng/ml cholera toxin, 10 ng/ml epidermal growth factor.

### 8.9.  High-salt Buffer

3.4 M NaCl, 50 mM Tris-HCl, 20 mM EDTA, 10 mM N-ethyl-maleimide (NEM) pH 7.4.

### 8.10.  Urea/guanidinium-HCl Extraction Buffer

6 M urea, 2 M guanidinium-HCl, 50 mM Tris-HCl, 20 mM EDTA, 10 mM NEM, pH 7.4, with added 2.0 mM dithiothreitol (DTT).

### 8.11.  Tris Buffered Saline (TBS)

0.15 M NaCl, 50 mM Tris-HCl, 20 mM EDTA, 10 mM NEM, pH 7.4.

## 9.  CULTURING TRANSFORMED AND NORMAL HUMAN MYOEPITHELIAL CELLS

### 9.1.  Transformed Myoepithelium

### Protocol 10.1.  Culture of Transformed Human Myoepithelial Cells

**Reagents and Materials**
*Sterile*
- Supplemented K-SFM: (see Section 8.1)
- Trypsin-EDTA: (see Section 8.5)
- Attachment medium: (see Section 8.2)
- CMF-HBSS: (see Section 8.3)
- Dimethylsulfoxide (DMSO)
- Culture dishes, 5 cm
- Culture flasks, 25 $cm^2$ Scalpels, #11 blade
- Cryovials

**Protocol**
(a) After human subject consent and approval of the institutional review board, obtain tissues following surgery.
(b) Mince portions of the surgical specimen to ~1 $mm^3$ under aseptic conditions.
(c) Transfer to dishes containing supplemented K-SFM. Have sufficient medium in the dishes so that the explants are in contact with medium but are able to adhere to the bottom of the dish without being dislodged.
(d) Culture the cells at 37°C in a humidified atmosphere at 5% $CO_2$ in air.
(e) Change the medium every third day. After 1–2 weeks, cells migrate out from the explants. The medium is suppressive to fibroblast growth but stimulatory to myoepithelial growth.

(f)   For subculture of cell monolayers, wash in CMF-HBSS, and detach with trypsin-EDTA.

(g)   Resuspend trypsinized cells in attachment medium and allow to attach overnight.

(h)   Replace medium with serum-free supplemented K-SFM and maintain thereafter in this serum-free growth medium. Cells are generally seeded in tissue culture flasks at 1/6–1/12 of the confluent density or $0.5–1 \times 10^4$ cells/cm$^2$.

(i)   Prepare frozen stocks in serum-free medium containing 10% DMSO.

If there is difficulty in establishing the cell line in this manner, you can serially transplant the initial tumoral explants subcutaneously in the flanks of female nude (nu/nu mutants on a BALB/c background) or SCID mice with a number 10 trochar and then attempt subsequent establishment of the cell line following steps (a)–(i). We have had a much higher success rate (virtually 100%) with xenograft establishment compared with cell line establishment (10% success rate). The xenografts are very slow growing, however, reaching approximately 1 cm in diameter after 6 months to 1 year (Fig. 10.3C). When the xenografts reach this size, they can be removed by sterile technique, minced, and transplanted to new mice. Portions can also be slow-frozen in serum-free medium containing 10% DMSO and stored in liquid nitrogen. At any point, attempts may again be made to establish a cell line by placing the xenograft explants in cell culture in a manner identical to that for the initial surgical specimen. Using this approach, the success rate for establishing a cell line can be doubled. These strategies are summarized in the Schematic of Figure 10.4.

The myoepithelial cells so obtained, though transformed, will serve as a normal myoepithelial surrogate and can be used in studies of tumor invasion and angiogenesis.

## 9.2.   Normal Myoepithelium

There is no doubt that the use of transformed myoepithelial cells, however convenient, has limitations. For one, these immortalized myoepithelial cells are highly selected and represent homogeneous cell populations, which may not accurately reflect the heterogeneous composition of myoepithelial cells within the normal human mammary gland. Also, transformed myoepithelial cell lines are likely to have acquired genetic changes in long-term culture that may confound analysis of the properties of the original myoepithelial cells of the breast or other primary organ. For these reasons, methods to obtain and culture primary normal myoepithelial cells would also be desirable.

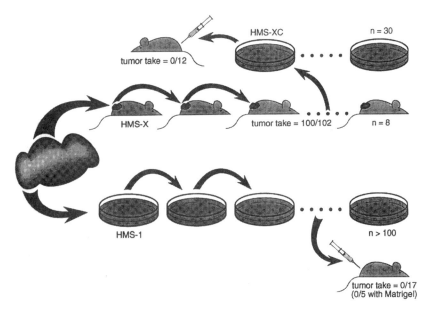

**Figure 10.4.** Schematic depicts successful method for obtaining myoepithelial cell lines (albeit transformed) from benign myoepithelial tumors of the salivary gland and breast.

Normal breast epithelium consists of different cell types, including luminal epithelial cells that line the ducts and alveoli and basally located myoepithelial cells. These two cell types can be distinguished on the basis of expression of distinct cell markers. The most commonly used phenotypic markers to identify luminal epithelial cells are the MUC-1 apical plasma membrane glycoprotein and cytokeratins 8, 18, and 19. For myoepithelial cells, markers include the common acute lymphocytic leukemia antigen (CALLA/CD10), smooth-muscle actin, cytokeratins 5, 14, and 17, S100, calponin, and maspin. Most studies examining mammary cells in primary culture have utilized the heterogeneous mixed population of epithelial cells (the total epithelial cell population) in the tissue sample. Studies indicate that this heterogeneous mixed population of epithelial cells really consists of luminal epithelial cells and basal myoepithelial cells (see also Chapter 9). Similarly, commercially available HMEC (Clonetics) really consist of both epithelial as well as myoepithelial cells.

### Protocol 10.2. Culture of Normal Human Myoepithelial Cells

#### *Reagents and Materials*

*Sterile*

❑ Disaggregation medium: (see Section 8.4)
❑ F12/DMEM/H (see Section 8.6) with 5% FBS

- Serum-free F12/DMEM/H (see Section 8.8) or supplemented K-SFM (se Section 8.1)
- Dispase medium: (see Section 8.7)
- Trypsin/EDTA (see Section 8.5)
- DMSO
- Nylon mesh, 20 μm
- Vitrogen
- *Humatrix*
- Collagen-coated tissue culture dishes or plastic flasks
- Cryovials

### Protocol

(a)  After human subject consent and approval of an institutional review board, obtain mammary gland tissues (mainly from reduction mammoplasties) following surgery.

(b)  After tissue is taken for adequate diagnosis, dissociate the remainder of the tissue and culture the epithelial cells [Emerman et al., 1990; Emerman et al., 1994] as follows.

(c)  Trim fat from tissue samples and mince the tissue.

(d)  Dissociate by shaking at 37°C for 18 h in disaggregation medium.

(e)  Collect the epithelial cell pellet by centrifuging the cell suspension at 80 g for 4 min.

(f)  If a single cell suspension is required; for example, for FACS-sorting, resuspend the cells in trypsin/EDTA for 4 min.

(g)  Add F12/DMEM/H containing 5% FBS to inhibit the enzymatic activity of trypsin.

(h)  Filter the resultant cell suspension through 20 μm nylon mesh

(i)  Centrifuge the filtrate at 100 g for 5 min, and resuspend the pellet in dispase medium to prevent reaggregation of the cells.

(j)  Count the number of viable cells, determined by Trypan Blue exclusion, on a hemocytometer.

(k)  Seed cells onto collagen-coated tissue culture dishes or plastic flasks.

(l)  Culture overnight in F12/DMEM/H containing 5% FCS to allow attachment of cells.

(m)  Switch the medium to serum-free F12/DMEM/H or supplemented K-SFM. The purpose of this initial phase of culture is to allow the "total epithelial cells" to adhere and to allow for the selective removal of non-attaching contaminating cells and debris.

(n)  At 80% confluence, harvest cells via trypsinization, wash, and freeze in liquid nitrogen as viable cell suspensions.

## 9.3. Recovery and Characterization

If you have performed steps (a)–(j) correctly the average epithelial cell numbers obtained from dissociated normal, dysplastic, and malignant human breast tissue should be: normal, $0.5 \times 10^6$ cells/g tissue; fibroadenoma, $1.5 \times 10^6$ cells/g tissue; fibrocystic, $0.7 \times 10^6$ cells/g tissue; carcinoma, $1.0 \times 10^6$ cells/g tissue. Cell numbers are increased by culturing. Cell numbers of $5 \times 10^6$–$1 \times 10^7$ can be obtained from primary and early passage cultures, providing sufficient material for experimental manipulations and analyses.

The harvested cells can be analyzed in either two-dimensional or three-dimensional culture systems, both of which reveal two populations of cells: epithelial cells with luminal characteristics and myoepithelial cells with characteristic myoepithelial markers [Yang et al., 1986; Rudland, 1991]. Using either two-dimensional or three-dimensional culture systems, we can detect clonal populations in cultures initiated at low cell densities ($<2500$ cells/cm$^2$) with colony yields linearly related to the concentration of cells initially plated. Using either system, we can obtain clonal populations of either cell type at an efficiency of 1–3%. Most colonies generated are composed of 2–10 cells after 14 days, but colonies $>50$ cells are observed occasionally.

In two-dimensional culture systems; for example, on plastic, the colonies exhibit two major morphologies, which include balls of tightly arranged cells and loosely arranged tear-drop shaped cells having very distinct cell borders. Immunocytochemical analysis (see Protocol 15.8) of the different colonies, using the monoclonal antibodies to MUC-1, keratins 8, 14, 18, 19, CALLA, smooth muscle actin, maspin, calponin and S-100 (see Sources of Materials), has shown that, in the colonies composed of tightly arranged cells, the cells express typical luminal epitopes (MUC-1, keratins 8, 18, and 19) but do not express the myoepithelial marker keratin 14 [Dairkee et al., 1988]. For the colonies composed of the loosely arranged tear-drop shaped cells, the cells express keratin 14, CALLA, smooth-muscle actin, maspin, calponin, S100, and other myoepithelial markers, but do not express the epithelial markers, MUC-1 or keratin 19. These dual populations of cells can be sorted by FACS to yield relatively pure populations of each cell type [O'Hare et al., 1991]. After sorting, the MUC-1$^+$, CALLA$^-$ subpopulation generates further colonies composed of tightly arranged cells. We believe that these are epithelial cells. After sorting, the MUC-1$^-$, CALLA$^+$ subpopulation generates colonies of dispersed teardrop-shaped cells. We think that these cells are myoepithelial cells. Sometimes, in the initial non-sorted population, cells expressing epithelial morphologies and cells expressing myoepithelial morphologies co-exist within a single colony (Fig. 10.5A), with the tightly arranged "luminal" cells in the center of the colony and the teardrop-shaped "myoepithelial" cells radiating out into the periphery of the colony analogous to the in vivo situation.

In three-dimensional culture systems; for example, Vitrogen and *Humatrix*, the mixed population of cells that gives rise to two populations on two-dimensional systems also gives rise to two populations in three-dimensional systems. The mixed population of cells is seeded at low cell density (<1000 cells/ml of gel) within the gel and is cultured in serum-containing medium for 10–14 days. Colonies of varying morphologies emerge with a frequency of ∼2%. Colony morphologies include small (<50 cells) spherical colonies composed of a simple cuboidal epithelium surrounding a central lumen, as well as larger (>100 cells) highly branched colonies composed of a solid cord of cells. Immunocytochemical analysis of the colonies generated in three-dimensional systems reveals that the small spherical colonies generally exhibit typical luminal epitopes (MUC-1$^+$, keratins 8/18/19$^+$, CALLA$^-$, keratin 14$^-$). The larger branched colonies, however, are mainly positive for myoepithelial markers (keratin 14$^+$, CALLA$^+$) and negative for luminal markers (MUC-1$^-$, keratin 19$^-$). Sorting by FACS of the mixed cell populations before seeding within gels (Vitrogen or *Humatrix*) demonstrates that the small spherical colonies derive from the MUC-1$^+$, CALLA$^-$ subpopulation (epithelial cells), and the large branched colonies derive from the MUC-1$^-$, CALLA$^+$ subpopulation (myoepithelial cells).

Interestingly, it should be recalled that our transformed myoepithelial cell lines; e.g., HMS-1 give evidence of branching and budding when grown in gels, evidence that they retain this basic myoepithelial characteristic. The schematic diagram (Fig. 10.5B) summarizes the method for obtaining and identifying myoepithelial cells from human mammary tissues with both two-dimensional and three-dimensional systems.

## 10. METHODS OF OBTAINING MYOEPITHELIAL MATRIX

Human myoepithelial matrix, termed *Humatrix*, can be obtained from our myoepithelial xenografts [Kedeshian et al., 1998]. This matrix is very rich in proteoglycans and hyaluronic acid (Fig. 10.6A). The method of extraction is as follows.

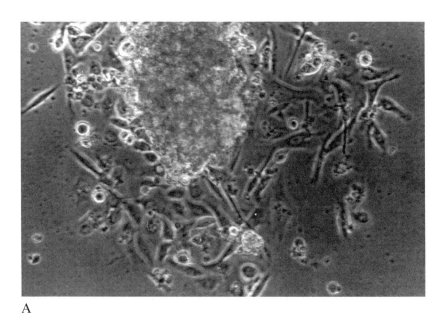

A

## MIXED CELL POPULATION GROWING IN KSFM

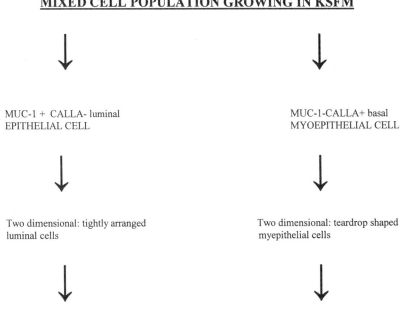

MUC-1 + CALLA- luminal
EPITHELIAL CELL

MUC-1-CALLA+ basal
MYOEPITHELIAL CELL

Two dimensional: tightly arranged
luminal cells

Two dimensional: teardrop shaped
myepithelial cells

Three dimensional: small spherical
alveolar-like colonies

Three dimensional: large branched
ductal-like colonies; homogeneous
cell types

B

## Protocol 10.3.  Preparation of Human Myoepithelial Matrix

### Reagents and Materials

*Sterile or aseptically prepared*
- ❏ HMS-X, HMS-3X, HMS-4X tumors (available to investigators from author upon request)

*Non-sterile*
- ❏ β-aminoproprionitrile (BAPN) fumarate fed mice
- ❏ High-salt buffer (see Section 8.9)
- ❏ Urea/guanidinium-HCl extraction buffer (see Section 8.10)
- ❏ Tris buffered saline (TBS) (see section 8.11)
- ❏ Dialysis membrane with a molecular weight cutoff of 5–10 kD
- ❏ Chloroform

### Protocol

(a) Grow the tumors in β-aminoproprionitrile (BAPN) fumarate fed mice, which are rendered lathyritic.

(b) Harvest HMS-X, HMS-3X, or HMS-4X tumors at 1–2 g.

(c) Homogenize 10 g of tumors in 2 ml/g , pH 7.4, at 4°C.

(d) Spin the homogenate for 15 min at 12,000 g at 4°C.

(e) Extract the pellet overnight at 4°C in 0.5 ml urea/guanidinium-HCl extraction buffer per g starting material with gentle stirring.

(f) Spin the extract for 30 min at 24,000 g at 4°C.

(g) Dialyze the supernatant against several changes of TBS at 4°C, followed by sequential dialyses against 0.5% chloroform in cell culture medium by using a dialysis membrane with a molecular weight cutoff of 5–10 kD. Using the urea/guanidinium-HCl method, 1 ml of *Humatrix* at a protein concentration of 1.5 mg/ml is obtained from each g of HMS-X.

(h) Concentrate the final solution to 3 mg/ml protein by ultra-filtration at 4°C.

(i) Store *Humatrix* at −20°C. At 4°C, *Humatrix* remains liquid; at 25 −37°C *Humatrix* undergoes gelation (Fig. 10.6B).

---

**Figure 10.5.** (A) Primary cultures of total breast epithelial cells reveal a clump of cuboidal luminal epithelial cells surrounded by more spindly or tear-dropped myoepithelial cells in periphery. These latter cells were CALLA+ and MUC1–. (B) Schematic depicts successful method for obtaining primary myoepithelial cells from human mammary explants and identifying myoepithelial cells out of a mixed cell population based on two- and three-dimensional appearances in tissue culture and confirmatory immunocytochemistry.

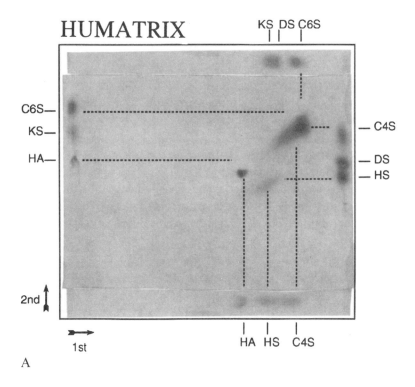

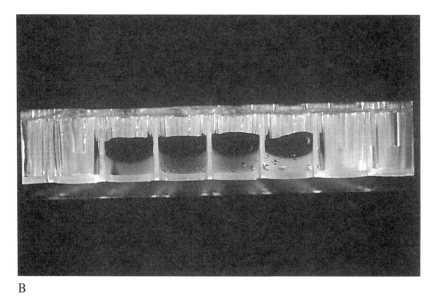

**Figure 10.6.** (A) Characteristics of the matrix extract from a human myoepithelial xenograft, HMS-X, is depicted in this 2-D cellulose acetate preparation. Large amounts of chondroitin-4-sulfate proteoglycan, heparan sulfate proteoglycan and hyaluronic acid are evident. (B) This matrix extract undergoes gelation at 25–37°C and excludes cells and large macromolecules. (C) Schematic depicts method of preparation of this human myoepithelial matrix gel, termed *Humatrix*.

## HMS-X Matrix Gel Preparation

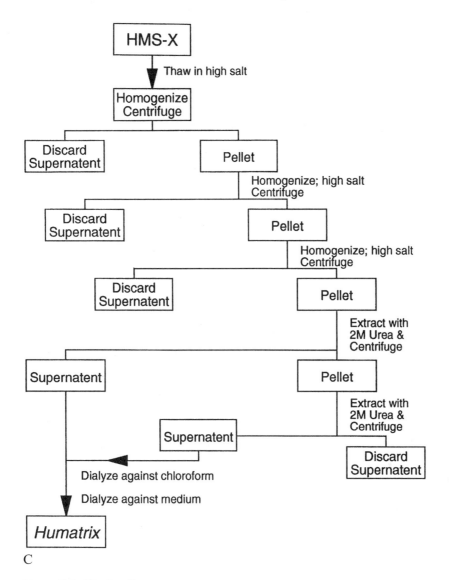

C

**Figure 10.6.** *(Continued)*

*Humatrix* can also be prepared by a pepsin hydrolysis method similar to that used in the preparation of Vitrogen 100. This method of human myoepithelial matrix extraction is summarized in the following schematic (Fig. 10.6C).

## II. FUTURE MYOEPITHELIAL RESEARCH DIRECTIONS

The human myoepithelial matrix, *Humatrix*, can be used as a gel to study morphogenesis, mitogenesis, invasion, and differentiation of tumor cells in a manner similar to the studies of these processes on *Matrigel*. Unlike *Matrigel*, *Humatrix* is human and myoepithelial-derived.

The observations that myoepithelial cells secrete suppressive gene products, such as maspin, in large quantities, whereas carcinoma cells do not, suggest that myoepithelial cells exert pleiotropic suppressive effects on tumor progression. Because these gene products are proteinase inhibitors, locomotion inhibitors, and angiogenesis inhibitors, their diverse actions may largely explain the pronounced anti-invasive and anti-angiogenic effects of myoepithelial cells on carcinoma and pre-carcinoma cells. Clearly, the gene products of myoepithelial cells have more than marker value. We need to better understand what it is about the myoepithelial phenotype that allows for high constitutive expression and secretion of tumor suppressive molecules. Studies of the maspin promoter and *cis/trans* interactions within the myoepithelial cell seem to be an attractive line of further research. We also need to understand better the mechanism by which certain pharmacological agents, such as PMA, and certain physiological agents, such as tamoxifen, bolster myoepithelial secretion of suppressive molecules, such as maspin. With this understanding we may be able to design smaller, less-toxic molecules that have the same effect. We need to exploit better the intricate paracrine and local relationships that exist between myoepithelial cells and epithelial cells (precancerous and cancerous) in the breast and other organs. This point is especially important and timely as intraductal approaches through the nipple are gaining in popularity as a means of screening women who are at risk for developing breast cancer. These intraductal approaches really exploit the local myoepithelial/epithelial relationships that exist. Screening for maspin levels as a surrogate end-point marker is only the beginning. One could envision delivering intraductal gene therapy designed to exploit the inherent differences between myoepithelial and epithelial cells. One could target and destroy the epithelial cells, selectively sparing the myoepithelium or alternately target the myoepithelial cells with a vector, which bolsters its secretion of suppressive molecules. If the myoepithelial defense can be bolstered in this manner, perhaps this natural barrier, which normally inhibits invasion for years, can be made into an impervious barrier, which inhibits invasion forever. At least that is one vision of scientists who are interested in myoepithelial cells.

# SOURCES OF MATERIALS

| Item | Supplier |
| --- | --- |
| Antibodies: | |
|   MUC-1 (Clone HMPV) | Pharmingen |
|   Keratins 8, 18, 19 (Clones 35BH11, DC10, BA17) | Dako |
|   Keratin 14 (Clone LL0002) | Research Diagnostics |
|   CALLA (Clone 56CC) | Zymed |
|   Smooth muscle actin (Clone 1A4) | DAKO |
|   Maspin (Clone EAW24) | Novocastra |
|   Calponin (Clone CALP) | DAKO |
|   S100 (Clone DAK-S100AY) | DAKO |
| Bovine pituitary extract | Gibco |
| BSA, fraction V | Gibco |
| Collagenase | Sigma |
| Collagen-coated tissue culture dishes or plastic flasks | Collaborative Biomedical Products (BD Biosciences) |
| Cryovials | Nalge Nunc |
| Culture dishes and flasks | BD Biosciences |
| Dialysis membrane (Spectra/Por) | Spectrum Labs |
| Dimethyl sulfoxide | Sigma |
| Dispase | Collaborative Biomedical Products (BD Biosciences) |
| EDTA | Sigma |
| EGF | Gibco |
| FBS | Gibco |
| Ham's F12: Dulbecco's modified Eagle's medium (F12:DMEM) | Stem Cell Technologies Inc. |
| Hanks' balanced salt solution, calcium- and magnesium-free | Gibco |
| HEPES | Sigma |
| *Humatrix* | Contact author |
| Hyaluronidase | Sigma |
| Insulin | Sigma |
| K-SFM keratinocyte serum-free medium | Gibco |
| Nylon mesh | BioDesign |
| Recombinant EGF | Gibco |
| Trypsin-EDTA | Gibco |
| Vitrogen | Nutacon |

# REFERENCES

Cavenee, W.K. (1993) A siren song from tumor cells. *J. Clin. Invest.* 91:3.

Cornil, I., et al. (1991) Fibroblast cell interactions with human melanoma cells affect tumor cell growth as a function of tumor progression. *Proc. Natl. Acad. Sci. USA* 88:6028–6032.

Cutler, L.S. (1990) The role of extracellular matrix in the morphogenesis and differentiation of salivary glands. *Adv. Dent. Res.* 4:27–33.

Dairkee, S.H., Puett, L., Hackett, A.J. (1988) Expression of basal and luminal epithelial-specific keratins on normal, benign and malignant breast tissue. *J. Natl. Cancer Inst.* 80:691–695.

Emerman, J.T., and Wilkinson, D.A. (1990) Routine culturing of normal, dysplastic and malignant human mammary epithelial cells from small tissue samples. *In Vitro Cell. Develop. Biol.* 26:1186–1194.

Emerman, J.T., and Eaves, C.J. (1994) Lack of effect of hematopoietic growth factors on human breast epithelial cell growth in serum-free primary culture. *Bone Marrow Transp.* 13:285–291.

Folkman, J., and Klagsbrun, M. (1987) Angiogenic factors. *Science* 235:442–447.

Guelstein, V.I., Tchypsheva, T.A., Ermilova, V.D., Ljubimov, A.V. (1993) Myoepithelial and basement membrane antigens in benign and malignant human breast tumors. *Int. J. Cancer* 53:269–277.

Kedeshian, P., Sternlicht, M.D., Nguyen, M., Shao, Z.M., Barsky, S.H. (1998) Humatrix, a novel myoepithelial matrical gel with unique biochemical and biological properties. *Cancer Lett.* 123:215–226.

Liotta, L.A., Steeg, P.S., Stetler-Stevenson, W.G. (1991) Cancer metastasis and angiogenesis: an imbalance of positive and negative regulation. *Cell* 64:327–336.

Nguyen, M., Lee, M.C., Wang, J.L., Tomlinson, J.S., Shao, Z.M., Alpaugh, M.L., Barsky, S.H. (2000) The human epithelial cell displays a multifaceted antigenic phenotype. *Oncogene* 19:3449–3459.

O'Hare, M.J, Ormerod, M.G., Monaghan, P., Lane, E.B., Gusterson, B.A. (1991) Characterization in vitro of luminal and myoepithelial cells isolated from the human mammary gland by cell sorting. *Differentiation* I46:209–221.

Rudland, P.S. (1991) Histochemical organization and cellular composition of ductal buds in developing human breast: evidence of cytochemical intermediate between epithelial and myoepithelial cells. *J. Histochem. Cytochem.* 39:1471–1484.

Safarians, S., Sternlicht, M.D., Freiman, C.J., Huaman, J.A., Barsky, S.H. (1996) The primary tumor is the primary source of metastasis in a human melanoma/SCID model: implications for the direct autocrine and paracrine epigenetic regulation of the metastatic process. *Int. J. Cancer* 66:151–158.

Shao, Z.M., Radziszewski, W.J., Barsky, S.H. (2000) Tamoxifen enhances myoepithelial cell suppression of human breast carcinoma progression by two different effector mechanisms. *Cancer Lett.* 157:133–144.

Sternlicht, M.D., Safarians, S., Calcaterra, T.C., Barsky, S.H. (1996) Establishment and characterization of a novel human myoepithelial cell line and matrix-producing xenograft from a parotid basal cell adenocarcinoma. *In Vitro Cell. Dev. Biol.* 32:550–563.

Sternlicht, M.D., Kedeshian, P., Shao, Z.M., Safarians, S., Barsky, S.H. (1997) The human myoepithelial cell is a natural tumor suppressor. *Clin. Cancer Res.* 3:1949–1958.

Yang, J., Balakrishnan, A., Hamamoto, S., Beattie, C.W., Das Gupta, T.K., Wellings, S.R., Nandi, S., Gupta, T.K. (1986) Different mitogenic and phenotypic responses of human breast epithelial cells grown in two versus three dimensions. *Exp. Cell Res.* 167:563–569.

Zhang, M., Volpert, O., Shi, Y.H., Bouck, N. (2000) Maspin is an angiogenesis inhibitor. *Nat. Med.* 6:196–199.

# Multistage Head and Neck Squamous Cell Carcinoma

Kirsten G. Edington[1], Isabella J. Berry[1], Margaret O'Prey[1],
Julie E. Burns[1], Louise J. Clark[1], Roy Mitchell[1], Gerry
Robertson[2], David Soutar[2], Lesley W. Coggins[1] and
E. Kenneth Parkinson[1]

[1]Beatson Institute for Cancer Research, Garscube Estate, Switchback Road,
Bearsden, Glasgow, Scotland; [2]Department of Plastic Surgery, Canniesburn
Hospital, Switchback Road, Bearsden, Glasgow, Scotland; and [3]Department
of Oral and Maxillofacial Surgery, City Hospital, Greenbank Road,
Edinburgh EH10 5SB, Scotland. Email: K.Parkinson@beatson.gla.ac.uk

*Culture of Human Tumor Cells*, Edited by Roswitha Pfragner and R. Ian Freshney.
ISBN 0-471-43853-7   Copyright © 2004 Wiley-Liss, Inc.

## I. INTRODUCTION

Squamous cell carcinomas of the head and neck (SCC-HN) constitute a significant percentage of all malignancies worldwide and are particularly common in parts of Southwest Asia and India [Pindborg, 1984; Million et al., 1989]. SCC-HN often arise without any obvious premalignant lesion, but the tumors also develop at low frequency (1–5%) from leukoplakias and at a much higher frequency (30–55%) from erythroplakias. On biopsy, the latter are much more often diagnosed as having dysplasia or carcinoma in situ than the former, and they often show signs of microinvasive carcinoma [see Pindborg, 1985, for a review]. However, very little is known about the sequential molecular and phenotypic changes that occur during the development and progression of the disease. In particular, although the loss of function of tumor suppressor genes has been implicated in the pathogenesis of many human tumor types [Fearon and Vogelstein, 1990], other than p53 [Maestro et al., 1992; Sakai and Tsuchida, 1992; Burns et al., 1993], little is known about tumor suppressor gene inactivation in SCC-HN [Latif et al., 1992]. Although it is now possible to study the molecular basis of tumor development from frozen or fixed tumor samples, it will be difficult to establish the functional significance of such molecular changes in the absence of genetically and phenotypically well-characterized cell culture models of human tumors [Huang et al., 1988; Tanaka et al., 1991]. Such in vitro models of multistage human colon carcinoma [Paraskeva et al., 1984; Williams et al., 1990] and melanoma [Mancianti and Herlyn, 1989] have been described, but little is known of the in vitro phenotypes that accompany human SCC-HN progression.

Cell cultures from malignant SCC-HN have frequently been reported to contain immortal variants, which give rise to continuous cell lines [Easty et al., 1981a, 1981b; Rheinwald and Beckett, 1981; Virolainen et al., 1984; Rupniak et al., 1985; Sacks et al., 1988]. These cells are aneuploid [Easty et al., 1981a; Rheinwald and Beckett, 1981; Sacks et al., 1988] and proliferate optimally in low levels of serum growth factors [Rheinwald and Beckett, 1981] and in the absence of added factors, such as hydrocortisone, cholera toxin, and epidermal growth factor. Some SCC-HN lines can also proliferate optimally in the absence of an irradiated Swiss 3T3 feeder layer [Rheinwald and Beckett, 1981] and most can be adapted to do so [Easty et al., 1981a, 1981b, Virolainen et al., 1984; Rupniak et al., 1985; Sacks et al., 1988]. Several SCC-HN lines are also defective in their ability to respond to terminal maturation signals [Rheinwald and Beckett, 1980; Parkinson et al., 1984], but few SCC-HN lines grow in suspension [Easty et al., 1981a, 1981b; Rheinwald and Beckett, 1981; Rupniak et al., 1985; Sacks et al., 1988], and not all can form progressively growing tumors when injected subcutaneously into immunosuppressed mice [Easty et al., 1981a; Rupniak et al., 1985; Sacks et al., 1988].

Despite extensive description of these SCC-HN phenotypes, nothing is known of when they arise during SCC-HN progression; there are few reports describing keratinocytes cultured from benign [Steinberg et al., 1982] or premalignant [Lindberg and Rheinwald, 1990; Chang et al., 1992] squamous lesions. Furthermore, most SCC-HN lines have been derived from recurrent or irradiated tumours [Easty et al., 1981a, 1981b; Rheinwald and Beckett, 1981; Virolainen et al., 1984] and have been, with one exception [Rheinwald and Beckett, 1981], cultured under conditions that are known to be suboptimal, rendering selection of fitter variants or more aggressive tumor phenotypes likely [see Rheinwald and Beckett, 1981]. Details of the tumor staging have sometimes not been given [Rheinwald and Beckett, 1981; Virolainen et al., 1984; Sacks et al., 1988], and none of the SCC-HN cell lines currently available are ideal for genetic analysis of tumor suppressor genes by allelic loss studies because samples of the normal patients' tissues were not retained [Latif et al., 1992].

We describe here the cultivation and characterization of keratinocytes from premalignant erythroplakias of the oral cavity and in addition keratinocytes from 21 SCC-HN at different stages of tumor progression.

## 2. PREPARATION OF MEDIA AND REAGENTS

### 2.1. Growth Media

**Fetal bovine serum (FBS):** The same batch of FBS, screened for its ability to support the clonal proliferation of normal human epidermal keratinocytes (HEK), was used to derive all cell cultures and was used in all experiments. Normal HEK cells were prepared from infant foreskins as described [Parkinson, 2001].

**Donor bovine serum:** Batch selected for growth and feeder capacity of 3T3 cells.

**DMEM/10DCS:** Dulbecco's modified Eagle's medium (DMEM) with 10% selected donor calf serum (DCS).

**Medium A:** DMEM with 20% v/v FBS, 0.4 µg/ml hydrocortisone, and 10 ng/ml ($1 \times 10^{10}$ M) cholera toxin [Green, 1978].

**Medium B:** DMEM with 10% v/v FBS, and 0.4 µg/ml hydrocortisone.

**Medium C:** DMEM with 5% v/v FBS, and 0.4 µg/ml hydrocortisone.

**Medium D:** DMEM with 2% v/v FBS, and 0.4 µg/ml hydrocortisone.

### 2.2. Trypsin/EDTA

Trypsin, 0.1%, mixed 1:1 with EDTA, 0.02% (5.4 mM), to give 0.05% trypsin in 2.7 mM EDTA

### 2.3. Preparation of Irradiated 3T3 (X3T3) Feeder Layers

(i) Trypsinize 3T3 cells for 3–4 min, and disperse the cells in DMEM/10DCS.

(ii) Centrifuge the cells, resuspend, count, and seed at a density of $1 \times 10^4$ cells per 9-cm plate to give confluence after 2 weeks.

(iii) Change the medium on day 11.

(iv) Trypsinize when confluent.

(v) Irradiate cells in suspension with 60 Gy of $\gamma$-irradiation from either a $^{60}$Co or a $^{37}$Cs source.

The cells are then either used immediately or stored for up to 48 h at 4°C without any appreciable loss of feeding capacity. If desired, the feeder cells can be plated in advance of the keratinocytes, as they are at their optimum 48 h after seeding. Normally for convenience we plate the lethally irradiated 3T3 feeder layer and the keratinocytes together.

Although 3T3 cells must not be allowed to reach confluence during serial propagation, it is important that they be allowed to reach confluence before being prepared for feeder layers and that they be checked routinely for mycoplasma contamination and replaced from frozen stock every 10–12 wk. Use selected donor calf serum and *not* fetal bovine serum or newborn calf serum to grow the 3T3 cells.

## 3. TUMOR SAMPLE COLLECTION, STAGING, AND PATHOLOGY

The following protocols were developed during the derivation of a number of cell lines from premalignant erythroplakias and SCC-HN. Tumor biopsies were taken from the center of each SCC-HN and bisected. Half of the sample was placed in DMEM/10DCS (see Section 2.1) before cell culture and the other half was snap-frozen in liquid nitrogen for histological examination and future DNA extraction. Blood (10–20 ml) from each patient was used as a source of normal lymphocyte DNA. All the SCC-HN were diagnosed as malignant squamous cell carcinomas, and their tumor-node-metastasis staging [UICC, 1987] is given in Table 11.1. The presence of malignant SCC cells in each biopsy was confirmed independently by two pathologists. All tumors were squamous cell carcinomas except BICR7, which had an adenosquamous histology. Three of the erythroplakia samples were taken when the patients were free of malignant disease; two were diagnosed as carcinoma in situ; and the other as severe dysplasia (Table 11.1). A fourth erythroplakia (BICRE5) was also diagnosed as severe dysplasia, but a developing SCC-HN was present in an adjoining area. Two of the patients had a previous history of malignant SCC-HN (BICRE4 and BICRE5), and three of the four patients developed SCC-HN within 12 months of the biopsy being taken. This finding demonstrated that these three erythroplakias were indeed premalignant and not benign lesions (Table 11.1). The fourth patient (BICRE1) had carcinoma in situ at the time of the biopsy, but this lesion, to date, has not progressed.

## 4. DERIVATION OF SCC-HN AND ERYTHROPLAKIA CULTURES

### Protocol 11.1. Explant Culture of Squamous Cell Carcinoma of the Head and Neck

#### Reagents and Materials

*Sterile or aseptically prepared*

❑ FBS
❑ DMEM/10FB: DMEM supplemented with 10% FBS
❑ DMEM/10FB/HC: DMEM/10FB with $8 \times 10^{-7}M$ (0.4 μg/ml) hydrocortisone hemisuccinate [Rheinwald and Green, 1975]
❑ Biopsies in DMEM/10FB
❑ X3T3: lethally irradiated Swiss 3T3 cells [Rheinwald and Green, 1975] (see Section 2.3)
❑ MRA: mycoplasma removal agent
❑ Scalpels, #11

#### Protocol

(a) Mince SCC-HN biopsies into 1-mm³-size fragments with cross scalpels.
(b) Transfer to a new culture dish.
(c) Rinse with FBS, and remove the serum.
(d) Incubate in a nonhumidified incubator at 37°C for 45 min to enable the explants to attach to the dish.
(e) Add 5 ml of DMEM/10FB/HC to the explants.
(f) Incubate for 5 d in a humidified incubator, by which time they should be firmly attached.
(g) At the first medium change, add 10 ml of the same medium containing $1 \times 10^6$ X3T3 cells
(h) At this time, also add MRA to a final concentration of 10 μg/ml. (Mycoplasma commonly infects the oropharynx of SCC-HN patients.)
(i) Following another 9–16 days growth with medium changes, the keratinocytes may be subcultured [Rheinwald and Green, 1975].

### Protocol 11.2. Subculture of SCC-HN Cells

#### Reagents and Materials

❑ Media A, B, C, D (see Section 2.1)
❑ Trypsin EDTA (see Section 2.2)
❑ MRA

### Protocol

(a) Incubate with trypsin EDTA at 37°C.

(b) Disaggregate into single cells after about 40 min of incubation [Rheinwald and Green, 1975; Rheinwald, 1980]; the exact length of time will depend on the extent of confluence of the cultures.

(c) Count the cells in a hemocytometer.

(d) Plate the keratinocyte suspensions in the four different growth media to determine the optimum growth conditions at the earliest possible stage and to avoid selection against the principal proliferative tumor cell populations.

(e) Omit the MRA 7 days after replating the keratinocytes, and test the cultures for mycoplasma contamination [Russell et al., 1986].

Due to the extremely small size of the erythroplakia biopsies, they were disaggregated by cold trypsinization [Parkinson et al., 1986].

## Protocol 11.3.  Cold Trypsinization of Erythroplakia Biopsies

### Reagents and Materials

*Sterile or Aseptically Prepared*

❑ DMEM/10FB: DMEM supplemented with 10% FBS
❑ DMEM/10FB/HC: DMEM/10FB with $8 \times 10^{-7}$M (0.4 μg/ml) hydrocortisone hemisuccinate [Rheinwald and Green, 1975]
❑ Biopsies in DMEM/10FB
❑ Media A, B, C, D
❑ MRA: mycoplasma removal agent
❑ Trypsin, 0.125% in PBSA
❑ Scalpels, #11

*Non-sterile*

❑ Ice bath

### Protocol

(a) Place the 1-mm$^3$ fragments in 0.125% trypsin.

(b) Place at 4°C for 16 h.

(c) Gently remove trypsin without removing the fragments.

(d) Incubate the fragments for 20 min at 37°C in the residue of the trypsin.

(e) Plate the disaggregated suspensions directly into Media A, B, C, or D (see Section 2.1).

By defining the optimum growth requirements of each cell culture immediately, cultures are more likely to be representative of the main population of the tumor biopsy than those described previously [Easty et al., 1981a, 1981b; Rheinwald and Beckett, 1981; Virolainen et al., 1984; Sacks et al., 1988].

## 5. CHARACTERIZATION OF THE CULTURES

### 5.1. Keratinocyte Identity

All four keratinocyte cultures prepared from premalignant erythroplakia biopsies had a quite distinctive morphology when compared with normal keratinocytes. The colonies were flatter at the edges and appeared much thinner, lacking the classic stratified cornified centers [Rheinwald and Green, 1975] of normal keratinocytes cultured in the absence of epidermal growth factor. However, this was most obvious when the colonies were small. Also, when the cells were stained by using a polyclonal antibody against the cornified envelope protein involucrin (data not shown), only a few flat cells in the centers of erythroplakia colonies reacted strongly with the antibody. This observation was consistent with the observation that all of these cultures formed far fewer cornified envelopes in surface culture than normal HEK, when plated at the same density (Table 10.2). All the malignant keratinocyte cultures showed markedly reduced stratification, regardless of the extent of progression of the SCC from which they were derived, and displayed much lower levels of involucrin expression (data not shown). In contrast, the large erythroplakia colonies were similar to large normal HEK colonies in appearance. All cultures studied reacted with monoclonal antibodies, which recognize stratified epithelial keratins (data not shown).

### 5.2. Transmission Electron Microscopy (TEM)

TEM was performed on Petri dish cultures, detached after fixation, to confirm the morphological characteristics of keratinocytes.

### Protocol 11.4. Transmission Electron Microscopy of Monolayer Keratinocytes

#### Reagents and Materials
*Non-sterile*
- ❑ Glutaraldehyde, 4% in 0.1 M phosphate buffer, pH 7.2, on ice
- ❑ Osmium tetroxide, 1% in 0.1 M PBS, pH 7.4
- ❑ Uranyl acetate, 1.5% in deionized water [Karnovsky, 1967].
- ❑ Lead citrate, 1.5% in deionized water
- ❑ Graded ethanol concentrations from 20% to absolute in stages of 20%

- Propylene oxide
- Araldite embedding resin
- Cell scraper
- Ultramicrotome
- Transmission electron microscope

### Protocol

(a) Grow the cultures to confluence in plastic Petri dishes.

(b) Fix in situ with cold glutaraldehyde for 30 min.

(c) Postfix with 1% osmium tetroxide for 1 h.

(d) Stain with 1.5% uranyl acetate.

(e) Dehydrate cells in a series of graded ethanol concentrations.

(f) Remove cell monolayer from the dish with propylene oxide [Easty et al., 1981a].

(g) Embed in Araldite resin.

(h) Cut thin sections on an ultramicrotome.

(i) Stain sections with uranyl acetate and lead citrate.

(j) Examine with a transmission electron microscope at 10,000–20,000 magnification.

The ultrastructure of lines BICRE5 (Fig. 11.1), BICRE4, BICR3, BICR6, BICR7, and BICR10 (data not shown) resembled that of normal cultured HEK except that the number of strata in the colonies was reduced (see above). All cultures were stratified however, and the cells contained tonofilaments and desmosomes. The BICRE5 cell culture (Fig. 10.1A,B) is composed of flattened cells that form up to three layers of stratification. The cell nucleus often has a convoluted profile, homogeneous nucleoplasm with a darker perinuclear region, and a large round nucleolus is sometimes visible. Sections cut perpendicular to the substratum showed that the cytoplasm is frequently organized into two layers. The majority of organelles (mitochondria, endoplasmic reticulum, Golgi complexes) are in the lower cytoplasmic layer, together with bundles of transversely cut tonofilaments that, in sections parallel to the substratum, appear as arrays radiating from the perinuclear area (data not shown). Centrioles and vacuoles, which contain an electron-lucent material, are sometimes observed in the perinuclear cytoplasm. In this layer, neighboring cells form interdigitating lamellipodia interspersed with desmosome connections. The upper cytoplasmic layer contains fine dispersed filaments and fewer cytoplasmic organelles, and the adjacent plasma membranes tend to be closely apposed. The free cell surface has spikes and areas with a slightly thickened plasmalemma, indicating a degree of cornification; numerous microvilli are present on the surface of some cells.

Taken together, the immunocytochemical and ultrastructural evi-

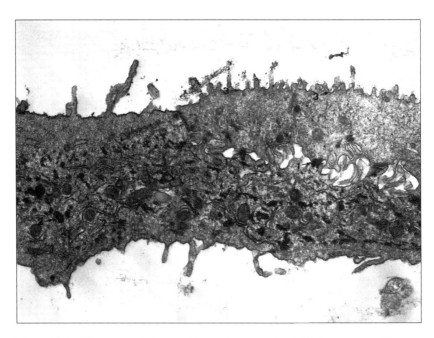

**Figure 11.1.** Ultrastructural features of premalignant erythroplakia keratinocytes. Electron micrographs of BICR E5 cells, sectioned perpendicular to the substratum, showing (**a**) perinuclear region, and (**b**) peripheral cytoplasmic regions of two adjacent cells, with hemidesmosomes and desmosomes (arrow heads), tonofilaments (t), cornified envelope (c), surface spikes (s), lamellipodia (l), and microvilli (m). Bar = 1 μm.

dence supports the view that our cultures are composed of keratinocytes [Rheinwald and Green, 1975; Rice and Thacher, 1986].

### 5.3. DNA Fingerprinting of Tumor Keratinocytes and Patient-Matched Lymphocytes and Fibroblasts

To confirm derivation, the cell lines were subjected to DNA fingerprinting and were compared with patient-derived lymphocyte DNA. DNA fingerprinting was performed as described by Jeffreys et al. [1985] by probing Southern blots of genomic DNA with minisatellite arrays 33.6 or 33.15 [Jeffreys et al., 1985]. Figure 11.2 shows typical DNA fingerprints of some of the premalignant and malignant keratinocytes used in the study. The fingerprints of all cell cultures were unique and were in all cases matched to the fingerprints of the patient's lymphocytes or fibroblasts. Occasionally a band present in the normal DNA sample was missing in the cultured tumor sample (see Fig. 11.2 lanes 3 and 4; 12, 13, and 14; 15 and 16). This finding has been reported previously for human kidney cell lines [Anglard et al., 1992] and may be related to allelic loss and/or chromosome abnormalities in the tumor cultures (see Table 11.2). The results nev-

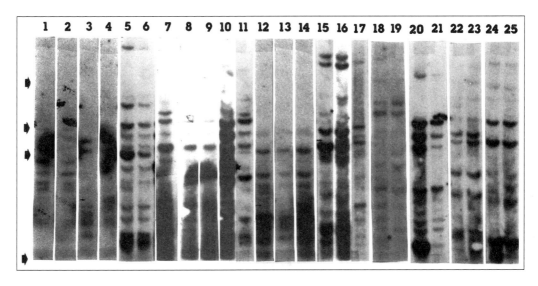

**Figure 11.2.** DNA fingerprinting of premalignant, malignant keratinocytes, and their patient-matched lymphocytes or fibroblasts. Lane 1: BICRE4. Lane 2: BICRE4 lymphocytes. Land 3: BICRES. Lane 4: BICRES lymphocytes. Lane 5: BICR3. Lane 6: BICR3 fibroblasts. Lane 7: BICR6. Lane 8: BICR6 fibroblasts. Lane 9: BICR6 lymphocytes. Lane 10: BICR18. Lane 11: BICR18 lymphocytes. Lane 12: BICR22. Lane 13: BICR22 fibroblasts. Lane 14: BICR22 lymphocytes. Lane 15: BICR31. Lane 16: BICR31 fibroblasts. Lane 17: BICR31 lymphocytes. Lane 18: BICR56. Lane 19: BICR56 fibroblasts. Lane 20: BICR63. Lane 21: BICR63 fibroblasts. Lane 22: BICR66. Lane 23: BICR66 fibroblasts. Lane 24: BICR68. Lane 25: BICR68 fibroblasts. The molecular weight markers are a XHind III digest, depicted by the large arrows, from top to bottom 23.1, 9.4, 6.6, and 4.4 kb. The small arrows show the positions of bands that are present in the normal tissue samples but absent in the tumor cultures.

ertheless confirmed the uniqueness of each tumor culture and within the limits of DNA fingerprint profiling [Jeffreys et al., 1985], suggesting that each culture was derived from the appropriate patient.

## 5.4. Proliferative Potential of Keratinocytes Isolated from Normal, Premalignant, and Malignant Squamous Epithelium

### 5.4.1. Serum and Feeder Layer Dependence

To determine the optimum growth requirements of each culture, we plated cells at clonal density ($500 - 5 \times 10^4$ cells per flask) with X3T3 feeder cells in 25 cm$^2$ flasks, with either Medium A, B, C, or D (see Section 2.1). The colonies were fixed, stained, and counted to determine cloning efficiency and the best conditions to culture and experiment with each line. Next, we cultured cells at $10\times$ clonal density ($1 \times 10^5$ cells/flask) in the optimum growth medium for each cell culture with and without X3T3 feeder cells to determine a given cell culture's dependence on growth factors and extracellular matrix molecules produced by the fibroblast feeder layer [Rheinwald and Green, 1975; Rheinwald and Beckett, 1981].

To test for X3T3-independent tumor proliferation, and also to obtain cultures of normal fibroblasts from each patient, some explants and cultures were plated in Medium A without X3T3 feeders. The growth requirements of each culture were quantitated and reassessed when the cells had completed 30–50 population doublings.

### 5.4.2. Anchorage-Independent Growth

To test for anchorage-independent growth, keratinocytes from each culture were cloned in suspension in agarose.

### Protocol 11.5.   Clonal Growth of Keratinocytes in Agar

#### Reagents and Materials
- Trypsin/EDTA (see Section 2.2)
- Agarose, 1.2% in ultrapure water (UPW), melted by autoclaving or boiling, and kept at 45°C
- Agarose, 1.0% in UPW, melted by autoclaving or boiling, and kept at 45°C
- 2× Growth medium: Growth medium at 20°C prepared at double strength
- Feeder cells: non-irradiated 3T3 cells (3T3 cells as a source of keratinocyte growth factors)
- DMEM/10DCS (see Section 2.1)
- Petri dishes, 5 cm diameter

#### Protocol
(a) Trypsinize 3T3 feeder cells and resuspend at $4 \times 10^5$ cells/ml in DMEM/10DCS.
(b) Plate out into dishes and incubate overnight at 37°C.
(c) Dilute 1.2% agarose 1:1 with 2× medium and dispense into dishes containing 3T3 monolayers, after removing the medium from the 3T3 cells.
(d) Place at 4°C to allow the agarose to gel.
(e) Trypsinize keratinocyte culture and resuspend cells at $5 \times 10^6$ cells/ml.
(f) Mix 1 part 0.66% agarose with 1 part 2× medium and add a total of $10^4$ or $10^5$ keratinocytes to 5ml of this mixture. Immediately plate the mixture on to preformed base layer of 0.6% agarose.

Using this procedure, we tested keratinocytes for anchorage independence in both their optimum attached monolayer culture medium, and in 20% v/v FBS with 0.4 μg/ml hydrocortisone if different from the former. However, colonies were observed only in cultures containing 20% serum. A colony was scored when it contained 50 cells or more. HeLa cells were used as a positive control; HEK, as a negative control. The absolute cloning efficiencies of the cultures under optimum conditions were as follows: HEK 34 (Pass 3) 0.3%; HEK 94 (Pass 3) 2.9%, (Pass 5) 1.4%; BICRE1, 0.8%; BICRE2, 0.6%; BICRE4, 0.5%; BICRE5, 0.7%; BICR3 39.1%; BICR63, 2.9%; BICR66, 2.5%; BICR6, 52.3%; BICR19, 39.8%; BICR31, 39.2%; BICR56, 9.5%; BICR68, 6.9%; BICR10, 31.5%; BICR16, 24.8%; BICR18, 20.5%; and BICR22, 8.8%. Results are from one representative experiment performed in duplicate.

### 5.4.3. Measurement of In Vitro Lifespan

The number of population doublings completed by each culture was measured at the earliest possible opportunity, usually after the first passage in optimum growth medium with X3T3 feeder cells. At each transfer, half the dishes were fixed and stained with Rhodamine B to determine cloning efficiency; a colony was defined as one that contained 50 cells or more. The remaining dishes were trypsinized, counted, and subcultured, and the number of population doublings was calculated [Rheinwald and Green, 1975; Stanley and Parkinson, 1979]. When a culture had completed 100 population doublings without any obvious reduction in cloning efficiency, it was considered to be immortal.

A culture was designated as senescent when no colonies of more than 30 cells appeared on the plates after 4 weeks and when any cells remaining appeared large, flat, and terminally differentiated. A culture was designated as being in crisis when the keratinocytes were vacuolated and cell proliferation appeared to be counterbalanced by death. Cells in crisis did not appear terminally differentiated. Normal HEKs from infants have a lifespan of 80 population doublings under the best conditions used here (Medium A) but attain only 20–30 population doublings in the absence of cholera toxin [Sun and Green, 1977]. All experiments were performed with cultures that had completed 30–50 population doublings.

Unlike most previous studies, the samples were shown to contain viable keratinocytes, and optimal culture conditions were selected at an early stage. Table 11.1 shows the details of all the tumour biopsies studied and their ultimate ability to form continuous cell lines. None of four keratinocyte cultures obtained from premalignant erythroplakia lesions formed continuous cell lines, whereas two out of seven

(28%) of those from early-stage (TNM stage 2) SCC-HN biopsies did so. These two cell lines, however, were highly vacuolated and slow growing at first, which suggests that they were perhaps emerging from crisis when placed in culture. In the case of BICR63, most explants produced keratinocytes of a more normal morphology, which senesced. Immortal variants, which proliferated rapidly in culture, were detected readily in keratinocyte cultures from late-stage T4 or recurrent SCC-HN (10 out of 14) and from lymph node metastases (2 out of 3). These cultures displayed no slowed growth when freshly explanted as BICR3 and BICR63 did. Cultures from two T4 tumors (BICR30 and BICR36) and the lymph node metastasis of the latter (BICR37) stratified and formed senescent cultures. However, keratinocytes from BICR7, another T4 tumor, formed abnormal, poorly stratified cultures. They did not senesce normally but instead appeared to enter crisis, and recently we have confirmed that this is the case.

Because SCC-HN cell lines had previously been reported to proliferate in the presence of low concentration of serum growth factors, and sometimes in the absence of anchorage to plastic [Easty et al., 1981a, 1981b; Rheinwald and Beckett, 1981; Virolainen et al., 1984; Sacks et al., 1988], or X3T3 [Rheinwald and Beckett, 1981], we tested for the acquisition of these phenotypes at different points during tumor progression.

All the premalignant and early-stage (T2) SCC tested required high levels of serum, hydrocortisone, cholera toxin, X3T3, and anchorage for optimum proliferation. In fact, no proliferation was possible in the absence of anchorage or an X3T3 feeder layer. A hallmark of the more advanced tumors was the ability to proliferate in the absence of cholera toxin and in low levels (2–5%) of serum. Indeed both cell lines obtained from lymph node metastases actually preferred 2% v/v serum. Only lines BICR6, BICR10, BICR18, and BICR31 formed colonies in the absence of an X3T3 feeder layer, and only BICR6 (efficiently) and BICR10, BICR16, and BICR31 (poorly) formed colonies in agar. BICR6 and BICR31 were derived from T4 SCCs of the hypopharynx and tongue, respectively, and from tumors that were very aggressive clinically. BICR 10 and BICR16 were isolated from recurrent SCCs and line BICR18 from a lymph node metastasis. Therefore, keratinocyte proliferation, which is anchorage-independent and does not require a 3T3 feeder layer, appears to be a property of aggressive late-stage, recurrent, or metastatic SCCs.

When we tested the premalignant and malignant keratinocytes for their ability to grow in serum-free MCDB 153 medium [Wille et al., 1984], proliferation was variable but generally very poor as has been noted previously [Rollins et al., 1989; Wise and Lechner, 2002]. These

**Table 11.1. Tumor staging, patient histories, and proliferative lifespan of keratinocytes obtained from biopsies.**

*(a) Erythroplakias*

| Patient/ Culture | Date of previous malignancy | Malignancy at time of biopsy | Biopsy histology | Date of Future malignancy | Proliferative fate of keratinocyte culture |
|---|---|---|---|---|---|
| BICR E1 | None | None; Nov. 1991 | | Carcinoma in situ | None at time of submission |
| Senescent | | | | | |
| BICR E2 | None | None; Nov. 1991 | | Carcinoma in situ | Nov. 1992 |
| Senescent | | | | | |
| BICR E4 | Sep. 1983 | None; Nov. 1991 | | Carcinoma in situ | Feb. 1992 |
| Senescent | | | | | |
| BICR E5 | Oct.1988 | SCC; Feb. 1992 | Severe dysplasia | N/A | Senescent |

*(b) Squamous cell carcinomas*

| Patient/ Culture | Site | Previous treatment | TNM staging[a] | Proliferative fate of keratinocyte culture |
|---|---|---|---|---|
| *Early-stage primary tumors* | | | | |
| BICR 1 | Tongue | None | $T_2N_0M_0$ | Senescent |
| BICR2 | Tongue | None | $T_2N_0M_0$ | Senescent |
| BICR3 | Alveolus | None | $T_2N_0M_0$ | Immortal |
| BICR25 | Floor of mouth | None | $T_2N_0M_0$ | Senescent |
| BICR63 | Tongue | None | $T_2N_{2b}M_0$ | Immortal |
| BICR66 | Retromolar trigone/tongue | None | $T_2N_0M_0$ | Senescent |
| BICR73 | Tongue | None | $T_2N_0M_0$ | Senescent |
| *Late-stage and recurrent primary tumors* | | | | |
| BICR6 | Hypopharynx | None | $T_4N_1M_0$ | Immortal |
| BICR7 | Tongue | | $T_4N_{2b}M_0$ | Crisis |
| BICR 10[b] | Buccal mucosa | Surgery and DXT | $T_4N_0M_0$[c] | Immortal |
| BICR16[b] | Tongue | " | $T_2N_0M_0$[c] | Immortal |
| BICR19[d] | Epidermis | None | [d] | Immortal |
| BICR30 | Larynx | None | $T_4N_1M_0$ | Senescent |
| BICR31 | Tongue | " | $T_4N_{2b}M_0$ | Immortal |
| BICR36 | Tongue | " | $T_4N_{2c}M_0$ | Senescent |
| BICR56 | Tongue | " | $T_4N_1M_0$ | Immortal |
| BICR59[b,e] | Oral cavity | " | $T_2N_0M_0C$ | Senescent |
| BICR68 | Tongue | " | $T_4N_0M_0$ | Immortal |
| BICR78 | Alveolus | None | $T_4N_1M_0$ | Immortal |
| BICR80 | Larynx | None | $T_4N_{2c}M_0$ | Senescent |
| BICR82 | Maxilla | Surgery and DXT | N.d. | Immortal |

Table 11.1. *(Continued)*

*(b)  Squamous cell carcinomas*

| Patient/ Culture | Site | Previous treatment | TNM staging[a] | Proliferative fate of keratinocyte culture |
|---|---|---|---|---|
| *Lymph node metastases* | | | | |
| BICRL8 | Larynx | None | $T_4N_1M_0$ | Immortal |
| BICR22 | Tongue | " | $T_4N_3M_0$ | Immortal |
| BICR37 | Tongue | " | $T_4N_{2c}M_0$ | Senescent |

[a] TNM = Tumor, node, metastasis staging (25). [b] Recurrent tumors. [e] Patient did not complete the course of radiotherapy. [c] Staging at recurrence. [d] Large and aggressive epidermal SCC, TNM staging not applicable. DXT = Deep X-ray therapy. N.d. = Not determined.

results were not affected by raising the calcium concentration of the medium or by omitting epidermal growth factor and/or bovine pituitary extract (data not shown). It is, in fact, unlikely that any of the erythroplakia cultures or most of the immortal SCC lines would have been isolated if MCDB 153 medium had been used from the beginning.

## 5.5.  Defective Response to Terminal Maturation Signals in Premalignant and Malignant Human Keratinocytes

Although many immortal SCC lines have been described, we thought it important to establish the frequency of immortal variants in tumor biopsies at different stages of progression.

### Protocol 11.6.  Assay of Terminal Maturation in Keratinocytes

#### Reagents and Materials
❑  Medium A (see Section 2.1) containing 0.15% w/v methyl cellulose
❑  Petri dishes coated with poly (2-hydroxyethyl) methacrylate (poly-HEMA) [Adams and Watt, 1989].

#### Protocol
(a)  Suspend keratinocytes at $1 \times 10^5$ cells/ml in Medium A containing methyl cellulose.
(b)  Seed into polyHEMA-coated dishes.
(c)  Recover keratinocytes from methyl cellulose after 0, 6, 12, 24, and 120 h.
   (i)  Dilute the cells from methyl cellulose 20-fold in Medium A.
   (ii)  Centrifuge at 2000 rpm (800–1000 $g$) for 10 min
   (iii)  Resuspend the cells in Medium A and count.

(d)  Replate on fresh X3T3 feeder layers in optimum growth medium to measure cell survival [Rheinwald and Beckett, 1980].

or

Treat with 5% w/v sodium dodecyl sulphate and 1% w/v 2-mercaptoethanol in Dulbecco's phosphate-buffered saline, solution A (PBSA) to measure the number of cornified envelopes [Sun and Green, 1977; Rheinwald and Beckett, 1980].

Some of the results described above suggested that all the cultured tumor keratinocytes studied were defective in their ability to stratify and hence undergo terminal maturation. Normal HEK commit to terminal maturation very rapidly when placed in suspension, by losing colony-forming ability and assembling cornified envelopes [Rheinwald

**Table 11.2.  Chromosome number and suspension-induced terminal maturation in premalignant and malignant human keratinocyte cultures.**

| Keratinocyte culture | Chromosome number | Relative colony-forming[a] efficiency (CFE) after 12 h in suspension | Half-life of CFE in suspension (T/2), h | Cornified envelopes (%)[d] In surface culture | After 5 days in suspension |
|---|---|---|---|---|---|
| Normal |||||| |
| HEK | 46 | $11.1 \pm 1.4$ | 3.2 | $6.1 \pm 4.3$ | $51.6 \pm 28.9$ |
| Erythroplakias ||||| |
| BICRE1 | N.d. | $47.2 \pm 19.6$ | 7.4 | $0.6 \pm 0.6$ | $2.3 \pm 0.5$ |
| BICRE2 | N.d. | N.d. | N.d. | $1.6 \pm 0.3$ | $37.0 \pm 11.7$ |
| BICRE4 | 46 | $90.4 \pm 7.8$ | 22.0 | $0.6 \pm 0.5$ | $4.0 \pm 3.2$ |
| BICRE5 | 46 | $33.0 \pm 4.0$ | 9.4 | $0.3 \pm 0.4$ | $11.6 \pm 2.2$ |
| *SCC (T2)* ||||| |
| BICR3 | 49 | $105.1 \pm 8.9$ | 34.0 | $0.6 \pm 0.9$ | $12.2 \pm 3.1$ |
| BICR63 | 137 | $58.0 \pm 4.1$ | 20.0 | $1.3 \pm 0.9$ | $10.0 \pm 0.8$ |
| BICR66 | 87 | $50.3 \pm 7.2$ | 18.0 | $0.9 \pm 0.4$ | $2.2 \pm 1.3$ |
| *SCC (T4 + Recurrent)* ||||| |
| BICR6 | 66 | $100.2 \pm 5.9$ | 31.8 | $0(<0.3)$ | $1.3 \pm 0.9$ |
| BICR10 | 78 | $105.9 \pm 10.7$ | [b] | $0.3 \pm 0.4$ | $1.6 \pm 0.4$ |
| BICR16 | 63 | $93.6 \pm 15.8$ | 29.0 | $0.3 \pm 0.4$ | $2.5 \pm 1.7$ |
| BICR19 | 44 | $42.6 \pm 13.3$ | 7.2 | $2.2 \pm 0.4$ | $20.6 \pm 2.6$ |
| BICR31 | 46 | $112.1 \pm 2.2$ | [b] | $0.3 \pm 0.4$ | $5.6 \pm 0.0$ |
| BICR56 | 86 | $102.4 \pm 8.8$ | 16.4 | $1.3 \pm 0.0$ | $26.9 \pm 1.8$ |
| BICR68 | 45 | $62.4 \pm 3.5$ | 28.0 | $1.6 \pm 1.3$ | $28.5 \pm 10.1$ |
| *SCC (Metastases)* ||||| |
| BICR18 | 58 | $71.0 \pm 12.6$ | 16.0 | $1.3 \pm 0.0$ | $2.8 \pm 0.4$ |
| BICR22 | 42 | $76.8 \pm 37.0$ | 18.8 | $1.6 \pm 0.4$ | $10.0 \pm 0.8$ |

[a] Relative to cloning efficiency at zero time, $\pm$ SD.  [b] No detectable loss of colony-forming ability.  [c] Determined from the cloning efficiency in surface culture after 0, 6, 12, and 24 h in suspension culture.  [d] Percentage of the total number of cells.

**Table 11.3. Tumor formation in nude mice by premalignant and malignant human keratinocyte cultures.**

| Keratinocyte culture | Latent period in days (to form a tumor of 1cm$^3$) | Latent period in days (to form tumor of 0.125 cm$^3$) | Number of mice with tumors/number of mice alive after 90 days |
|---|---|---|---|
| Normal Keratinocytes | | | |
| HEK34 | NT216 | NT216 | 0/6 |
| *Erythroplakia keratinocytes* | | | |
| BICRE4 | NT216 | NT216 | 0/4 |
| BICRE5 | NT209 | NT209 | 0/3 |
| SCC (T2) | | | |
| BICR3 | NT163 | NT163 | |
| BICR63 | NT108 | 34 | 0$^c$/3 |
| BICR66 | NT90 | NT90 | 0/1 |
| *SCC (T4 and Recurrent)* | | | |
| BICR6 | 101 | N.d. | 5/6 |
| BICR10 | 98 | 16 | 3/3 |
| BICR16 | 84 | 16 | 3/3 |
| BICR19 | 78$^a$ | 13 | 4/4 |
| BICR31 | 71 | 20 | 2/2 |
| BICR56 | 77$^a$ | 35 | 2/2 |
| BICR68 | NT108 | 34 | 0$^c$/5 |
| *SCC (Metastases)* | | | |
| BICR18 | 90 | N.d. | 2/4 |
| BICR22 | 102 | 80 | 2/4 |

NT = No tumors after no. of days indicated.
[a] Mice died bearing tumors >0.125 cm$^3$ but <1 cm$^3$. [b] Maximum size of tumor 0.064 cm$^3$, but it regressed after 134 days. [c] Maximum size of tumors 0.27 cm$^3$, but they regressed after 70 days. [d] Maximum size of tumors 0.72 cm$^3$, but they regressed after 70 days.

and Beckett, 1980]. When we tested BICRE4, BICRE5, BICR66, and all the immortal malignant SCC lines, we found that all of the tumor cultures tested committed to terminal maturation more slowly than normal as assessed by loss of colony-forming ability (see Table 11.2). All tumor keratinocytes tested showed a lower frequency of cornified envelopes than normal in surface culture and, when placed in suspension culture, most lines (except BICR19, BICR56, and BICR68) did not assemble cornified envelopes in more than 20% of the cells (see Table 11.2). Significantly, BICR19 and BICR56 formed slow-growing tumors and BICR68 regressing tumors in nude mice (Table 11.3). The results in Table 11.2 support the argument that blocked terminal maturation is an early and ubiquitous event in squamous tumor progression, which is furthermore not necessarily a property of continuous cell lines because all the erythroplakia and the BICR66 carcinoma cultures senesced (see Table 11.1).

## 5.6. Tumorigenicity of Premalignant and Malignant Human Keratinocytes

### Protocol 11.7. Tumorigenicity in Nude Mice

***Reagents and Materials***

*Sterile*

❑ Serum-free DMEM

❑ Trypsin/EDTA (see Section 2.2)

*Non-sterile*

❑ Nude mice

❑ Calipers

***Protocol***

(a) Harvest keratinocyte cultures with trypsin and EDTA when 50–75% confluent.

(b) Wash twice with serum-free DMEM.

(c) Resuspend at $1 \times 10^7$ cells per 200 μl.

(d) Inject subcutaneously into the flank of a 4-week-old nude mouse [Rheinwald and Beckett, 1981].

(e) Examine mice weekly for the presence of tumors and record the tumor volumes by caliper measurements.

(f) A mouse developing a progressively growing tumor that remains for 3 months is scored as positive.

(g) Record the latent periods and time taken for the tumors to reach 0.125 cm$^3$ and 1 cm$^3$ in volume.

(h) Snap-freeze in liquid nitrogen to preserve tumors for future immunocytochemical and histological analysis.

Several SCC cell lines previously studied form tumors when injected into the flanks of immunosuppressed mice [Easty et al., 1981a, 1981b; Rheinwald and Beckett, 1981; Rupniak et al., 1985; Sacks et al., 1988], but generally their growth rate in mice has not been quantitated nor has the relationship to the stage of tumor progression in humans been investigated. We assessed the tumorigenicity of our cultured tumor keratinocytes and showed that only keratinocytes from malignant lesions form progressively growing tumors in nude mice (see Table 11.3). Furthermore, keratinocytes cultured from early-stage carcinomas were either non-tumorigenic (BICR66) or formed slow-growing tumors, which eventually regressed (BICR3, BICR63). One late-stage tumor culture also formed tumors that regressed in mice, and two late-stage tumor cultures BICR19 and BICR56 formed slow growing tumors, which were still expanding in size when the animals died. All the other T4, recurrent, and metastatic lines formed large progressively growing tumors that reached a volume of 1 cm$^3$ within 71 to

102 days. The normal HEK, BICRE4, and BICRE5 formed only squamous cysts, which disappeared within 2 weeks. Mice injected with cells from these cultures failed to develop tumors over periods of up to 216 days (see Table 10.3).

## 5.7. Cytogenetics

### Protocol 11.8. Chromosomal Analysis of Premalignant and Malignant Human Keratinocytes

#### Reagents and Materials

*Sterile*
- ❑ Colchicine, 50 μM, dilute to 0.5 μM for use
- ❑ Trypsin/EDTA (see Section 2.2)

*Non-sterile*
- ❑ Hypotonic solution: 0.056 M KCl, 8.5 mM sodium citrate
- ❑ Fixative: 1 part, v/v, glacial acetic acid to 3 parts anhydrous methanol, freshly prepared and kept on ice
- ❑ Giemsa stain

#### Protocol

(a) Culture keratinocytes in their optimum growth medium.

(b) Check for presence of mitoses, and treat with 0.5 μM colchicine for 30 min.

(c) Remove medium, rinse gently with PBSA, and add trypsin/EDTA, 0.2 ml/cm$^2$.

(d) When cells have detached, disperse the cells in the trypsin and centrifuge at 100 g for 2 min.

(e) Resuspend the pellet in hypotonic solution and incubate at 37°C for 20 min.

(f) Place the cells on ice and add the fixative directly to the hypotonic solution (1 part fixative to 4 parts hypotonic solution). This helps to prevent the keratinocytes from clumping.

(g) Centrifuge at 100 g for 2 min and discard supernate.

(h) Flick tube to disperse pellet and resuspend pellet in fresh fixative, and place on ice for 30 min.

(i) Repeat centrifugation and fixation two more times, 30 min each time.

(j) After the last centrifugation, discard the fixative, flick the tube to disperse the pellet, and resuspend the pellet in 200 μl fresh fixative, depending on the cell concentration.

(k) Drop the cells on to ice-cold microscope slides and dry rapidly.

(l) Check density of cells in spread by phase-contrast microscopy and dilute suspension if necessary before preparing more slides.

(m) Stain with Giemsa.

(n)   Count the chromosomes in well-spread metaphases under bright field optics.

BICRE4, BICRE5 and BICR6 were karyotyped in detail by Mrs. M. Fitchett at the Oxford Medical Genetics Laboratory, Churchill Hospital, Oxford. The karyotypes of BICRE4 and BICRE5 were examined in detail, and BICRE4 and most (more than 90%) of the cells of BICRE5 had normal karyotypes. BICRE5 did, however, contain a clone of cells that had lost a Y chromosome (data not shown). The SCC-HN cultures all had numerical chromosome alterations in most metaphases counted, although lines BICR19, BICR68, and BICR22 were hypodiploid or diploid in the majority of their divisions.

## 6.  DISCUSSION AND APPLICATIONS

Only a few previous reports have focused on the culture of benign laryngeal papillomas [Steinberg et al., 1982] and oral leukoplakias [Lindberg and Rheinwald, 1990; Chang et al., 1992] before the present study. Furthermore, only one study has detailed the growth properties of the isolated cells and their tumorigenicity in nude mice [Chang et al., 1992]. We have described here for the first time the cultivation and characterization of keratinocytes from premalignant erythroplakias of the oral cavity. These cells have normal keratinocyte growth requirements in that they require high levels of serum, hydrocortisone, cholera toxin, anchorage, and an X3T3 feeder layer for optimum growth. They also have normal diploid karyotypes, senesce, and do not form tumors in nude mice.

However, all the erythroplakia cultures were weakly stratified, had low levels of cornified envelopes in surface culture, and were resistant to suspension-induced terminal maturation (see Table 11.2). Because this phenotype was shared by all the SCC-HN cultures, regardless of the stage of progression of the original lesion, blocked terminal maturation may be an early marker of human squamous neoplasia, which is also essential for tumor progression. Resistance to inducers of keratinocyte terminal maturation has also been reported to be an early event in the induction and progression of mouse epidermal squamous cell carcinoma [Yuspa and Morgan, 1981; Parkinson et al., 1984]. However, the identification for the first time that blocked terminal maturation is an early marker of human SCC-HN development will facilitate the identification of the genetic initiation events of this disease, which so far have remained elusive.

Human premalignant keratinocytes, such as normal keratinocytes [Rheinwald and Green, 1975; Wille et al., 1984] and those cultured from benign laryngeal papillomas [Steinberg et al., 1982] have a finite proliferative lifespan in vitro (see Table 10.1), as do the majority of cells cultured from human premalignant lesions [Paraskeva et al., 1984; Mancianti and Herlyn, 1989]. There is only one report of an established cell line arising from a premalignant oral lesion adjacent to a malignant SCC [Chang et al., 1992]; furthermore this cell line was clearly a subpopulation of the keratinocytes biopsied, because most of the keratinocytes cultured from the lesion senesced. In our experience, immortal cells were detectable only in carcinomas and, even then, they were less frequent than in more advanced lesions. Also, the two immortal cell lines obtained from T2 tumors did not form progressively growing tumors in nude mice (see Table 11.3). The results may be a reflection of fewer genetic changes in the T2 carcinomas than in the more advanced lesions. Alternatively, or in addition, early-stage carcinomas may contain small, expanding clones of immortal cells, which are not always present in the part of the tumor put into culture. This possibility is supported by the observation that line BICR63 was obtained from a single explant, which produced abnormal cells; all the others produced cells that appeared morphologically normal. On transfer, all the morphologically normal cells senesced and were outgrown by the abnormal ones. This type of observation was also made by Rheinwald and Beckett [1981], but the tumor staging was not published in their study. The appearance of the keratinocytes from later stage tumors was usually abnormal, and all the cells had a uniform and distinctive morphology. The frequency of immortal variants in advanced tumors was very high, but, nevertheless, their presence in some advanced tumors could not be demonstrated (see Table 11.1). Either the keratinocytes in these tumors had still failed to accumulate the appropriate genetic changes necessary for immortalization or there is a subgroup of SCC-HN that can form local metastases in vivo, even though they are not immortal, at least not in vitro. Rheinwald and Beckett also reported that 7 out of 13 SCC-HN placed into culture senesced and, although it is difficult to relate their results to ours in the absence of their tumor staging data, it is notable that overall our observed frequency of senescent SCC-HN cultures (47%) is similar to theirs (54%). Our data suggest that the phenotype of in vitro immortality is normally a late event in the development of human SCC-HN, similar to the conclusions drawn in investigations of colon cancer [Paraskeva et al., 1984, Paraskeva et al., 1988] and melanoma [Mancianti and Herlyn, 1989]. However, not all SCC-HN arise in the same manner, and 10–15% of these lesions arise from a clearly discernable premalignant lesion, such as leukoplakia or erythroplakia. When a much larger series of these lesions was examined by our colleagues at the Beatson Institute [F. McGreggor, A. Mutoni, and P. Harrison; unpublished data] and when the results were combined with ours, it was found that around 8 out of 18 were immortal. These results contrast with our study of carcinomas and suggest that immortality may sometimes be an early event in SCC-HN that arises from premalignant lesions. All the cultures obtained from SCC-HN were aneuploid (see Table 11.2), even BICR66, which had a limited lifespan in vitro, which suggests that aneuploidy per se is an insufficient condition for immortality, although more senescent SCC-HN need to be examined to confirm this.

All the cultures were tested to determine their requirements for optimum clonal proliferation, and most of the immortal cell lines from late-stage SCC-HN required much lower levels of serum growth factors for optimal proliferation than normal HEK (see also Table 11.4). Our results also indicate that many tumors did not grow optimally under conditions where normal HEK grew well. The results with BICR18 and BICR22 were particularly interesting, as their growth requirements strongly favored low levels of serum and no cholera toxin. Indeed, these cell lines might not have been derived at all if their in vivo growth requirements had not been defined early. The observa-

**Table 11.4. Summary of phenotypes seen throughout the development and progression of human SCC.**

| Tumor stage | Cell line | Blocked terminal[c] maturation | Aneu-ploidy | Immortal | Multi-plication in low serum[d] | Multi-plication in agar[e] | Multi-plication without X3T3 | Tumor formation in nude mice[f] |
|---|---|---|---|---|---|---|---|---|
| Normal | HEK | − | − | − | − | − | − | − |
| Premalignant | BICRE1 | + | N.d. | − | − | − | − | N.d. |
| Premalignant | BICRE2 | +/− | N.d. | − | − | − | − | N.d.. |
| Premalignant | BICRE4 | + | − | − | − | − | − | − |
| Premalignant | BICRE5 | + | − | − | − | − | − | − |
| $T_2N_0N_0$ | BICR66 | + | + | − | − | − | − | − |
| $T_2N_0M_0$ | BICR3 | + | + | + | +/− | − | − | − |
| $T_2N_{2B}M_0$ | BICR63 | + | + | + | +/− | − | − | +/− |
| N/A[a] | BICR19 | +/− | + | + | + | − | − | +/− |
| $T_4N_0M_0$ | BICR68 | +/− | + | + | +/− | | | |
| $T_4N_1M_0$ | BICR56 | +/− | + | + | +/− | − | − | +/− |
| $T_4N_1M_0$ | BICR6 | + | + | + | + | + | + | + |
| $T_4N_{2B}M_0$ | BICR31 | + | + | + | + | +/− | + | + |
| N/A[b] | BICR10 | + | + | + | + | +/− | | |
| N/A[b] | BICR16 | + | + | + | + | +/− | + | + |
| $T_4N_1M_0$ | BICR18 | + | + | + | + | − | + | + |
| $T_4N_3M_0$ | BICR22 | + | + | + | + | − | − | + |

[a] Tumor-node-metastasis staging not applicable to this tumor, which was derived from epidermis (it was T2/T3 size). [b] Recurrent tumor; therefore staging inappropriate or not known. [c] ± /− = only weakly blocked terminal maturation—either poor survival or >20% form cornified envelopes in suspension culture (see Table 11.2). [d] + = Optimum growth in 5% serum versus 20% serum or better. ±/− = less-than-optimum growth in 5% serum versus 20% serum. − = No growth in 5% serum. [e] + = Large colonies at a frequency >1% cloning efficiency. ±/− = small colonies at a frequency of less than 1% cloning efficiency;. − = no colonies. [f] + = Large, progressively growing tumors. ±/− = Slow growing, highly differentiated, or regressing tumors. − No tumors. N.d. = Not determined.

tion that metastatic SCCs have completely different growth requirements from primary and recurrent tumors might be a reflection of the special adaptation required for keratinocytes to proliferate in the lymph node, but greater numbers of metastatic cell lines must be studied to confirm this. Metastatic SCC-HN have been reported to grow in higher levels of serum when cultured without X3T3 [Easty et al., 1981a; Virolainen et al., 1984], but the optimum growth requirements of these lines were not defined early. In our study, only certain aggressive T4 tumors, recurrent tumors, or metastases gave rise to cultures that could proliferate at low density without an X3T3 feeder layer. As pointed out by Rheinwald and Beckett [1981], the practice of deriving and cultivating SCC lines without fibroblast feeders may lead to fitter variant lines that have deviated phenotypically from the original tumor cells. Furthermore, because we have now shown that an X3T3-independent phenotype is associated with the later stages of tumor progression, the case for optimizing the culture conditions for each individual tumor as early as possible assumes even greater importance. Similarly only four SCC-HN lines, all from advanced tumors, could form colonies to any great extent in agar, and only one formed large colonies (see Table 11.4).

The cell lines from the most advanced tumors also formed the most rapidly growing tumors in nude mice (see Table 11.3) and, when the lines are ranked according to the tumor-node-metastasis stage of the tumor, they were derived from (see Table 11.4), there was a good relationship between the extent of progression of the original tumor and the ability of the resultant cell line to form experimental tumors. Furthermore, when all our results are summarized (see Table 10.4), it is clear that four in vitro phenotypes are essential to the formation of experimental tumors in nude mice. These are immortalization, aneuploidy, blocked terminal maturation, and the ability to proliferate in low levels of serum growth factors. Also five out of six of the cell lines, which formed rapidly growing tumors in mice, could grow either in suspension or in the absence of an X3T3 feeder layer, and three of them could do both. Anchorage-independent growth has been used by many investigators to study keratinocyte progression to SCC in vitro, usually starting with cells immortalized with DNA tumor viruses, and rendering them independent of 3T3-derived factors [Rhim et al., 1985]. Although such models are of undoubted value, our results would suggest that some of the genetic changes observed in such models [Stacey et al., 1990] might not occur in humans until a very late stage of SCC progression, if at all.

It can also be seen from Table 11.4 that, although the number of phenotypic changes increases with the stage of SCC-HN progression, these phenotypes are not necessarily acquired in the same sequence in every tumor. However, certain trends are obvious: Resistance to terminal maturation signals appears to be an early and ubiquitous event; aneuploidy and immortalization are associated with malignancy, whereas independence from anchorage and serum- or 3T3-derived growth factors increases at a later stage of SCC progression.

The availability of a series of SCC-HN cultures and cell lines from early as well as late stages of tumor progression has presented us with a powerful human model system for studying the development and progression of the disease. Furthermore, because we have retained samples of normal material from all the patients described here, we have been able to use our in vitro system to study the role of tumor suppressor genes, as well as oncogenes in the pathogenesis of SCC-HN. As our cell lines are phenotypically characterized and can be genetically characterized, they should, in particular, make excellent targets for the introduction of candidate tumor suppressor genes [Huang et al., 1988] or the chromosomes carrying such suppressor genes [Tanaka et al., 1991].

## ACKNOWLEDGMENTS

The authors would like to thank Margaret Fitchett, Maud Clark, and the Oxford Regional Cytogenetics service for help with karyotyping, Ian Macmillan of the Veterinary Pathology Department, Glasgow University, for histology, Professor Gordon Macdonald of the Dental Pathology Department, Glasgow University and Dr. Peter Johnson of the Pathology Department, Aberdeen Royal Infirmary, for pathological Assessment of tumors, Tom Hamilton and Steven Bell for help with the animal experiments, Dr. Fiona Watt for the gift of the rabbit anti-involucrin antibody and advice concerning keratinocyte suspension in methocel, David Tallach for artwork, Dr. John Wyke, Dr. Christos Paraskeva, and Dr. Brad Ozanne for helpful comments on the manuscript, and, finally, we wish to thank the Cancer Research Campaign (now Cancer Research UK) for continued financial support.

## SOURCES OF MATERIALS

| Material | Supplier |
| --- | --- |
| Cholera toxin | Sigma |
| Fetal bovine serum (FBS) | Globepharm |
| Hydrocortisone hemisuccinate | Sigma |
| Mycoplasma removal agent (MRA) | ICN |
| Trypsin, crystalline | Worthington |
| Cell scraper | Nunc |
| Rhodamine B | Sigma |
| Methyl cellulose | Sigma |
| Poly (2-hydroxyethyl) methacrylate | Sigma |

# REFERENCES

Adams, J.C., and Watt, F.M. (1989) Fibronectin inhibits the terminal differentiation of human keratinocytes. *Nature (London)* 340:307–309.

Anglard, P., Trahan, E., Llu, S., Latif, F., Merino, M.J., Lerman, M.I., Zbar, B. Linehan, W.M. (1992) Molecular and cellular characterization of human renal cell carcinoma cell lines. *Cancer Res.* 52:348–356.

Burns, J.E., Baird, M.C., Clark, L.J., Burns, P.A., Edington, K., Chapman, C., Mitchell, R., Robertson, G., Soutar, D. Parkinson, E.K. (1993) Gene mutations and increased levels of p53 protein in human squamous cell carcinomas and their cell lines. *Br. Cancer* 67:1274–1284.

Chang, S.E., Foster, S., Betts, D., Marnock, W.E. (1992) DOK, a cell line established from human dysplastic oral mucosa, shows a partially transformed non-malignant phenotype. *Int. J. Cancer* 52:896–902.

Easty, D.M., Easty, G.C., Carter, R.L., Monaghan, P., Butler, L.J. (1981a) Ten human carcinoma cell lines derived from squamous carcinomas of the head and neck. *Br. J. Cancer* 43:772–785.

Easty, D.M., Easty, G.C., Carter, R.L., Monaghan, P., Pittam, M.R., James, I. (1981b) Five human tumour cell lines derived from a primary squamous carcinoma of the tongue, two subsequent local recurrences and two nodal metastases. *Br. J. Cancer* 44:363–370.

Fearon, E., and Vogelstein, B. (1990) A genetic model of colorectal tumorigenesis. *Cell* 61:759–767.

Green, H. (1978) Cyclic AMP in relation to proliferation of the epidermal cell: a new view. *Cell* 15:801–811.

Huang, H-J. S., Yee, J-K., Shew, J-Y., Chen, P-L., Bookstein, R., Friedmann, T., Lee, E. Y-H. P., Lee, W-H. (1988) Suppression of the neoplastic phenotype by replacement of the RB gene in human cancer cells. *Science* 242:1563–1566.

Jeffreys, A.J., Wilson, V., Thein, S.L. (1985) Hypervariable "minisatellite" regions in human DNA. Nature *(London)* 314:67–73.

Karnovsky, M.J. (1967) The ultrastructural basis of capillary permeability studied with peroxidase as a tracer. *J. Cell Biol.* 35:213–236.

Latif, F., Fivash, M., Glenn, G., Tory, K., Orcutt, M.L., Hampsch, K., Delisio, J., Lerman, M., Cowan, J., Beckett, M., Weichselbaum, R. (1992): Chromosome 3p deletions in head and neck carcinomas: statistical ascertainment of allelic loss. *Cancer Res.* 52:1451–1456.

Lindberg, K., and Rheinwald, J.G. (1990) Three distinct keratinocyte subtypes identified in human oral epithelium by their patterns of keratin expression in culture and in xenografts. *Differentiation* 45:230–241.

Maestro, R., Dolcetti, R., Gasparotto, D., Dogljoni, C., Pelucchi, S., Barzan, L., Grandi, E., Boiocchi, M. (1992) High frequency of p53 gene alterations associated with protein overexpression in human squamous cell carcinoma of the larynx. *Oncogene* 7:1159–1166.

Mancjanti, M-L., and Herlyn, M. (1989) Tumor progression in melanoma: the biology of epidermal melanocytes in vitro. In Conti, C.J., Slaga, T.J., Klein-Szanto, A.J.P. (eds) *Skin Tumors: Experimental and Clinical Aspects.* Raven Press: New York, NY, 369–383.

Million, R.R., Cassisi, N.J., Clark, J.R. (1989) Cancer of the head and neck. In DeVita, Jr., V.T., Hellman, S., Rosenberg, S.A. (eds) *Cancer: Principles and Practice of Oncology.* J B Lippencott Company: Philadelphia, PA, 488–590.

Paraskeva, C., Buckle, B.C., Sheer, D., Wigley C.B. (1984) The isolation and characterisation of colorectal epithelial cell lines at different stages in malignant transformation from familial polyposis coli patients. *Int. J. Cancer* 34:49–56.

Paraskeva, C., Finerty, S., Powell, S. (1988) Immortalisation of a human colorectal adenoma cell line by continous in vitro passage: possible involvement of chromosome I in tumour progression. *Int. J. Cancer* 41:908–912.

Parkinson, E.K., Hume, W.J., Potten, C.S. (1986) The radiosensitivity of cultured human and mouse keratinocytes. *Int. J. Rad. Biol.* 50:717–726.

Parkinson, E.K., Pera, M.F., Emmerson, A., Gorman, P.A. (1984) Differential effects of complete and second-stage tumour promoters in normal but not transformed human and mouse keratinocytes. *Carcinogenesis* 5:1071–1077.

Parkinson, E.K. and Yeudall W.A. (2002) The epidermis. In Freshney, R.I, and Freshney, M.G.. (ed) *Culture of Epithelial Cells*, 2nd Edition. John Wiley & Sons: New York, NY, pp 65-94..

Pindborg, J.J. (1984) Control of oral cancer in developing countries. *Bull WHO* 62:817–830.

Pindborg, J.J. (1985) Oral precancer. In Barnes, L. (ed) *Surgical Pathology of the Head and Neck*, Vol. 1. Marcel Dekker: New York, 279–331.

Rheinwald, J.G., and Beckett, M.A. (1980) Defective terminal differentiation in culture as a consistent and selectable character of malignant human keratinocytes. *Cell* 22:629–632.

Rheinwald, J.G., and Beckett, M.A. (1981) Tumorigenic keratinocyte lines requiring anchorage and fibroblast support cultured from human squamous cell carcinomas. *Cancer Res.*, 41:1657–1663.

Rheinwald, J.G., and Green, H. (1975) Serial cultivation of strains of human epidermal keratinocytes: the formation of keratinizing colonies from single cells. *Cell* 6:331–343.

Rhim, J.S., Jay, C., Arnstein, P., Price, F.M., Sanford, K.K., Aaronson, S.A. (1985) Neoplastic transformation of human epidermal keratinocytes by Ad12-SV40 and Kirsten Sarcoma viruses. *Science* 227:1250–1252.

Rice, R.M., and Thacher, S.M. (1986) Involucrin: a constituent of cross-linked envelopes and marker of squamous maturation, In Bereiter-Hann, J., Maltoltsy, A.G., Richards, K.S. (eds) *Biology of the Integument 2.* Springer-Verlag: Berlin, 752–761.

Rollins, B., O'Connell, T.M., Bennett, C., Burton, L.E., Stiles, C.D., Rheinwald, J.G. (1989) Environmental dependent growth inhibition of human epidermal keratinocytes by recombinant human transforming growth factor-beta. *J. Cell Physiol.* 139:455–462.

Rupniak, H.T., Rowlatt, C., Lane, E.B., Steele, J.G., Trejdosiewicz, L.K., Lakiewicz, B., Povey, S., Hill, B.T. (1985) Characteristics of four new human cell lines derived from squamous cell carcinomas of the head and neck. *J. Natl. Cancer Inst.* 75:621–635.

Russell, W.C., Newman, C., Williamson, D.H. (1986) A simple cytochemical demonstration of DNA in cells infected with mycoplasma and viruses. *Nature (London)* 253:461–462.

Sacks, P.G., Parnes, S.M., Gallick, G.E., Mansouri, Z., Lichtner, R., Satya-Prakash, K.L., Pathak, S., Parsons, D.F. (1988) Establishment and characterization of two new squamous cell carcinoma cell lines derived from tumours of the head and neck. *Cancer Res.* 48:2858–2866.

Sakai, E., and Tsuchida, N. (1992) Most human squamous cell carcinomas in the oral cavity contain mutated p53 tumor-suppressor genes. *Oncogene* 7:927–933.

Stacey, M., Gallimore, P.H., Meconville, C., Taylor, A.M.R. (1990) Rearrangement of the same chromosome regions in different SV40 transformed human skin keratinocyte lines is associated with tumorigenicity. *Oncogene* 5:727–739.

Stanley, M.A., and Parkinson, E.K. (1979) Growth requirements of human cervical epithelial cells in culture. *Int. J. Cancer* 24:407–414.

Steinberg, B.M., Abramson, A.L., Meade, R.P. (1982) Culture of human laryngeal papilloma cells in vitro. *Otolaryngol. Head Neck Surg.* 90:728–735.

Sun, T-T., and Green, H. (1977) Cultured epithelial cells of cornea, conjunctiva and skin: absence of marked intrinsic divergence of their differentiated states. *Nature (London)* 269:489–493.

Tanaka, K., Oshimura, M., Kikuchi, R., Seki, M., Hayashi, T., Miyaki, M. (1991)

Suppression of tumorigenicity in human colon carcinoma cells by introduction of normal chromosome 5 or 18. *Nature* (*London*) 349:340–342.

UICC (1987) TNM classification of malignant tumours. In Hermande, P., Fobin, L. (eds) *Union Internationale Contre le Cancer* Springer Verlag: Berlin.

Virolainen, E., Janharanta, R., Carey, T.E. (1984) Steroid hormone receptors in human squamous carcinoma cell lines. *Int. J. Cancer* 33:19–25.

Wille, J.J., Jr., Pittelkow, M.R., Shipley, G.D., Scott, R.E. (1984) Integrated control of growth and differentiation of normal prokeratinocytes cultured in serum-free medium: clonal analyses, growth kinetics and all cycle studies. *J. Cell Physiol.* 121:31–44.

Williams, A.C., Harper, S.J., Paraskeva, C. (1990) Neoplastic transformation of a human colonic epithelial cell line: in vitro evidence for the adenoma to carcinoma sequence. *Cancer Res.* 50:4724–4730.

Wise, J. and Lechner, J.F. (2002) Normal human bronchial epithelial cell culture. In Freshney, R.I. and Freshney, M.G. (eds) *Culture of Epithelial Cells.* John Wiley and Sons: New York, pp 257–276.

Yuspa, S.H., and Morgan, D. (1981) Mouse skin cells resistant to terminal differentiation associated with the initiation of carcinogenesis. *Nature* (*London*) 293:72–74.

# 12

# Culture of Melanocytes from Normal, Benign, and Malignant Lesions

Ruth Halaban

*Department of Dermatology, Yale University School of Medicine, New Haven, Connecticut. ruth.halaban@yale.edu*

*Culture of Human Tumor Cells,* Edited by Roswitha Pfragner and R. Ian Freshney.
ISBN 0-471-43853-7   Copyright © 2004 Wiley-Liss, Inc.

## I. INTRODUCTION

1.1. Pure Cultures of Normal Melanocytes

### 1.1.1. Melanocyte Mitogens: Historical Overview

Attempts to establish normal melanocytes in culture by using the standard serum-supplemented media that support melanoma cell growth have failed. Seminal observations by three groups provided the basis for the current medium supplements used by most investigators. Eisinger and Marko [1982] were the first to identify phorbol esters (such as TPA, 12-O-tetradecanoyl phorbol-13-acetate) and cholera toxin (CT) (a stimulator of intracellular cyclic adenosine monophosphate, cAMP), as the two chemical mitogens for normal human melanocytes derived from newborn foreskins. Three years later, it was shown, independently, that a melanocyte growth factor(s) exists in bovine brain [Wilkins et al., 1985], and in cultured melanoma cells, as well as other cell types (astrocytomas and the human embryonic lung WI-38)

[Eisinger et al., 1985]. This ubiquitously expressed mitogen was identified shortly after as bFGF (basic fibroblast growth factor, FGF-2) [Halaban et al., 1987]. Basic FGF was demonstrated also as the melanocyte mitogen in keratinocytes, melanoma cells, and other tissues [Halaban et al., 1988a; Halaban et al., 1988b]. Since then, the list of growth factors that stimulate melanocyte proliferation in culture has expanded to include hepatocyte growth factor/scatter factor (HGF/SF), mast/stem-cell Factor (M/SCF, known also as the KIT ligand and steel factor), and the neuropeptides endothelin-1, 2 and 3 (ET-1, ET-2, ET-3) [Matsumoto et al., 1991; Yada et al., 1991; Funasaka et al., 1992; Halaban et al., 1992; Imokawa et al., 1992; Halaban et al., 1993a; Böhm et al., 1995]. These studies demonstrated that normal human melanocytes, unlike melanoma cells from advanced lesions, require a selective combination of growth factors in order to proliferate in culture.

### 1.1.2. Synergistic Stimulation

Supplementation with synergistic growth factors is the key for successful establishment of normal melanocytes in culture. Mass population of human melanocytes can be grown relatively easily in the serum and chemical supplemented medium, termed TICVA (Table 12.1). The synergism here is between the protein kinase C stimulator TPA (T), three agents that increase intracellular levels of cAMP; that is, IBMX (3-isobutyl-1-methyl xanthine, I), CT (C), and dbcAMP (N6, 2′-O-dibutyryladenosine 3:5-cyclic monophosphate, A), and sodium orthovanadate ($Na_3VO_4$, V), an inhibitor of tyrosine phosphatases that prolongs the activity of endogenous tyrosine kinases. Any one of these classes of compounds does not support melanocyte proliferation when supplemented alone. Furthermore, each one of these reagents has to be provided at optimal concentrations in order to elicit rapid proliferation (Fig. 12.1) [Halaban et al., 1986]. Although optimal levels of intracellular cAMP can be achieved by high concentrations of CT in

**Table 12.1. Components of TICVA medium.**

| Reagent | Amount | Final Concentrations |
|---|---|---|
| Ham's F10 or Ham's F12 | 500 ml | |
| Glutamine | 5 ml | 2 mM |
| Pen/strep | 5 ml | 100 U/ml; 100 μg/ml |
| Fetal bovine serum | 35 ml | 7% |
| TPA (100 μg/ml DMSO stock) | 250 μl | 50 ng/ml |
| IBMX (5 mM stock solution) | 10 ml | 0.1 mM |
| $Na_3VO_4$ (1 mM stock solution) | 500 μl | 1.0 μM |
| DbcAMP (stock solution 0.1 M) | 5 ml | 1 mM |
| Cholera toxin (2.5 μM stock solution) | 500 μl | 2.5 nM |

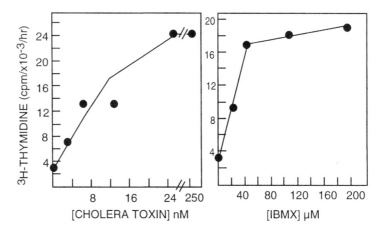

**Figure 12.1.** Synergistic stimulation of normal human melanocytes by IBMX or CT in TPA and serum supplemented medium. Melanocytes from newborn foreskins were grown for 3 months in the presence of TPA, IBMX and CT. They were then transferred to 4-cm² cluster-well plates and incubated in medium supplemented with TPA and the indicated concentrations of CT (left) or IBMX (right). The figure is taken from [Halaban et al., 1986].

the medium (Fig. 12.1) [Halaban et al., 1986], it is too costly and economically unsound.

Likewise, none of the peptide growth factors (Table 12.2) supports melanocyte proliferation when supplemented individually, even in the presence of the obligatory mammalian mitogens, such as insulin and transferrin, or serum (Fig. 12.2) [Bhargava et al., 1992; Funasaka et al., 1992; Halaban et al., 1992; Böhm et al., 1995; Swope et al., 1995]. Peptide growth factors are synergistic with dbcAMP, TPA, or with each other [Halaban et al., 1987; Medrano and Nordlund, 1990; Funasaka et al., 1992; Halaban et al., 1992; Böhm et al., 1995]. Interestingly, among the peptide growth factors, only the endothelins, such as ET-1, induce extended long dendrites and maintain viability over several days when provided as the sole growth factor (Fig. 12.2) [review by Halaban, 1994; Böhm et al., 1995]. The four peptides bFGF,

**Table 12.2. Peptide growth factor supplemented medium.**

| Reagent | Amount | Final Concentrations |
| --- | --- | --- |
| Opti-MEM | 500 ml | |
| Fetal bovine serum | 15 ml | 3% |
| Pen/strep | 2.5 ml | 100 U/ml/100 µg/ml |
| Hepatocyte growth factor/SF | 500 µl | 40 ng/ml |
| Endothelin 1 | 500 µl | 10 nM ($1 \times 10^{-8}$ M) |
| bFGF | 500 µl | 10 ng/ml |
| Heparin | 500 µl | 1 ng/ml |
| $Na_3VO_4$ | 0.5 ml | 1.0 µM |

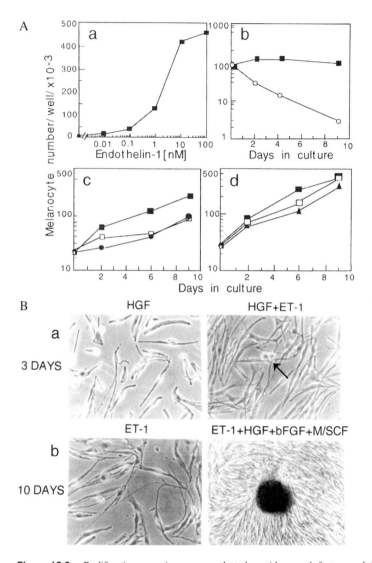

**Figure 12.2.** Proliferative synergism among selected peptide growth factors and the effect on melanocyte morphology. Normal neonatal human melanocytes were grown in chemically defined medium in the presence or absence of growth factors. A. Proliferative responses of melanocytes to growth factors: a, Melanocytes were incubated for 10 days in the presence of HGF/SF (40 ng/ml) and the indicated concentrations of ET-1; b, no additions (o), or 10 nM ET-1 (■); c, 10 nM ET-1 plus 40 ng/ml HGF/SF (■), or plus 200 ng/ml M/SCF (□), or plus 10 ng/ml bFGF (●); d, ET-1 plus HGF/SF plus M/SCF plus bFGF (■), or ET-1 plus M/SCF plus HGF/SF (□), or ET-1 plus HGF/SF plus bFGF (▲). Data in "a" are plotted on an arithmetic scale and in "c–d" on a logarithmic scale. They represent means from duplicate wells. All experiments were performed at least twice with similar results. Standard error did not exceed 12%. B. Morphology of melanocytes under different culture conditions: The photomicrographs show morbid melanocytes cultured for 3 days in the presence of HGF/SF alone (a, left), in contrast to those proliferating in the presence of HGF/SF plus ET-1 (a, right). Melanocytes are maintained healthy with ET-1 alone over 10 days (b, left) and are thriving in the presence of four growth factors (b, right) [from Böhm et al., 1995].

HGF/SF, M/SCF and ET-1 co-operate with each other when supplied in pairs, but proliferation is best promoted when three or more growth factors are added to the culture medium (Fig. 12.2) [Böhm et al., 1995].

Heparin is required to stabilize bFGF and to facilitate its interaction with the FGFR1 [Ogura et al., 1999]. A ratio of 1:10 heparin: bFGF (w:w) provides the needed protection from degradation and facilitates receptor ligand interaction. Higher levels of heparin relative to bFGF can sequester the peptide from its receptor and suppress mitogenic activity.

Melanotropin (or melanocyte stimulating hormone, MSH), is a potent mitogen for mouse melanocytes in culture [Tamura et al., 1987], but less so for human melanocytes in vitro [Hunt et al., 1994; Swope et al., 1995]. Its main role in the skin is likely to be the modulation of pigmentation in general, and in response to UVB irradiation in particular [Valverde et al., 1995; Im et al., 1998; Tada et al., 1998; Abdel-Malek et al., 2000].

The TICVA medium is usually supplemented with 7–10% fetal borine serum (FBS). However, serum proteases shorten the life span of peptide growth factors. For this reason, commercially available low-serum media (which included insulin and transferrin and possibly other components) are better suited for peptide-growth factor supplemented medium.

### 1.1.3. Basic Medium

Although melanocytes originate from the skin, unlike keratinocytes, they do not need low calcium to proliferate. In fact, they require the optimal level of calcium supplied in standard media (0.2–1 mM $CaCl_2$), and they proliferate at a slower rate when incubated in the low-calcium medium formulated for keratinocytes [unpublished observations].

Different basal media are used routinely to grow melanocytes by various investigators (such as Ham's F10, RPMI 1640, MCDB 153). Although side by side comparisons have not been reported, specific attention should be made to the tyrosine concentration and its effect on the cells. Melanocytes synthesize pigment by converting tyrosine to melanin [Lerner et al., 1949], a reaction that involves the production of toxic quinones and semiquinone that can accumulate in the medium under certain conditions, such as infrequent feeding and high ratio of cells/medium [Halaban and Lerner, 1977]. In fact, the production of tyrosinase-driven toxic compounds was considered a possible way to arrest melanoma growth in patients [Naish-Byfield et al., 1991; Naish-Byfield and Riley, 1992; Cooksey et al., 1995]. For this reason, the Ham's F10 or Ham's F12 medium with low-tyrosine concentration (10 µM tyrosine) has been originally selected to grow mouse melanoma cells [Wong et al., 1974] and currently also normal melanocytes. However, other media, such as Dulbecco's modified Eagle's medium (DMEM) and RPMI 1640, containing high levels of tyrosine (200–400 μM) are also in use. The accumulation of toxic tyrosinase reaction products is particularly prominent in cultures of melanocyte expressing high levels of the enzyme tyrosinase, and can be observed by the darkening of the medium due to the accumulation of melanin. Under these circumstances, it is advisable to change the medium more frequently (3×/week), to increase the volume of the medium in the flask or to switch to low tyrosine-containing medium (i.e., 100 μM or less).

### 1.1.4. Differentiation

The expression of differentiated functions and proliferation are not mutually exclusive in melanocytes. Melanocytes isolated from newborn or adult skin are already fully differentiated as far as expression of several pigmentation genes, such as tyrosinase, the tyrosinase-related proteins Tyrp1 and Dct (previously known as TRP1/gp75 and TRP2, respectively), silver/Pmel 17 and P-protein. Furthermore, the growth factors, especially M/SCF, ET-1, and MSH, enhance the expression of genes controlling pigmentation and the production of melanin [Kwon et al., 1987; Halaban et al., 1993b; Tada et al., 1998]. The growth factors can also increase pigmentation of adult skin when injected subcutaneously, as shown for M/SCF [Costa et al., 1996; Grichnik et al., 1998], or when taken orally, as in the case of MSH [Lerner and McGuire, 1961; Hadley et al., 1993; Hadley et al., 1998]. Studies with murine or avian neural crest derived cells demonstrated that TPA, bFGF, M/SCF, endothelin-3, and HGF/SF have also a profound effect on differentiation of melanocyte progenitors; that is, melanoblasts into melanocytes [Stocker et al., 1991; Sherman et al., 1993; Reid et al., 1995; Lahav et al., 1996; Guo et al., 1997; Lahav et al., 1998; Ito et al., 1999; Kos et al., 1999]. These agents are likely to participate in the proliferation and differentiation of human melanoblasts as well, because inherited disorders that suppress the production of a specific ligand or disable the respective receptor are the cause of reduced pigmentation. The classical example is piebaldism, a condition characterized by white patches devoid of melanocytes due to inactivation mutation in the KIT receptor [reviewed by Spritz, 1994]. Likewise, mutations in the endothelin-B receptor (EDNRB) and ET-3 have been identified in patients with Hirschsprung's disease (HSCR) and Waardenburg syndrome (WS), disorders involving neural crest-derived cells associated with piebaldism [Puffenberger et al., 1994; Attie et al., 1995; Amiel et al., 1996; Edery et al., 1996].

### 1.1.5. Aging

Normal human melanocytes, unlike melanoma cells from advanced lesions (or normal mouse melanocytes), have a definite life span in culture. As much as the chemical supplemented medium (TICVA) is a convenient and relatively inexpensive way of growing large quantities of human melanocytes, it accelerates the onset of growth arrest and senescence. In this medium melanocytes show signs of senescence during the fifth passage in culture [Medrano et al., 1994; Halaban et al., 2000a]. The peptide growth factor supplemented medium (bFGF, HGF/SF, and ET-1), facilitates a faster growth rate and a longer life span compared with TICVA-supplemented medium, although fibroblasts are also stimulated by the presence of bFGF and can rapidly overgrow the melanocytes (see below how to eliminate fibroblasts).

The observation that TPA stimulates melanocyte proliferation paved the way for the establishment of melanocytes from other species as well, such as mice and chicken [Boissy et al., 1987; Tamura et al., 1987]. Unlike their human counterparts, melanocytes from murine or avian origin become immortal [Dotto et al., 1989]. These melanocytes maintain all melanocytic characteristics, are stable, grow easily in culture with a single growth factor, and serve as a continuous supply of normal melanocytes. Furthermore, the vast repertoire of transgenic and knockout mice is a rich resource for mutant melanocytes to study cancer-related genes [see for example Halaban et al., 1998).

### 1.1.6. Sources of Normal Melanocytes

The commonest source is newborn foreskin, which is continuously available, and easily accessible with no restrictions by human investigative committees, as it would otherwise be discarded. The foreskins provide melanocytes with relative long life span, up to 18 passages in culture. Other possibilities are adult skins removed for cosmetic or health reasons. In our experience, skins from adults below age 30 are adequate and can generate melanocytes with short life span, such as five passages [Lerner et al., 1987; Abdel-Malek et al., 1994]. Skin from individuals older than 30 years of age could generate melanocytes that last for about two passages, which then stop proliferating [Lerner et al., 1987].

The normal melanocytes exist in the basal layer of the epidermis interspersed among the keratinocytes at a ratio of ∼1:15 to 1:30 melanocytes:keratinocytes, depending on the body site. Proper trimming of the tissue and enzymatic digestion are critical for dissociating the melanocytes and increase yield. The dermal tissue has to be removed so digestion enzymes can diffuse easily into the dermal/epidermal junction. There should be a good balance of trimming, otherwise the tissue is over or under-digested. The enzymatic digestion preferably splits the epidermis from the dermis at the basal layer in such a way that the melanocytes remain attached to the epidermis when the two sections are peeled apart. Once this is accomplished, the epidermis can be exposed to a brief treatment with trypsin/EDTA solution, after which the melanocytes are easily detached into a cell suspension and plated culture flasks.

### 1.1.7. Elimination of Other Skin Cells

The primary cultures can contain keratinocytes and fibroblasts, in addition to melanocytes. Keratinocytes do not pose a problem, as they become differentiated and growth-arrested in the presence of high calcium. Most of the times, the keratinocytes do not survive the first round of trypsinization. Furthermore, keratinocytes round up in response to TPA and do not attach to the culture flask. Fibroblasts, however, proliferate in the TPA-supplemented medium, even in the presence of CT. They particularly flourish in the peptide growth factor supplemented medium, in response to bFGF. Because fibroblasts assume different shapes, and in mixed cultures contain melanin due to transfer from melanocytes, they can be easily mistaken for melanocytes. Furthermore, the presence of fibroblasts suppresses the growth of melanocytes [unpublished observations].

Treatment with geneticin (G418) is the best way to eliminate fibroblasts from melanocyte cultures. It was discovered serendipitously that melanocytes are more resistant to G418 (100 μg/ml) during a short-term exposure (2 days) [Halaban and Alfano, 1984]. This treatment is particularly effective in TICVA-supplemented medium. Unfortunately, melanocytes are more susceptible to G418 when grown in peptide growth factor-supplemented medium. The elimination of fibroblasts by geneticin is not immediate but becomes apparent only 4–5 days after treatment. In fact, at the end of the two-day treatment, the cultures seem unaffected. The cultures should be inspected for leftover fibroblasts 7–10 days after the G418 treatment, as very often, one or two additional cycles of exposure to G418 is needed. A good test for surviving fibroblasts is to grow a portion in medium without growth factors. The cells that survive and proliferate under these conditions are likely to be fibroblasts.

## 1.2. Melanocytes from Nevi, Primary, and Advanced Lesions

As in the case of foreskins, the key factors for growing melanocytes from various lesions are: a) proper dissociation; b) optimal balance of growth factors; and c) elimination of fibroblasts. However, paradoxically the procedures to accomplish the three conditions are not as well-defined as for normal melanocytes derived from foreskins. Mechanical and enzymatic dissociation is particularly critical for nevi and primary melanoma cells because frequently the cells are embedded in the stroma and are not easily accessible. Cells that remain in chunks of tissues do not attach to the culture flasks and eventually die.

Melanocytes from nevi, such as congenital or acquired nevi, require the full complement of growth factors listed for normal melanocytes. These melanocytes also possess finite life span. Proliferation is improved in the peptide-growth factor-supplemented medium because the TICVA induce an early growth arrest.

The acquisition of independence from external growth factors, characteristic to melanoma cells from advanced lesions, is not accomplished in a single step. Our experience has shown that melanocytes from early primary thin melanomas (up to Breslow depth $\sim$1 mm) are similar to normal melanocytes and nevus cells in that they require the full spectrum of growth factors for proliferation in culture and they resemble normal melanocytes in their morphology and high levels of pigmentation (Fig. 12.3). Melanocytes from slightly more advanced primary tumors (Breslow depth $\sim$2 mm) still require at least one exogenous mitogen (such as agents that elevate intracellular cAMP concentrations) (Fig. 12.4). Transformed melanocytes cultured from invasive primary tumors (nodular melanomas) or from metastases, with some exceptions, proliferate vigorously when supplied with serum alone. Most of the cultured melanoma cells used in laboratory studies are derived from tumors at these advanced stages. However, the in vitro growth of cells from these lesions sometimes requires the addition of growth factors, especially during the early stages of establishment in culture.

The tailoring of growth factor supplements to melanoma cells is still an "art form" because different molecular changes that lead to transformation may impose different requirements. The precise identification of optimal conditions is hampered, in large part, by the scarcity of the tissue and the need to use most of it for pathological assessment. Under ideal circumstances, the cells should be distributed immediately into media supplemented with different combination of growth factors. If left unattended, the melanocytes from early lesions die before the growth factor requirement has been established.

The TICVA medium is not the best choice for culturing primary or metastatic melanoma cells. Paradoxically, various components, especially TPA, can inhibit growth [Halaban et al., 1986]. Among the peptide growth factors, M/SCF was also shown to inhibit the proliferation of melanoma cells that express the cognate KIT receptor

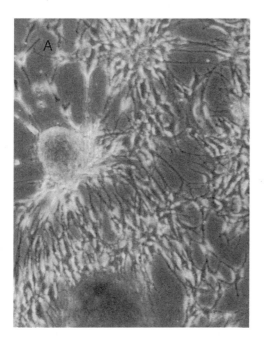

A

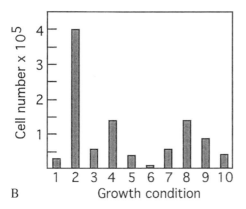

B

**Figure 12.3.** Primary melanoma cells WW295 in culture. A. Phase microscope photograph of melanocytes isolated from the superficial component of primary melanoma (0.82 mm) and cultured with multiple growth factors (bFGF, HGF/SF, ET-1, dbcAMP and heparin) for 2 weeks. Pigmentation can be easily observed in the proliferating melanocytes in the foci of piled up cells. B. Histogram showing that the primary melanoma cells WW295 required several growth factors for optimal proliferation. Cell proliferation was tested in PC1 chemical defined medium (Hycor Biomedical Inc. Portland, ME) [Böhm et al., 1995] supplemented as follows: 1: no further additions; 2) bFGF, heparin, HGF/SF, ET-1, dbcAMP, Na$_3$VO$_4$ and cholera toxin; 3: bFGF, heparin, HGF/SF, dbcAMP; 4: bFGF, heparin, HGF/SF; 5: bFGF, heparin; 6: HGF/SF; 7:HGF/SF, dbcAMP; 8: bFGF, heparin, dbcAMP; 9: M/SCF, dbcAMP; or 10: Hams' F10 supplemented with serum, TPA, IBMX, Pituitary extract (40 µg/ml) and Na$_3$VO$_4$. Concentrations of individual components are as given in tables 1 and 2. The cells were grown for 4 days under the experimental condition with medium change after 2 days, detached by trypsinization, and counted in the Coulter counter.

Culture of Melanocytes from Normal, Benign, and Malignant Lesions

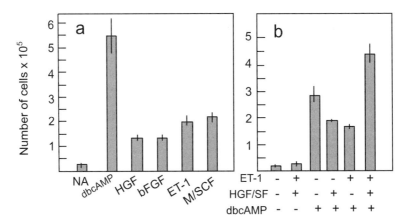

**Figure 12.4.** Growth response of melanocytes from primary melanomas WW165. Melanocytes were cultured from the more advanced primary melanoma lesion (WW165, 2.25 mm).
Proliferation assays were performed by incubating the cells in 6 cluster-well plates ~80,000
cells/well) in PC1 defined medium plus the indicated growth factors for 8 days and number of
cells determined with the Coulter counter. The concentration of each component was as listed
in tables I and II. The histogram shows that WW165 proliferate well in the presence of only
dbcAMP as a supplement. Each value is an average of two wells and bars indicate standard
deviation of the mean.

[Zakut et al., 1993; Huang et al., 1996]. HGF/SF, ET-1 and IBMX or
dbcAMP are the best combination when in doubt. As mentioned
above, bFGF should be used with caution, because it induces the
proliferation of fibroblasts. The latter might be confused with melanoma cells that often become amelanotic and appear fibroblastic. In
addition, fibroblasts suppress the proliferation of melanoma cells from
early lesions due to the production of interleukin-6 [Cornil et al., 1991;
Lu et al., 1992; Lu and Kerbel, 1993].

Fibroblasts can be eliminated from melanocytes cultured from nevi
and, to some degree, from early primary lesions by short exposure (2
days) to G418 supplemented to TICVA medium. However, this procedure cannot be used for advanced primary and especially metastatic
lesions because the malignant cells are no longer resistant to G418.
One way to overcome the presence of fibroblasts is essentially to allow
the "natural selection" of malignant cells. The fibroblasts are contact-
inhibited and have a limited life span, whereas the melanoma cells are
not contact inhibited and are likely to be immortal or to possess prolonged life span. Therefore, maintaining the cultures without splitting
into fresh culture dishes can force the fibroblasts out. The melanoma
cells can be recognized in this mix culture population by their size,
shape, and presence of pigment, if the original tumor was pigmented.
In this respect, medium with high tyrosine can enhance pigment formation, aiding the recognition of melanoma cells.

## 1.3. Validation of the Melanocytic Origin

Under most circumstances it is very easy to recognize normal melanocytes and malignant cells from early lesions, on the basis of melanin production. Highly pigmented cells contain brown melanin when observed under phase or light microscopy. However, very often the melanin is finely dispersed and cannot be ascertained by microscopical examination. An easy alternative is look for pigment in the cell pellet. However, pigmentation is dependent on tyrosinase activity, which is often down-regulated in melanoma cells [Halaban et al., 1997; Halaban et al., 2001a; Halaban et al., 2001b]. Pigmentation can be enhanced in amelanotic cells by increasing the tyrosine concentration in the medium, adjusting the extracellular pH to 7.4–7.8 and/or adding the tyrosinase co-factor DOPA (dihydroxyphenylalanine, 50 µM) to the culture medium [Halaban et al., 2001a; Halaban et al., 2001b]. Alternatively, the cells can be stained for melanocytic markers, such as gp100 (also known as silver or Pmel 17), Dct/TRP2, MITF (microphthalmia transcription factor), melan A and others [Blessing et al., 1998; Kaufmann et al., 1998; Orosz, 1999; Orchard, 2000; Busam et al., 2001; Clarkson et al., 2001; Juergensen et al., 2001; King et al., 2001; Koch et al., 2001; Miettinen et al., 2001; Zavala-Pompa et al., 2001]. Low levels of tyrosinase can also be detected, even in amelanotic melanoma [Halaban et al., 1997; Halaban et al., 2000b; Halaban et al., 2001a]. Western blotting or immunofluorescence can be applied to detect melanocyte-specific markers by using pigmented normal melanocytes and fibroblasts or any other non-melanocytic cell type as positive and negative controls, respectively.

## 2. PREPARATION OF STOCK SOLUTIONS AND MEDIA

### 2.1. Stock Solutions

#### 2.1.1. Chemical Growth Factors

**TPA** (12-$O$-tetradecanoyl phorbol-13-acetate, 616.8 d): Prepare $2000 \times 85$ µM stock solution (100 µg/ml) in dimethyl sulfoxide (DMSO). TPA in DMSO is sterile, and there is no need to pass through a filter. Keep at $-20°$C (not in self-defrost freezer).

**IBMX** (222.3 d): Prepare 10 mM solution (0.556 g/450 ml in double-distilled $H_2O$), autoclave, cool to $\sim 60°$C, then add 50 ml of $10\times$ basic medium (without serum or antibiotics) to make $50\times$ solution. Keep at room temperature in the dark. IBMX crystallizes if refrigerated, and your final concentration will not be the 0.1 mM required for growth.

**dbcAMP** (491.4 d): Prepare 0.1 M stock solution (1 g in 20 ml $H_2O$) and filter-sterilize. Keep at $-20°$C (not in self-defrost freezer).

**Cholera toxin** (84 kd): Prepare 2.5 µM stock (1000×). Add 4.8 ml distilled sterilized water into 1 mg vial cholera toxin (CT). Do not be filter because CT is supplied sterile. Keep at 4°C.

**Na₃VO₄** (183.9 d): Prepare 1 mM stock in $H_2O$ (18.4 mg/100 ml), filter-sterilize. Keep at room temperature.

### 2.1.2. Peptide Growth Factors

To preserve the integrity and stability of peptide growth factors, dilute in sterile phosphate buffered saline without $Ca^{2+}$ and $Mg^{2+}$ (PBSA) supplemented with 0.5% bovine serum albumin (BSA) and 0.1% CHAPS (a nondenaturing zwitterionic detergent sterilized by filtration). Most growth factors are provided "tissue culture-ready;" that is, they are sterile and there is no need for further filtration. For convenience, all the peptide growth factors are prepared as 1000× stock solutions. Keep at −20°C (not in self-defrost freezer).

**bFGF:** Fibroblast growth factor-basic, human, recombinant; 10 µg/ml.
**Heparin:** 1 µg/ml ($H_2O$).
**ET-1:** Endothelin 1, human, or porcine; 10 µM stock solution
**HGF/SF:** Hepatocyte growth factor/Scatter factor, human, recombinant, (Sigma, Cat. # H 1404); 40 µg/ml
**M/SCF:** Stem cell factor/mast cell factor, 100 µg/ml

### 2.1.3. Digestion Enzymes

**Trypsin 0.25%:** Aliquot 20 ml of 2.5% stock trypsin solution, and bring up to 200 ml with S-MEM ($Ca^{2+}$-$Mg^{2+}$-free cell suspension medium).
**Trypsin 0.25% + 2 mM EDTA:** Prepare 100 mM stock solution of EDTA. Add 2 ml of the 100 mM stock EDTA to 200 ml of 0.25% trypsin in S-MEM. Alternatively, use Gibco ready-made trypsin/ EDTA solution. Keep refrigerated at 4°C.
**Trypsin/TPA:** 0.25% trypsin in S-MEM supplemented with 50 ng/ml TPA and antibiotics (2.5 µg/ml Fungizone, 100 U/ml penicillin and 100 µg/ml streptomycin).
**Dispase:** Prepare at 25.4 mg/ml in 635 mg/25 ml aliquots in low $Ca^{2+}$ Eagle's MEM (S -MEM). Keep at −20°C (not in self-defrost freezer).
**Collagenase:** Final concentration, 200 U/ml. Example: If collagenase stock is 144 U/mg, use 265.3 mg/200 ml of S-MEM. Note the concentration of collagenase (units/mg) varies from lot to lot, therefore it is necessary to recalculate the amount needed. Sterilize by filtration. Keep at −20°C (not in self-defrost freezer).
**Dispase/collagenase:** Dilute Dispase 1:25 in collagenase solution.

### 2.1.4. Other Solutions

**FBS + 9.5% DMSO:** In a 15 ml sterile test tube, aliquot 10 ml FBS and add plus 95 µl DMSO. Mix well.

**Geneticin (G418):** Stock solution, 200 mg/ml. Notice that the specific activity depends on the lot. Example: If Geneticin is supplied at 1/2 activity, weigh out 400 mg/ml to generate stock solution of 200 mg/ml of 2× distilled water. Sterilize by filtration. Keep at −20°C (not in self-defrost freezer).

**70% EtOH:** Mix in 1 L bottle 700 ml absolute ethanol plus 300 ml $H_2O$. Keep at room temperature.

**PBSA:** Phosphate-buffered saline (without $MgCl_2$ or $CaCl_2$) 2.68 mM KCl, 1.47 mM $KH_2PO_4$, 136.8 mM NaCl, 8.0 mM $Na_2HPO_4$ $7H_2O$. Adjust pH to 7.0 to 7.2 with HCl, $H_2O$ to 1000 ml. Autoclave for 30 min. Store at room temperature or 4°C. PBSA is also commercially available as ready-made solution from Gibco.

**Dissection PBS:** PBSA with 2.5 µg/ml Fungizone, 100 U/ml penicillin and 100 µg/ml streptomycin.

## 2.2. Preparation of Media

### 2.2.1. Media Additives for Normal Human Melanocytes

The potency of chemical and peptide growth factors decay with time; therefore, prepare medium to be used within 7–10 days. Alternatively, supplement the complete medium that has been sitting in the refrigerator for more than a week with half the amount of TPA needed for final concentration, because it is the most sensitive component in the TICVA medium. Tables 12.1 and 12.2 are examples, and changes should be made according to the final volume of basic medium used. Feed melanocytes 2–3×/week.

### 2.2.2. Basal Medium for Melanomas

This is the routine medium for melanoma cells from metastatic lesions. Example of components is given in Table 12.3.

**Table 12.3. Components of basal medium.**

| Reagent | Amount (concentration) |
| --- | --- |
| Ham's F10 (or Ham's F12) | 400 ml |
| Glutamine | 4 ml |
| Pen/strep | 4 ml |
| Calf serum | 28 ml (7%) |
| Fetal bovine serum | 20 ml (2%), add only if cells do no look good |

## 2.3.  Tissue Preparation

Melanocytes can be obtained from neonatal human foreskins. Foreskins (and regular skins) should be immersed immediately after excision in sterile collection medium (any base medium), supplemented with antibiotics penicillin–streptomycin (Pen/Strep) without serum. The nursery in the hospital should be supplied with several 15 ml tubes containing ~5 ml of sterile medium. These tubes can be kept at room temperature adjacent to the operating bench for the convenience of the person involved in the operation. However, once excised, the foreskins, each in individual tube, should be kept in the refrigerator until use. Do not freeze.

Then the foreskins (and regular skins) can be digested with Dispase (the preferred method) or trypsin overnight at 4°C. The following is the general method of handling the skins, following with either the Dispase or trypsin treatment, splitting epidermis from the dermis and culturing the cells.

**Melanocytes from Neonatal Human Foreskins**

### Protocol 12.1.   Digestion of Normal Human Foreskins with Dispase

#### Reagents and Materials
*Sterile*
- ❑  Collection medium (see Section 2.3, above)
- ❑  Resuspension medium: Ham's F10 + 3% FBS
- ❑  TICVA medium: see Table 12.1
- ❑  Dissection PBS: PBSA with 2.5 µg/ml Fungizone, 100 U/ml penicillin and 100 µg/ml streptomycin
- ❑  Dispase (see Section 2.1.3)
- ❑  Trypsin/EDTA: 0.25% trypsin in 0.3 mM EDTA in PBSA
- ❑  70% ethanol (EtOH)
- ❑  Petri dishes, 10 cm
- ❑  Flasks, 75 cm$^2$
- ❑  Sterile tubes, 15 ml
- ❑  Forceps
- ❑  Spring scissors

#### Protocol
(a)  Use fresh newborn foreskins kept in collection medium at 4°C for up to 24 h.

(b)  Prepare a 10 cm (55 cm$^2$) Petri dish with ~10 ml of 70% EtOH, and prepare several Petri dishes with ~4 ml sterile dissection PBS (1 to 2 foreskins or adult skin/dish).

(c) Using sterile forceps, transfer and immerse the foreskins in the 70% EtOH Petri dish. Make sure all surfaces of the skins are covered and transfer the skins to the dissection PBS dishes for dissection.

(d) Trim off dermis under dissecting microscope or magnifying glass by using sterile forceps and spring scissors. Keep instruments sterile by immersing in 70% EtOH. Dip in 70% ethanol and shake off surplus between specimens.

(e) In cell culture hood, transfer to 6 cm Petri dishes, epidermis side up.

(f) Add 3 ml Dispase/6 cm dish, and stretch the skin by using forceps ensuring that the skin is flat with edges uncurled. This allows the Dispase to penetrate evenly.

(g) Incubate at 4°C overnight.

(h) Separate the epidermis from the dermis, and discard the dermis.

(i) Place 2–3 pieces of epidermis in 3 ml solution of trypsin/EDTA in a 15 ml sterile tube, and incubate for 15 min at 37°C.

(j) Add 3 ml of resuspension medium to the tube, and gently disperse the cells by shaking until the solution becomes cloudy.

(k) Remove the remainder of the epidermis with a pipette and discard.

(l) Spin in a bench-top centrifuge at 100–200 $g$ for 5 min to pellet the suspended cells.

(m) Remove supernatant carefully by suction without disturbing the pellet, and resuspend the cell pellet in 5 ml TICVA medium (Table 12.1).

(n) Triturate well with a 10 ml pipette, and transfer the dissociated cells from 2–3 foreskins into a 75 cm$^2$ flask.

(o) Add 5 ml TICVA medium and incubate at 37°C in a $CO_2$ incubator (5% $CO_2$).

(p) On the following day, add 2 ml of TICVA to the flask and triturate well; notice that cell aggregates float and do not attach. The efficiency of cell attachment is improved when these aggregates are dissociated.

(q) Change the medium 3 × /week.

## Digestion of Skins with Trypsin

The trypsin digestion method is similar to Protocol 12.1, with slight modification.

## Protocol 12.2.  Digestion of Human Foreskins with Trypsin

### Reagents and Materials

*Sterile*

- ❑ Collection medium (see Section 2.3, above)
- ❑ Resuspension medium: Ham's F10 + 3% FBS
- ❑ TICVA medium: see Table 12.1
- ❑ Dissection PBS: PBSA with 2.5 μg/ml Fungizone, 100 U/ml penicillin and 100 μg/ml streptomycinrypsin/TPA (see Section 2.1.3)
- ❑ Trypsin/EDTA: 0.25% trypsin in 0.3 mM EDTA in PBSA
- ❑ 70% ethanol (EtOH)
- ❑ Petri dishes, 10 cm, 6 cm
- ❑ Flasks, 75 cm$^2$
- ❑ Sterile tubes, 15 ml
- ❑ Forceps
- ❑ Spring scissors

### Protocol

(a) Follow steps (a–d) as above (Protocol 12.1).

(b) In Step (e), add 2.5 ml solution of trypsin/TPA.

(c) Incubate overnight at 4°C (in a refrigerator).

(d) Prepare 15 ml sterile tubes with 3 ml of TICVA medium.

(e) Using fine forceps, separate dermis from epidermis, put both tissues into the 15 ml test tube containing ~3 ml TICVA medium, and Vortex for 1 min at top speed to dissociate cells. Epidermis and dermis is used because the melanocytes are present on both tissues.

(f) Remove large pieces of tissues by suction with a glass pipette.

(g) Seed directly into 25 cm$^2$ or larger flasks, depending on the number of foreskins. using this method fibroblasts are also included. They have to be eliminated at a later time.

### Melanocytes from Adult Shave Biopsy

Using 1 or 2 cm$^2$ shave biopsies, incubate without trimming (unless thick) for 1–2 h at 37°C or overnight in Dispase as in Protocol 12.1, depending on the thickness and size of the skin. Proceed through steps (i)–(l) as in Protocol 12.1. Adult skins are usually kept in individual flasks. Size of flask depending on the size of the biopsy.

## Protocol 12.3.  Melanocytes from Nevi and Primary Melanomas

### Reagents and Materials

As for Protocol 12.1 plus the following

*Sterile*

❑ Dispase/collagenase (see Section 2.1.3)

### Protocol

(a) Follow steps (a)–(f) as in Protocol 12.1. The Dispase treatment can be performed for a short period at 37°C in case the tissue is small. If tissue is large, incubate overnight in a refrigerator.

(b) Proceed with step (g), but do not discard the dermis. Dissociate the cells by shaking both tissues, because a large portion of melanocytes may reside in the dermis. Remove the dermis and epidermis, and place in a 6 cm fresh Petri dish.

(c) Harvest the loose cells by centrifugation, and seed in flasks with medium supplemented with growth factors (see Tables 12.1, 12.2).

(d) Wash the remaining tissue with PBSA, add 1–2 ml Dispase/collagenase solution for about 1 h at 37°C. The exact timing depends on the size of the tissue. Transfer to a 15 ml test tube, and gently agitate it over a vortex. At the end of an adequate digestion, the tissue is mushy and well dissociated.

(e) Add 9 ml PBSA to the test tube, and centrifuge for 10 min. Make sure cells are sedimented at the bottom of the tube. The collagenase treatment causes viscosity, and sometimes further dilution with PBSA in 50 ml sterile tube is needed to allow sedimentation. This is a critical and sometimes tricky step, but final yield is dependent on proper digestion and sedimentation of dissociated cells.

(f) Remove Dispase/collagenase gently without agitating the pellet, resuspend in growth medium, and seed in flasks.

(g) The following day, triturate the medium in the flasks vigorously for at least 10 times with a 5 ml or 10 ml pipette.

(h) The suspended cells in clumps should be dissociated to settle down. Keep triturating for four consecutive days, adding a little bit of medium on the second day (0.5 ml). In some cases, the suspended clumped cells have to be centrifuged and re-suspended in fresh medium to settle down.

### Protocol 12.4.  Melanoma Cells from Advanced Lesions

### Reagents and Materials

*Sterile*

❑ Dissection PBS or medium without serum
❑ Growth medium (see Tables 12.1, 12.2)
❑ Petri dishes, 10 cm
❑ Centrifuge tube, 15 ml
❑ Culture flasks, 25 cm²

## Protocol

(a) Specimens should be placed in sterile saline or medium without serum as above, immediately after surgery, and processed for cell culture as soon as possible.

(b) Place the specimens and medium in a 10 cm Petri dish, and tease the tissue with fine sterile forceps. Malignant cells ooze out very often, clouding the medium. Their presence can be verified when observed under the microscope.

(c) Collect the dissociated cells into a sterile tube, centrifuge, and plate in culture flasks.

(d) Treat the remaining intact tissue with collagenase at 37°C for 1–3 h until digested. Test for digestion every 30 min to 1 h by transferring the tissue to a test tube and shaking the tube over a vortex until the medium becomes turbid.

(e) Plate in fresh flask.

## Protocol 12.5. Elimination of Fibroblasts by Geneticin (G418) Treatment

### Reagents and Materials

❑ Melanocyte medium: TICVA or peptide growth factor supplemented medium (see Table 12.1, 12.2)
❑ Geneticin, 200 mg/ml
❑ Melanocyte medium with Geneticin, 100 μg/ml (100 μl Geneticin, 200 mg/ml in 200 ml medium)

### Protocol

(a) Incubate normal human melanocytes in melanocyte medium (TICVA or peptide growth factor supplemented medium) containing 100 μg/ml Geneticin for 2–3 days.

(b) At the end of 2 day's incubation, replace with regular medium and continue with regular feeding schedule.

(c) Observe under a microscope for residual fibroblasts; if any remain, repeat the treatment 10–14 d later.

## Protocol 12.6. Maintaining Melanocyte Cultures

### Reagents and Materials

❑ Melanocyte medium (see Table 12.1, 12.2)
❑ Trypsin/EDTA (see Section 2.1.2)

### Protocol

(a) Feed the cells every 3 days, preferably 3× /week.

(b) Trypsinize the cultures when confluent to prepare for a split at a ratio of 1:3.

(i)   Remove the entire medium by suction; add ~1–3 ml of trypsin/EDTA solution, depending on the size of the flask, to cover the surface.

(ii)  Rock the flask gently for few seconds just to make sure that the solution is well-spread (avoid the neck of the flask).

(iii) Remove all the trypsin/EDTA solution by suction.

(iv)  Let sit for 5–10 min, shake the flasks gently to test for detachment. Confirm cell detachment by looking at the culture under the microscope.

(c)  Add medium, triturate (~5 times), and transfer to a larger flask or any other culture dish. Keep one-third of the cells in the original flask; it is still "good" for cell culture. Notice that there is no need to centrifuge the cells to get rid of the trypsin/EDTA solution. However, make sure that all the trypsin/EDTA solution is removed before the cells are detached. Leftover EDTA in the growth medium may kill the cells.

*Note:* Melanocytes should be kept at 30% confluence. Sparsely seeded cells do not recover well from the splitting.

(d)  Freeze cells in 0.5–1 ml fetal bovine serum (FBS) plus 95 µl DMSO. Place on ice for 30 min, and transfer to −80°C freezer. After 1 week, frozen cells can be transferred to liquid nitrogen (−175°C).

## 2.4.  Variations on the Method

Various investigators use a slight modification in the components described above for normal human melanocyte culture. For example, some investigators use MCDB 153 as the basic medium with 5% heat-inactivated FBS, bFGF (0.6 ng/ml), TPA (8 nM), and bovine pituitary extract (BPE, 13 µg/ml) as growth supplements [Medrano and Nordlund, 1990]. BPE was originally included as the source for bFGF, but its presence is no longer required when a pure source of bFGF is available. Another alternative is a similar medium in which TPA and BPE are replaced by endothelin-1 (1 nM) and alpha-MSH (1 nM) [Swope et al., 1995].

## 3.  SAFETY PRECAUTIONS

There is nothing inherently dangerous in handling melanoma cells, as the cause of malignant transformation is not of viral origin. Common sense has to be applied in handling human tissues. This includes wearing gloves and using instruments to transfer tissues from one dish to another. Immersing the foreskin in 70% ethanol before any further manipulations is likely to eliminate viruses and any other superficial contaminants.

Although TPA is known as tumor promoter due to its enhancement of skin cancer in mouse skin after treatment with a carcinogen, by itself it cannot cause any malignant transformation. In over 25 years of research with normal human melanocytes, no one so far has generated even an immortal line after prolonged exposure to TPA. Nevertheless, TPA should be handled with gloves, as one's exposure to unknown carcinogens in the environment cannot be ruled out. The use of gloves is also recommended while preparing TPA and cholera toxin and handling normal human melanocytes in culture.

The routine feeding of cells should be in laminar flow biohazard hoods (Class II). All surfaces should be wiped out with 70% ethanol in the beginning and the end of the workday. The use of individually wrapped pipettes reduces the chances for contamination. Media should be removed from flasks by suction into a large Erlenmeyer flask separated from the vacuum system by an additional small Erlenmeyer flask. The used medium removed from flasks should be decontaminated with Chlorox before its discharge into the sink. Rinse the Erlenmeyer flask, attach to the vacuum, and pass again some Chlorox to maintain sterility.

## 4. MAIN APPLICATIONS

Pure cultures of melanocytes from normal skin and various stages of transformation are a resource for basic and applied research, diagnostic, and clinical applications. It is not the scope of this chapter to describe all these applications in detail, but some examples are briefly listed.

### 4.1. Genetic Alterations

The melanoma cells from various lesions have been used for chromosomal analysis, which resulted in the identification of chromosome 9p21 as the site of melanoma tumor suppressor [Cowan et al., 1986; Cowan et al., 1988; Cannon-Albright et al., 1992; Holland et al., 1994; Walker et al., 1994]. Subsequently, the INK4a locus (also known as CDKN2, MTS1), encoding the CDK (cyclin dependent kinase) inhibitor (CKI) p16$^{INK4a}$, has been identified as one of the genes residing in chromosomal band 9p21 [Kamb et al., 1994a; Kamb et al., 1994b]. With the advance in molecular genetic techniques, the normal and malignant melanocytes are currently a critical source for mutational analysis and differential gene expression by using cDNA microarray [Bittner et al., 2000; Su et al., 2000]. Although fresh tumors and paraffin-embedded tissues are also used, the pure culture is important in validating changes that occur in the malignant cells versus the stroma. Its is hoped that molecular characterization of various melanomas may indicate a pattern that can be used in diagnosis and treatment.

## 4.2. Signal Transduction and Autonomous Growth

The normal melanocytes are routinely used as the "yard stick" for evaluating malignant changes in melanoma cells. For example, the knowledge of cell surface receptors involved in normal melanocyte proliferation served to identify bFGF as the aberrantly produced ligand in melanoma cells that confers autonomous growth [Halaban et al., 1988a]. It was shown that the synergistic effect of growth factors serves to elevate and maintain the levels of signal transducers, such as the MAPK cascade, critical for regulating genes involved in cell cycle progression [Böhm et al., 1995]. Side-by-side analyses of normal and malignant melanocytes demonstrated that the regulators of cycle progression, a group of kinases collectively known as CDKs, are aberrantly active in melanoma cells leading to the inactivation of the retinoblastoma tumor suppressors pRb, p107, and p130. Components of this pathway are good targets for novel drug design, as was shown in the case of flavopiridol, the inhibitor of CDK activity [Halaban et al., 2000a].

## 4.3. In Vivo and In Vitro Models for Malignant Progression

Normal and malignant melanocytes have been used to develop models for malignant progression by using the skin equivalent procedure [Meier et al., 2000]. Alternatively, these cells have been used in in vivo models, such as xenografts in nude mice, to test various effects of therapeutic drugs, such as toxicity, modulation of specific gene expression, and suppression of receptor kinases [see for example Trotta and Harrison, 1987; Dykes et al., 1995; Plowman et al., 1995; Oku et al., 1998; Davol et al., 1999].

## 4.4. Melanoma Vaccines

Melanoma cells serve as a source for tumor-associated antigens, which can be used as vaccines for active immunotherapy. They were used in vitro for the large-scale expansion of melanoma-reactive TIL (tumor infiltrating lymphocytes) or directly as immunogen derived from autologous (AVAX dinitrophenyl-modified) tumor cells or allogenic melanoma cell lysates (Melacin) to boost patient response to melanoma [Kang et al., 1997; Maguire et al., 1998; Mitchell, 1998; Minev et al., 1999; Berd, 2000; Berd, 2001; Dillman et al., 2001].

## 4.5. Transplantation

Cultured normal human melanocytes have been used to re-pigment white patches of skin in vitiligo patients [Lerner et al., 1987; Lontz et al., 1994].

## ACKNOWLEDGMENT

I thank Ms. Donna LaCivita for carefully reading the manuscript and for helpful suggestions.

## SOURCES OF MATERIALS

| Item | Catalog no. | Supplier |
|---|---|---|
| Bovine serum | 100-506 | Gemini |
| BSA fraction V | 100 018 | Roche |
| CHAPS | 220201 | Calbiochem |
| Cholera toxin | C-3012 | Sigma |
| Collagenase, Type IA | C-9891 | Sigma |
| dbcAMP, $N^6,2'$-O-dibutyryladenosine 3:5-cyclic monophosphate | D-0627 | Sigma |
| Dispase, Dispase II | 165-859 | Roche |
| DMSO | D-8779 | Sigma |
| Donor horse serum | 100-508 | Gemini |
| EDTA | E6511 | Sigma |
| Endothelin 1, human, porcine (ET-1) | E-7764 | Sigma |
| FCS | 100-106 | Gemini |
| Fibroblast growth factor-basic (bFGF), human, recombinant | F-0291 | Sigma |
| Fungizone | 15295-017 | Gibco |
| Geneticin | 11811-031 | Gibco |
| L-Glutamine, 200 mM | | |
| Ham's F-10 (10 µM tyrosine) , powder | 81200 | Gibco |
| Ham's F12 medium (30 µM tyrosine), powder | 21700 | Gibco |
| Hepatocyte growth factor/scatter factor (HGF/SF), human, recombinant | H-1404 | Sigma |
| IBMX, 3-isobutyl-1- methyl xanthine | I-5879 | Sigma |
| Illuminated magnifying glass (for dissection, better than dissecting microscope) | | |
| M/SCF (see Stem cell factor) | 25030-081 | Gibco |
| Opti-MEM, reduced serum medium | 31985-070 | Gibco |
| Penicillin/streptomycin | 15140-122 | Gibco |
| S-MEM, minimal essential medium without calcium or magnesium for cell suspension, powder | 21800-081 | Gibco |
| Sodium orthovanadate, $Na_3VO_4$, | S-6508 | Sigma |
| Spring scissors, curved, Noyes 12 cm, or slightly curved, Westcott 11 cm | 15011-12, 15015 | Fine science Tools Inc |
| | 18-1516 | VWR |
| Stem cell factor/mast cell factor (M/SCF) | S-7901 | Sigma |
| TPA, 12-O-tetradecanoyl phorbol-13-acetate | P8139 | Sigma |

## SOURCES OF MATERIALS *(Continued)*

| Item | Catalog no. | Supplier |
| --- | --- | --- |
| TRP1 (C-16) goat polyclonal IgG against the carboxy terminus of TRP1 of human origin | sc-10445 | Santa Cruz |
| TRP1 (G-17) goat polyclonal IgG against an internal region of TRP1 of human origin | sc-10443 | Santa Cruz |
| TRP2 (D-18) goat polyclonal IgG against the amino terminus of TRP2 of human origin | sc-10451 | Santa Cruz |
| Trypsin, 0.25%, with 1 mM EDTA | 25200-056 | Gibco |
| Trypsin, 2.5% | 15090-046 | Gibco |
| Tyrosinase (C-19) goat polyclonal IgG against carboxy terminus of tyrosinase of human origin | sc-7833 | Santa Cruz |

## REFERENCES

Abdel-Malek, Z.A., Scott, M.C., Suzuki, I., Tada, A., Im, S., Lamoreux, L., Ito, S., Barsh, G., Hearing, V.J. (2000) The melanocortin-1 receptor is a key regulator of human cutaneous pigmentation [In Process Citation]. *Pigm. Cell Res.* 8:156–162.

Abdel-Malek, Z.A., Swope, V.B., Nordlund, J.J., Medrano, E.E. (1994) Proliferation and propagation of human melanocytes in vitro are affected by donor age and anatomical site. *Pigm. Cell Res.* 7:116–122.

Amiel, J., Attie, T., Jan, D., Pelet, A., Edery, P., Bidaud, C., Lacombe, D., Tam, P., Simeoni, J., Flori, E., Nihoul-Fekete, C., Munnich, A., Lyonnet, S. (1996) Heterozygous endothelin receptor B (EDNRB) mutations in isolated Hrischsprung disease. *Hum. Mol. Genet.* 5:355–357.

Attie, T., Till, M., Pelet, A., Amiel, J., Edery, P., Boutrand, L., Munnich, A., Lyonnet, S. (1995) Mutation of the endothelin-receptor B gene in Waardenburg-Hirschsprung disease. *Hum. Mol. Genet.* 4:2407–2409.

Berd, D. (2000) Melanoma vaccines as a therapeutic option. *South Med. J.* 93:439–40.

Berd, D. (2001) Autologous, hapten-modified vaccine as a treatment for human cancers. *Vaccine* 19:2565–2570.

Bhargava, M., Joseph, A., Knesel, J., Halaban, R., Li, Y., Pang, S., Goldberg, I., Setter, E., Donovan, M.A., Zarnegar, R., Michalopoulos, G.A., Nakamura, T., Faletto, D., Rosen E. (1992) Scatter factor and hepatocyte growth factor: activities, properties, and mechanism. *Cell Growth Diff.* 3:11–20.

Bittner, M., Meltzer, P., Chen, Y., Seftor, E., Hendrix, M., Radmacher, M., Simon, R., Yakhini, Z., Ben-Dor, A., Sampas, N., Dougherty, E., Wang, E., Marincola, F., Gooden, C., Lueders, J., Glatfelter, A., Pollock, P., Carpten, J., Gillander, E.S., Lejy, D., Dietrich, K., Beaudry, C., Berens, M., Alberts, D., Sondak, V., Hayward, N., Trent, J. (2000) Molecular classification of cutaneous malignant melanoma by gene expression profiling. *Nature* 406:536–540.

Blessing, K., Sanders, D.S., Grant, J.J. (1998) Comparison of immunohistochemical staining of the novel antibody melan-A with S100 protein and HMB-45 in malignant melanoma and melanoma variants. *Histopathology* 32:139–146.

Böhm, M., Moellmann, G., Cheng, E., Alvarez-Franco, M., Wagner, S., Sassone-Corsi, P., Halaban, R. (1995) Identification of p90[RSK] as the probable CREB-Ser[133] kinase in human melanocytes. *Cell Growth Diff.* 6:291–302.

Boissy, R.E., Moellmann, G.E., Halaban, R. (1987) Tyrosinase and acid phosphatase activities in melanocytes from avian albinos. *J. Invest. Dermatol.* 88:292–300.

Busam, K.J., Iversen, K., Coplan, K.C., Jungbluth, A.A. (2001) Analysis of microphthalmia transcription factor expression in normal tissues and tumors, and comparison of its expression with S-100 protein, gp100, and tyrosinase in desmoplastic malignant melanoma. *Am. J. Surg. Pathol.* 25:197–204.

Cannon-Albright, L.A., Goldgar, D.E., Meyer, L.J., Lewis, C.M., Anderson, D.E., Fountain, J.W., Hegi, M.E., Wiseman, R.W., Petty, E.M., Bale, A.E. et al. (1992) Assignment of a locus for familial melanoma, MLM, to chromosome 9p13–p22. *Science* 258:1148–1152.

Clarkson, K.S., Sturdgess, I.C., Molyneux, A.J. (2001) The usefulness of tyrosinase in the immunohistochemical assessment of melanocytic lesions: a comparison of the novel T311 antibody (anti-tyrosinase) with S-100, HMB45, and A103 (anti-melan-A). *J. Clin. Pathol.* 54:196–200.

Cooksey, C.J., Land, E.J., Ramsden, C.A., Riley, P.A. (1995) Tyrosinase-mediated cytotoxicity of 4-substituted phenols—quantitative structure thiol-reactivity relationships of the derived o-quinones. *Anti-Cancer Drug Des.* 10:119–129.

Cornil, I., Theodorescu, D., Man, S., Herlyn, M., Jambrosic., Kerbel, R.S. (1991) Fibroblast cell interactions with human melanoma cells affect tumor cell growth as a function of tumor progression. *Proc. Natl. Acad. Sci. USA* 88:6028–6032.

Costa, J.J., Demetri, G.D., Harrist, T.J., Dvorak, A.M., Hayes, D.F., Merica, E.A., Menchaca, D.M., Gringeri, A.J., Schwartz, L.B., Galli, S.J. (1996) Recombinant human stem cell factor (kit ligand) promotes human mast cell and melanocyte hyperplasia and functional activation in vivo. *J. Exp. Med.* 183:2681–2686.

Cowan, J.M., Halaban, R., Francke, U. (1988) Cytogenetic analysis of melanocytes from premalignant nevi and melanomas. *J. Natl. Cancer Inst.* 80:1159–1164.

Cowan, J.M., Halaban, R., Lane, A.T., Francke, U. (1986) The involvement of 6p in melanoma. *Cancer Genet. Cytogenet.* 20:255–261.

Davol, P.A., Garza, S., Frackelton, A.R. Jr. (1999) Combining suramin and a chimeric toxin directed to basic fibroblast growth factor receptors increases therapeutic efficacy against human melanoma in an animal model. *Cancer* 86:1733–1741.

Dillman, R.O., De Leon, C., Beutel, L.D., Barth, N.M., Schwartzberg, L.S., Spitler, L.E., Garfield, D.H., O'Connor, A.A., Nayak, S.K. (2001) Short-term tumor cell lines for the active specific immunotherapy of patients with metastatic melanoma. *Crit. Rev. Oncol. Hematol.* 39:115–123.

Dotto, G.P., Moellmann, G., Ghosh, S., Edwards, M., Halaban, R. (1989) Transformation of murine melanocytes by basic fibroblast growth factor cDNA and oncogenes and selective suppression of the transformed phenotype in a reconstituted cutaneous environment. *J. Cell Biol.* 109:3115–3128.

Dykes, D.J., Bissery, M.C., Harrison, S.D. Jr., Waud, W.R. (1995) Response of human tumor xenografts in athymic nude mice to docetaxel (RP 56976, Taxotere). *Invest. New Drugs* 13:1–11.

Edery, P., Attie, T., Amiel, J., Pelet, A., Eng, C., Hofstra, R.M., Martelli, H., Bidaud, C., Munnich, A., Lyonnet, S. (1996) Mutation of the endothelin-3 gene in the Waardenburg-Hirschsprung disease (Shah-Waardenburg syndrome). *Nat. Genet.* 12: 442–444.

Eisinger, M., and Marko, O. (1982) Selective proliferation of normal human melanocytes in vitro in the presence of phorbol ester and cholera toxin. *Proc. Natl. Acad. Sci. USA* 79:2018–2022.

Eisinger, M., Marko, O., Ogata, S., Old, L.J. (1985) Growth regulation of human melanocytes: mitogenic factors in extracts of melanoma, astrocytoma, and fibroblast cell lines. *Science* 229:984–986.

Funasaka, Y., Boulton, T., Cobb, M., Yarden, Y., Fan, B., Lyman, S.D., Williams, D.E., Anderson, D.M., Zakut, R., Mishima, Y., Halaban, R. (1992) c-Kit kinase induces a cascade of protein tyrosine-phosphorylation in normal human melanocytes in response to mast cell growth factor and stimulates mitogen-activated protein kinase but is down regulated in melanomas. *Mol. Biol.Cell* 3:197–209.

Grichnik, J.M., Burch, J.A., Burchette, J., Shea, C.R. (1998) The SCF/KIT pathway plays a critical role in the control of normal human melanocyte homeostasis. *J. Invest. Dermatol.* 111:233–238.

Guo, C.S., Wehrle-Haller, B., Rossi, J., Ciment, G. (1997) Autocrine regulation of neural crest cell development by steel factor. *Dev. Biol.* 184:61–69.

Hadley, M.E., Hruby, V.J., Blanchard, J., Dorr, R.T., Levine, N., Dawson, B.V., al-Obeidi, F., Sawyer, T.K. (1998) Discovery and development of novel melanogenic drugs. Melanotan-I and -II. *Pharm. Biotechnol.* 11:575–595.

Hadley, M.E., Sharma, S.D., Hruby, V.J., Levine, N., Dorr, R.T. (1993) Melanotropic peptides for therapeutic and cosmetic tanning of the skin. [Review]. *Ann. NY Acad. Sci.* 680:424–439.

Halaban, R. (1994) Signal transduction in normal and malignant melanocytes. *Pigm. Cell Res.* 7:89–95.

Halaban, R., and Alfano, F.D. (1984) Selective elimination of fibroblasts from cultures of normal human melanocytes. *In Vitro* 20:447–450.

Halaban, R., Cheng, E., Smicun, Y., Germino, J. (2000) Deregulated E2F transcriptional activity in autonomously growing melanoma cells. *J. Exp. Med.* 191: 1005–1015.

Halaban, R., Cheng, E., Svedine, S., Aron, R., Hebert, D.N. (2001) Proper folding and ER to Golgi transport of tyrosinase are induced by its substrates, DOPA and tyrosine. *J. Biol. Chem.* 276:11933–11938.

Halaban, R., Cheng, E., Zhang, Y., Mandigo, C.E., Miglarese, M.R. (1998) Release of cell cycle constraints in mouse melanocytes by overexpressed mutant E2F1$_{E132}$, but not by deletion of p16$^{INK4A}$ or p21$^{WAF1/CIP1}$. *Oncogene* 16:2489–24501.

Halaban, R., Cheng, E., Zhang, Y., Moellmann, G., Hanlon, D., Michalak, M., Setaluri, V., Hebert, D.N. (1997) Aberrant retention of tyrosinase in the endoplasmic reticulum mediates accelerated degradation of the enzyme and contributes to the dedifferentiated phenotype of amelanotic melanoma cells. *Proc. Natl. Acad. Sci. USA* 94:6210–6215.

Halaban, R., Ghosh, S., Baird, A. (1987) bFGF is the putative natural growth factor for human melanocytes. *In Vitro Cell. Dev. Biol.* 23:47–52.

Halaban, R., Ghosh, S., Duray, P., Kirkwood, J.M., Lerner, A.B. (1986) Human melanocytes cultured from nevi and melanomas. *J. Invest. Dermatol.* 87:95–101.

Halaban, R., Kwon, B.S., Ghosh, S., Delli Bovi, P., Baird, A. (1988) bFGF as an autocrine growth factor for human melanomas. *Oncogene Res.* 3:177–186.

Halaban, R., Langdon, R., Birchall, N., Cuono, C., Baird, A., Scott, G., Moellmann, G., McGuire, J. (1988) Paracrine stimulation of melanocytes by keratinocytes through basic fibroblast growth factor. *Ann. NY Acad. Sci.* 548:180–190.

Halaban, R., and Lerner, A.B. (1977) Tyrosinase and inhibition of proliferation of melanoma cells and fibroblasts. *Exp. Cell Res.* 108:119–125.

Halaban, R., Patton, R.S., Cheng, E., Svedine, S., Trombetta, E.S., Wahl, M.L., Ariyan S, Hebert D.N. (2001) Abnormal acidification of melanoma cells induces tyrosinase retention in the early secretory pathway. *J Biol Chem*–277:14821–8.

Halaban, R., Rubin, J.S., Funasaka, Y., Cobb, M., Boulton, T., Faletto, D., Rosen, E., Chan, A., Yoko, K., White, W., Cook, C., Moellmann, G. (1992) Met and hepatocyte growth factor/scatter factor signal transduction in normal melanocytes and melanoma cells. *Oncogene* 7:2195–2206.

Halaban, R., Rubin, W., White, W. (1993) *Met* and *HGF/SF* in normal melanocytes and melanoma cells. In Goldberg, I.D. (ed) *Hepatocyte Growth Factor-Scatter Factor and the C-Met Receptor.* Birkhäuser Verlag: Basel, Switzerland, 329–339.

Halaban, R., Svedine, S., Cheng, E., Smicun, Y., Aron, R., Hebert, D.N. (2000) Endoplasmatic reticulum is a common defect associated with tyrosinase-negative albinism. *Proc. Natl. Acad. Sci. USA* 97:5889–5894.

Halaban, R., Tyrrell, L., Longley, J., Yarden, Y., Rubin, J. (1993) Pigmentation and proliferation of human melanocytes and the effects of melanocyte-stimulating hormone and ultraviolet B light. *Ann. NY Acad. Sci.* 680:290–301.

Holland, E.A., Beaton, S.C., Edwards, B.G., Kefford, R.F., Mann, G.J. (1994) Loss of heterozygosity and homozygous deletions on 9p21-22 in melanoma. *Oncogene* 9:1361–1365.

Huang, S.Y., Luca, M., Gutman, M., McConkey, D.J., Langley, K.E., Lyman, S.D., Bar-Eli, M. (1996) Enforced C-Kit expression renders highly metastatic human melanoma cells susceptible to stem cell factor-induced apoptosis and inhibits their tumorigenic and metastatic potential. *Oncogene* 13:2339–2347.

Hunt, G., Todd, C., Cresswell, J.E., Thody, A.J. (1994) Alpha-melanocyte stimulating hormone and its analogue Nle4DPhe7 alpha-MSH affect morphology, tyrosinase activity and melanogenesis in cultured human melanocytes. *J. Cell Sci.* 107:205–211.

Im, S., Moro, O., Peng, F., Medrano, E.E., Cornelius, J., Babcock, G., Nordlund, J.J., Abdel-Malek, Z. A. (1998) Activation of the cyclic AMP pathway by alpha-melanotro pin mediates the response of human melanocytes to ultraviolet B radiation. *Cancer Res.* 58:47–54.

Imokawa, G., Yada, Y., Miyagishi, M. (1992) Endothelins secreted from human keratinocytes are intrinsic mitogens for human melanocytes. *J. Biol. Chem.* 267:24675–24680.

Ito, M., Kawa, Y., Ono, H., Okura, M., Baba, T., Kubota, Y., Nishikawa, S.I., Mizoguchi, M., (1999) Removal of stem cell factor or addition of monoclonal anti-c-KIT antibody induces apoptosis in murine melanocyte procursors. *J. Invest. Dermatol.* 112:796–801.

Juergensen, A., Holzapfel, U., Hein, R., Stolz, W., Buettner, R., Bosserhoff, A. (2001) Comparison of two prognostic markers for malignant melanoma: MIA and S100 beta. *Tumour Biol.* 22:54–58.

Kamb, A., Gruis, N.A., Weaver-Feldhaus, J., Liu, Q., Harshman, K., Tavtigian, S.V., Stockert, E., Day, R.S., R., Johnson, B.E., Skolnick, M.H. (1994) A cell regulator potentially involved in genesis of many tumor types [see comments]. *Science* 264:436–440.

Kamb, A., Shattuck-Reidens, D., Eeles, R., Liu, Q., Gruis, N.A., Ding, W., Hussey, C., Tran, T., Weaver-Feldhaus, J., McClure, M., Aitken, J.F., Anderson, D.E., Bergman, W., Frantz, R., Goldger, D.E., Green, A., MacLennan, R., Martin, N.G., Meyer, L.J., Youl, P., Zone, J.J., Skolnick, M.H., Cannon-Albright, A.A. (1994) Alnalysis of p16 gene (CDKN2) as a candidate for the chromosome 9p melanoma susceptibility locus. *Nat. Genet.* 8:23–26.

Kang, X., Robbins, P.F., Fitzgerald, E.B., Wang, R., Rosenberg, S.A., Kawakami, Y. (1997) Induction of melanoma reactive T cells by stimulator cells expressing melanoma epitope-major histocompatibility complex class I fusion proteins. *Cancer Res.* 57:202–205.

Kaufmann, O., Koch, S., Burghardt, J., Audring, H., Dietel, M. (1998) Tyrosinase, melan-A, and KBA62 as markers for the immunohistochemical identification of metastatic amelanotic melanomas on paraffin sections. *Mod. Pathol.* 11:740–746.

King, R., Googe, P.B., Weilbaecher, K.N., Mihm, M.C., Fisher, D.E. (2001) Microphthalmia transcription factor expression in cutaneous benign, malignant melanocytic, and nonmelanocytic tumors. *Am. J. Surg. Pathol.* 25:51–57.

Koch, M.B., Shih, I.M., Weiss, S.W., Folpe, A.L. (2001) Microphthalmia transcription factor and melanoma cell adhesion molecule expression distinguish desmoplastic/spindle cell melanoma from morphologic mimics. *Am. J. Surg. Pathol.* 25:58–64.

Kos, L., Aronzon, A., Takayama, H., Maina, F., Ponzetto, C., Merlino, G., Pavan, W. (1999) Hepatocyte growth factor/scatter factor-MET signaling in neural crest-derived melanocyte development. *Pigm. Cell. Res.* 12:13–21.

Kwon, B.S., Halaban, R. , Kim, G.S., Usack, L., Pomerantz, S., Haq, A.K. (1987) A melanocyte-specific complementary DNA clone whose expression is inducible by melanotropin and isobutylmethyl xanthine. *Mol. Biol. Med.* 4:339–355.

Lahav, R., Dupin, E., Lecoin, L., Glavieux, C., Champeval, D., Ziller, C., Le Douarin, N.M. (1998) Endothelin 3 selectively promotes survival and proliferation of neural crest-derived glial and melanocytic precursors in vitro. *Proc. Natl. Acad. Sci. USA* 95:14214–14219.

Lahav, R., Ziller, C., Dupin, E., LeDouarin, N.M. (1996) Endothelin 3 promotes neural crest cell proliferation and mediates a vast increase in melanocyte number in culture. *Proc. Natl. Acad. Sci. USA* 93:3892–3897.

Lerner, A.B., Fitzpatrick, T.B., Calkins, E., Summerson, W.H. (1949) Mammalian tyrosinase: preparation and properties. *J. Biol. Chem.* 178:185–195.

Lerner, A.B., Halaban, R., Klaus, S.N., Moellmann, G.E. (1987) Transplantation of human melanocytes. *J. Invest. Dermatol.* 89:219–224.

Lerner, A.B., and McGuire, J. (1961) Effect of a- and b-melanocyte-stimulating hormones on the skin colour of man. *Nature (London)* 189:176–179.

Lontz, W., Olsson, M.J., Moellmann, G., Lerne, A.B. (1994) Pigment cell transplantation for treatment of vitiligo: a progress report. *J. Am. Acad. Dermatol.* 30:591–597.

Lu, C., and Kerbel, R.S. (1993) Interleukin-6 undergoes transition from paracrine growth inhibitor to autocrine stimulator during human melanoma progression [published erratum appears in *J. Cell Biol.* 1993 Apr; 121(2): following 477]. *J. Cell Biol.* 120:1281–1288.

Lu, C., Vickers, M.F., Kerbel, R.S. (1992) Interleukin 6: a fibroblast-derived growth inhibitor of human melanoma cells from early but not advanced stages of tumor progression. *Proc. Natl. Acad. Sci. USA* 89:9215–9219.

Maguire, H.C. Jr., Berd, D., Lattime, E.C., McCue, P.A., Kim, S. Chaptman, P.B., Mastrangelo, M.J. (1998) Phase I study in patients with metastatic melanoma including evaluation of immunology parameters. *Cancer Biother. Radiopharm.* 13:13–23.

Matsumoto, K., Tajima, H., Nakamura, T. (1991) Hepatocyte growth factor is a potent stimulator of human melanocyte DNA synthesis and growth. *Biochem. Biophys. Res. Commun.* 176:45–51.

Medrano, E.E., and Nordlund, J.J. (1990) Successful culture of adult human melanocytes obtained from normal and vitiligo donors. *J. Invest. Dermatol.* 95:441–445.

Medrano, E.E., Yang, F., Boissy, R., Farooqui, J., Shah, V., Matsumoto, K., Nordlund, J.J., Park, H.Y. (1994) Terminal differentiation and senescence in the human melanocyte: repression of tyrosine-phosphorylation of the extracellular signal-regulated kinase 2 selectively defines the two phenotypes. *Mol. Biol. Cell* 5:497–509.

Meier, F., Nesbit, M., Hsu, M.Y., Martin, B., Van Belle, P., Elder, D.E., Schaumburg-Lever, G., Garbe, C., Walz, T.M., Donatien, P., Crombleholme, T.M., Herlyn, M. (2000) Human melanoma progression in skin reconstructs: biological significance of bFGF. *Am. J. Pathol.* 156:193–200.

Miettinen, M., Fernandez, M., Franssila, K., Gatalica, Z., Lasoto, J., Sarlomo-Rikala, M. (2001) Microphthalmia transcription factor in the immunohistochemical diagnosis of metastatic melanoma: comparison with four other melanoma markers. *Am. J. Surg. Pathol.* 25:205–211.

Minev, B.R., Chavez, F.L., Mitchell, M.S. (1999) Cancer vaccines: novel approaches and new promise. *Pharmacol. Ther.* 81:121–139.

Mitchell, M.S. (1998) Perspective on allogeneic melanoma lysates in active specific immunotherapy. *Semin. Oncol.* 25:623–635.

Naish-Byfield, S., Cooksey, C.J., Latter, A.M., Johnson, C.I., Riley, P.A. (1991) In vitro assessment of the structure-activity relationship of tyrosinase-dependent cytotoxicity of a series of substituted phenols. *Melanoma Res.* 1:273–287.

Naish-Byfield, S., and Riley, P.A. (1992) Oxidation of monohydric phenol substrates by tyrosinase. An oximetric study [see comments]. *Biochem. J.* 288:63–67.

Ogura, K., Nagata, K., Hatanaka, H., Habuchi, H., Kimata, K., Tate, S., Ravera, M.W., Jaye, M., Schlessinger, J., Inagaki, F. (1999) Solution structure of human acidic fibroblast growth factor and interaction with heparin-derived hexasaccharide. *J. Biomol. NMR* 13:11–24.

Oku, T., Tjuvajev, J.G., Miyagawa, T., Sasajima, T., Joshi, A., Joshi, R., Finn, R., Claffey, K.P., Blasberg, R.G. (1998) Tumor growth modulation by sense and antisense vascular endothelial growth factor gene expression: effects on angiogenesis, vasular permeability, blood volume, blood flow, fluorodeoxyglucose uptake, and proliferation of human melanoma intracerebral xenografts. *Cancer Res.* 58:4185–4192.7

Orchard, G.E. (2000) Comparison of immunohistochemical labelling of melanocyte differentiation antibodies melan-A, tyrosinase and HMB 45 with NKIC3 and S100 protein in the evaluation of benign naevi and malignant melanoma [In Process Citation]. *Histochem. J.* 32:475–481.

Orosz, Z. (1999) Melan-A/Mart-1 expression in various melanocytic lesions and in non-melanocytic soft tissue tumours. *Histopathology* 34:517–525.

Plowman, J., Dykes, D.J., Narayanan, V.L., Abbott, B.J., Saito, H., Hirata, T., Grever, M.R. (1995) Efficacy of the quinocarmycins KW2152 and DX-52-1 against human melanoma lines growing in culture and in mice. *Cancer Res.* 55:862–867.

Puffenberger, E.G., Hosoda, K., Washington, S.S., Nakao, K., Dewit, D., Yanagisawa, M., Chakravarti, A. (1994) A missense mutation of the endothelin-B receptor gene in multigenic Hirschsprung's disease. *Cell* 79:1257–1266.

Reid, K., Nishikawa, S., Bartlett, P.F., Murphy, M. (1995) Steel factor directs melanocyte development in vitro through selective regulation of the number of c-kit+ progenitors. *Dev. Biol.* 169:568–579.

Sherman, L., Stocker, K.M., Morrison, R., Ciment, G. (1993) Basic Fibroblast Growth Factor (bFGF) acts intracellularly to cause the transdifferentiation of avian neural crest-derived Schwann cell precursors into melanocytes. *Development* 118:1313–1326.

Spritz, R.A. (1994) Molecular basis of human piebaldism. [Review]. *J. Invest. Dermatol.* 103:137S–140S.

Stocker, K.M., Sherman, L., Rees, S., Ciment, G. (1991) Basic FGF and TGF-b1 influence commitment to melanogenesis in neural crest-derived cells of avian embryos. *Development* 111:635–641.

Su, Y.A., Bittner, M.L., Chen, Y., Tao, L., Jiang, Y., Zhang, Y., Stephan, D.A., Trent, J.M. (2000) Identification of tumor-suppressor genes using human melanoma cell lines UACC903, UACC903(+6), and SRS3 by comparison of expression profiles. *Mol. Carcinog.* 28:119–127.

Swope, V.B., Medrano, E.E., Smalara, D., Abdel-Malek, Z.A. (1995) Long-term proliferation of human melanocytes is supported by the physiologic mitogens alpha-melanotropin, endothelin-1, and basic fibroblast growth factor. *Exp. Cell Res.* 217:453–459.

Tada, A., Suzuki, I., Im, S., Davis, M.B., Cornelius, J., Babcock, G., Nordlund, J.J., Abdel-Malek, Z.A. (1998) Endothelin-1 is a paracrine growth factor that modula tes melanogenesis of human melanocytes and participates in their responses to ultraviolet radiation. *Cell Growth Diff.* 9:575–584.

Tamura, A., Halaban, R., Moellmann, G., Cowan, J.M., Lerner, M.R., Lerner, A.B., (1987) Normal murine melanocytes in culture. *In Vitro Cell. Dev. Biol.* 23:519–522.

Trotta, P.P., and Harrison, S.D. Jr. (1987) Evaluation of the antitumor activity of recombinant human gamma-interferon employing human melanoma xenografts in athymic nude mice. *Cancer Res.* 47:5347–5353.

Valverde, P., Healy, E., Jackson, I., Rees, J.L., Thody, A.J. (1995) Variants of the melanocyte-stimulating hormone receptor gene are associated with red hair and fair skin in humans. *Nat. Genet.* 11:328–330.

Walker, G.J., Palmer, J.M., Walters, M.K., Nancarrow, D.J., Parsons, P.G., Hayward, N.K. (1994) Refined localization of the melanoma (MLM) gene on chromosome 9p by analysis of allelic deletions. *Oncogene* 9:819–824.

Wilkins, L., Gilchrest, B.A., Szabo, G., Weinstein, R., Maciag, T. (1985) The stimulation of normal human melanocyte proliferation in vitro by melanocyte growth factor from bovine brain. *J. Cell. Physiol.* 122:350–361.

Wong, G., Pawelek, J., Sansone, M., Morowitz, J. (1974) Response of mouse melanoma cells to melanocyte stimulating hormone. *Nature (London)* 248:351–354.

Yada, Y., Higuchi, K., Imokawa, G. (1991) Effects of endothelins on signal transduction and proliferation in human melanocytes. *J. Biol. Chem.* 266:18352–18357.

Zakut, R., et al. (1993) KIT ligand (mast cell growth factor) inhibits the growth of KIT-expressing melanoma cells. *Oncogene* 8:2221–2229.

Zavala-Pompa, A., Folpe, A.L., Jimenez, R.E., Lim, S.D., Cohen, C., Eble, J.N., Amin, M.B. (2001) Immunohistochemical study of microphthalmia transcription factor and tyrosinase in angiomyolipoma of the kidney, renal cell carcinoma, and renal and retroperitoneal sarcomas: comparative evaluation with traditional diagnostic markers. *Am. J. Surg. Pathol.* 25:65–70.

# 13

# Establishment and Culture of Human Leukemia-Lymphoma Cell Lines

Hans G. Drexler

*DSMZ-German Collection of Microorganisms and Cell Cultures, Braunschweig, Germany. hdr@dsmz.de*

*Culture of Human Tumor Cells,* Edited by Roswitha Pfragner and R. Ian Freshney.
ISBN 0-471-43853-7   Copyright © 2004 Wiley-Liss, Inc.

# I. INTRODUCTION

Along with other improvements, the advent of continuous human leukemia-lymphoma (LL) cell lines as a rich resource of abundant, accessible, and manipulable living cells has contributed significantly to a better understanding of the pathophysiology of hematopoietic tumors [Drexler et al., 2000b]. The first LL cell lines, Burkitt's lymphoma-derived lines, were established in 1963 [Drexler and Minowada, 2000]. Since then, large numbers of LL cell lines have been described, though not all of them in full detail [Drexler, 2000].

## 1.1. Background

The major advantages of continuous LL cell lines are the unlimited supply and worldwide availability of identical cell material and the infinite viable storability in liquid nitrogen. LL cell lines are characterized generally by monoclonal origin and differentiation arrest, sustained proliferation in vitro under preservation of most cellular features, and specific genetic alterations (Table 13.1) [Drexler et al., 2000b].

Truly malignant cell lines must be discerned from Epstein-Barr virus (EBV)-immortalized normal cells. However, the picture is not quite so straightforward, as some types of LL cell lines are indeed EBV$^+$, and some EBV$^+$ normal cell lines carry also genetic aberrations and may mimic malignancy-associated features [Drexler and Matsuo, 2000; Drexler et al., 2000b].

By opening new avenues for investigation, studies of LL cell lines have provided seminal insights into the biology of hematopoietic neoplasia. Over nearly four decades, these cell cultures, which were initially known only to a few specialists, have become ubiquitous powerful research tools, available to every investigator.

## 1.2. Principles of Methodology

It is extremely difficult to establish continuous LL cell lines, and most attempts fail. Seeding of neoplastic cells directly into suspension

**Table 13.1. Major advantages and common characteristics of LL cell lines.**

- Unlimited supply of cell material
- Worldwide availability of identical cell material
- Indefinite storability in liquid nitrogen and recoverability
- Monoclonal origin
- Differentiation arrest at a discrete maturation stage
- Sustained proliferation in culture
- Stability of most features in long-term culture
- Specific genetic alterations

Modified from ref. [Drexler et al., 2000b].

cultures is a common procedure in attempts to establish cell lines. The success rate for the establishment of continuous cell lines is low and also varies strongly between different reports. However, the malignant nature of the derived LL cell lines and the immortalization of the cell lines have not been demonstrated in all instances.

It has been purported that B-cell precursor (BCP)-cell lines derived from patients at relapse or from cases with poor prognostic features have an enhanced growth potential in vitro compared with samples obtained at presentation or from children with good prognostic parameters [Schneider et al., 1977; Zhang et al., 1993]. However, this notion is not supported by the results of other studies [Lange et al., 1987; Kees et al., 1996]. A recent review reported that on aggregate the success rate for BCP-cell lines established from patients at diagnosis was 6%, whereas the success rate was 29% for relapse samples [Matsuo and Drexler, 1998]. There seems to be a higher success rate in cases that have certain primary chromosomal aberrations or gene mutations. This notion has been confirmed in T-cell lines carrying an (8;14) (q24;q11) translocation and in a variety of cell lines carrying alterations of the *P53*, *P15INK4B*, or *P16INK4* genes [O'Connor et al., 1991; Lange et al., 1992; Drexler, 1998; Drexler et al., 2000a].

One particular cell culture method promoted by Smith and colleagues deserves further comment: when leukemic T-cells were cultured in wells with a feeder layer consisting of complete medium, human serum, and agar in an hypoxic environment with supplemental insulin-like growth factor (IGF)-I, 10 out of 12 frozen samples and five out of six fresh samples grew and were established as cell lines [Smith et al., 1984, 1989]. However, when BCP-type acute lymphoblastic leukemia (ALL)-derived cells were cultured under these conditions, rapid cell death occurred concomitant with the outgrowth of normal mononuclear cells. On applying the monocyte toxin L-leucine methyl ester and insulin instead of IGF-I, the success rate increased from 11% to 40% [Zhang et al., 1993]. It should be noted that some of these BCP-cell lines have extremely long doubling times (10–14 days), which clearly limits their usefulness. The reproducibility of this method in other hands and the long-term growth (immortalization) of these cultures are not known.

The reasons for the frequent failure to establish cell lines remain unclear. According to general experience, the major causes appear to be culture deterioration with cessation of multiplication of the neoplastic cells and overgrowth by normal fibroblasts, macrophages, or lymphoblastoid cells. Although the lymphoblastoid cells may give rise to a continuous cell line (an EBV$^+$ B-lymphoblastoid cell line, B-LCL), human fibroblast and macrophage cultures are commonly not immortalized.

Despite the fact that the proliferation of malignant hematopoietic cells in vivo seems to be independent of the normal regulatory mechanisms, these cells usually fail to proliferate autonomously in vitro even for short periods of time. In vivo, at least initially, these cells seem to require one, or probably several, hematopoietic growth factors for proliferation. The addition of regulatory proteins (for example, so-called hematopoietic growth factors, such as erythropoietin (EPO), granulocyte colony-stimulating factor (G-CSF), granulocyte-macrophage CSF (GM-CSF), interleukin-2 (IL-2), IL-3 or IL-6, stem cell factor (SCF) or thrombopoietin, mitogens such as phytohemagglutinin (PHA), or conditioned medium (CM) secreted by certain tumor cell lines, which often contains various factors) is a culturing technique that appears to increase the frequency of success by overcoming the "crisis" period in which the neoplastic cells cease proliferating. These molecules enable the LL cells from the majority of patients to multiply for about 2–4 weeks. Out of these short-term cultures, continuous cell lines, which are derived from the malignant cells, can be established.

Taken together, the efficiency of cell line establishment is still rather low, and the deliberate establishment of LL cell lines remains by and large an unpredictable random process. Clearly, difficulties in establishing continuous cell lines may be due to the inappropriate selection of nutrients and growth factors for these cells. Thus, a suitable microenvironment for hematopoietic cells, either malignant or normal, cannot yet be created in vitro. Further work is required to achieve significant improvements in the success rates of LL cell line immortalization.

In the following, some of the more promising techniques for establishing new LL cell lines are described. Table 13.2 provides a statistical overview on three clinical and cell culture parameters of the well-characterized LL cell lines as published in the literature, which might have some influence on the success rate: i) choice of culture medium, ii) specimen site of the primary cells, and iii) status of the patient at the time of sample collection.

## 2. PREPARATION OF MEDIA AND REAGENTS

### 2.1. Culture Media

RPMI 1640, IMDM, MEM-α, or McCoy's 5A, with 20% fetal bovine serum (FBS), plus additional 10% CM from cell line 5637 (see Section 2.2) or with an appropriate concentration of purified or recombinant growth factors (e.g., 10 ng/ml of GM-CSF, IL-3 or SCF). Antibiotics (commonly penicillin-streptomycin) may be added as ready-for-use commercially available 100× solutions. Although this does not appear

**Table 13.2. Statistics on establishment of continuous LL cell lines: media, specimen sites, treatment status.**

| Variable | No. of lines/ No. of attempts | Success rate |
|---|---|---|
| *Culture Medium:* | | |
| RPMI 1640 | 335/454 | 74% |
| IMDM | 53/454 | 12% |
| McCoy's 5A | 30/454 | 7% |
| MEM-$\alpha$ | 21/454 | 5% |
| Others[a] | 15/454 | 3% |
| *Specimen Site:* | | |
| Peripheral blood | 210/439 | 48% |
| Bone marrow | 106/439 | 24% |
| Pleural effusion | 60/439 | 14% |
| Ascites | 24/439 | 5% |
| Tumor | 15/439 | 3% |
| Lymph node | 12/439 | 3% |
| Cerebrospinal fluid | 4/439 | 1% |
| Pericardial effusion | 3/439 | 1% |
| Spleen | 2/439 | 1% |
| Liver | 1/439 | |
| Meninges | 1/439 | |
| Tonsil | 1/439 | |
| *Treatment Status:* | | |
| At relapse/refractory/terminal | 226/413 | 55% |
| At diagnosis/presentation | 141/413 | 34% |
| At blast crisis | 39/413 | 9% |
| During therapy | 7/413 | 2% |

Modified from ref. [Drexler, 2000] and updated. Literature data were not available in all categories for all cell lines. Counted were only data on well-characterized cell lines (without sister cell lines or subclones).
[a] Other media used: Cosmedium, DMEM, Eagle's MEM, Fischer's, Ham's F10, Ham's F12, L-15.

to be harmful, it is also clearly not necessary if proper cell culture techniques ("good culture practices") are used. l-Glutamine (at 2 mM and 2-mercaptoethanol (at 0.05 mM) may also be added, but their possible effects, if any at all, on the success rate of establishment are questionable. We strongly discourage the routine addition of antimycoplasmal reagents, as this practice would lead very quickly to the development of resistant mycoplasma strains [Drexler and Uphoff, 2000].

FBS is the standard supplement in the suspension culture system of LL cell lines. It is commonly used in concentrations of 5–20%. LL cell lines that do not require FBS in the culture medium have been described; however, they appear to be rather rare [Drexler, 2000]. Prior to usage, it is recommended that batches of serum be pretested for their ability to support vigorous cell growth and for viral (in particular for bovine viral diarrhea virus), mycoplasmal, and other bacterial contamination. If possible, a large supply of the FBS from a pretested batch that supports cell growth well and has no contamination should be purchased and stored at $-20°C$ for future use. Alternatives are newborn calf serum (usually at only 25% of the price of the expensive FBS) or serum-free media. However, not all LL cell lines will grow in newborn calf serum as well as in FBS. Although serum-free media do not provide financial advantages, they do allow for certain experimental manipulations, which are not possible with FBS as one can control all the substances to which the cells are exposed. FBS is known to contain many unidentified ingredients at variable concentrations.

## 2.2. Conditioned Medium

The adherent bladder carcinoma cell line 5637 is known to produce and secrete large quantities of various cytokines, including G-CSF, GM-CSF, SCF and others (but not IL-3!) [Quentmeier et al., 1997]. Conditioned medium should be made in large volumes in order to guarantee continuity over a period of time; for example, six months to one year.

## Protocol 13.1.  Preparation of Conditioned Medium from 5637 Cells

### Reagents and Materials
*Sterile or Aseptically Prepared*
- ❑  5637 cells
- ❑  Growth medium: RPMI 1640 or IMEM with 10% FBS
- ❑  Trypsin, 0.05%, EDTA, 0.5 mM

### Protocol
(a)  At initiation of culture, seed out at $\sim 2 \times 10^6$ 5637 cells/80 cm$^2$ in 10 ml of complete medium.

(b)  Incubate at 37°C in 5% $CO_2$ in a humidified atmosphere, with the flask tops slightly loosened, until the cell monolayer becomes almost confluent.

(c)  Replace the medium and grow for another 4 days.

(d)  Collect this medium and store it at 4°C.

(e)  Add fresh complete medium to the 5637 culture.

(f)  Repeat this process at 3- to 5-day intervals, pooling the culture supernatants.

(g)  After 1–2 weeks, split the confluent 5637 cultures 1:4 to 1:5 by using trypsin/EDTA for 5 min and reseed the cells in fresh flasks. 5637 cells have a doubling time of about 24 h, leading to a cell harvest of $8 \times 10^6 - 1.0 \times 10^7$ cells/80 cm$^2$ [Drexler et al., 2001].

(h)  Spin down the pooled conditioned medium at 200 g for 5–10 min before filtering through sterile filters at 0.2 µm to exclude any contaminating nonadherent cells.

(i)  Aliquot into 30 or 50 ml tubes and store at −20°C.

(j)  Test the 5637 CM in proliferation assay at 5%, 10%, and 20% v/v, using specific growth factor-dependent cell lines; for example, cell line M-07e [Drexler et al., 2001], to determine the optimum volume necessary for maximal stimulation. Alternatively, the exact growth factor concentrations in the 5637 CM can be determined with enzyme-linked immunoassays [Quentmeier et al., 1997].

## 3.  ESTABLISHMENT OF CELL LINES

### 3.1.  Acquisition of Cells

The most commonly used specimens are peripheral blood or bone marrow samples, as these are relatively easily obtained from patients (see Table 13.2). Other liquid specimens (such as pleural effusion, ascites, cerebrospinal fluid, etc.) and solid tissue samples (lymph node, tonsil, spleen, etc.) are used less often but can be processed similarly to peripheral blood or bone marrow samples. Solid tissues require a prior step, whereby the tissue is dissociated mechanically. All solutions and utensils must be sterile. Work performed in a laminar flow cabinet (class II biological safety cabinet) under sterile conditions is recommended.

### Protocol 13.2.  Collection of Samples for Culture of Leukemia/Lymphoma Cells

#### Reagents and Materials

*Sterile*

- ❏  Tubes containing anticoagulant (e.g., heparin)
- ❏  Universal containers
- ❏  Collection medium (for solid tissues): culture medium supplemented with a high concentration of antibiotics (e.g., penicillin, 250 U/ml, streptomycin 250 µg/ml, gentamycin 50 µg/ml, amphotericin B, 25 µg/ml)

#### Protocol

(a)  Collect specimens of peripheral blood or bone marrow or other samples in sterile heparinized tubes. Place lymph nodes and other solid tissues in sterile containers.

(b)  Ideally, specimens should be processed as soon as possible after receipt, but may be stored overnight at room temperature (peripheral blood and bone marrow should remain undiluted, solid tissues should be placed in culture medium).

(c) For attempts to establish cell lines, cryopreserved samples can also be used, but it appears to be of advantage to isolate the mononuclear cells prior to cryopreservation.

## 3.2. Isolation of Cells

Single-cell suspensions are prepared from patient-derived samples. To that end, the mononuclear cells are isolated by density gradient centrifugation.

## Protocol 13.3. Isolation of Mononuclear Cells by Density Gradient Centrifugation

### Reagents and Materials
*Sterile*
- ❑ Culture medium: (see Section 2.1 and Table 13.2)
- ❑ FBS, heat-inactivated at 56°C for 30–45 min
- ❑ Ficoll-Hypaque density 1.077 g/L
- ❑ Pasteur pipettes
- ❑ Scissors
- ❑ Metal gauze or screen, ~100 μm mesh

### Protocol

(a) Cut solid tissue specimens (e.g., lymph nodes) with scissors, and force the particles through a fine metal mesh.

(b) Suspend the cells in 50–100 ml culture medium (possibly also more depending on the size of the tissue specimen). The most commonly used media are listed in Table 13.2.

(c) Dilute the blood and bone marrow samples 1:2 with culture medium. Isolation of cells from a leucophoresis collection requires dilution of the sample with culture medium at 1:4.

(d) Pipette 5 ml Ficoll-Hypaque density gradient solution into a 15 ml or 10 ml into a 30 ml conical centrifuge tube.

(e) Slowly layer an equal volume of the mixture of medium and sample over the Ficoll-Hypaque solution. Do not disturb the Ficoll-Hypaque/sample interface. It is helpful to hold the centrifuge tube at a 45° angle.

(f) Centrifuge for 20–30 min at 450 g at room temperature (with the centrifuge brakes turned off).

(g) After centrifugation, a layer of mononuclear cells should be visible on top of the Ficoll-Hypaque phase, as they have a lower density than the Ficoll-Hypaque solution. The anucleated erythrocytes and the polynucleated granulocytes are concentrated as pellet below the Ficoll-Hypaque layer.

(h)　Using a Pasteur pipette, transfer the interface layer containing the mononuclear cells to a centrifuge tube.

(i)　Wash the cells by adding culture medium plus 2% FBS (add about 5 times the volume of the mononuclear cell solution) and centrifuge for 10 min at 200 g at room temperature.

(j)　Discard the supernate, resuspend the cell pellet in culture medium plus 2% FBS, and repeat the washing procedure.

(k)　Finally, resuspend the cells in 20 ml of culture medium with 20% FBS.

(l)　Count the cells and determine their viability.

The washing steps described above are performed to remove the acidic heparin or other anti-coagulants that may harm the cells; to remove the patient's serum, which may inhibit cell growth; and to remove the Ficoll-Hypaque, which is hypertonic for the cells. Two washes will generally suffice. FBS is included in order to prevent cells from adhering to one another.

In general, more than 90% of the mononuclear cells can be recovered by this procedure. Cell yields depend on the number of malignant cells in the specimen and are highly variable from patient to patient. On average, each ml of leukemia bone marrow yields $1.5 \times 10^7$–$3.0 \times 10^7$ mononuclear cells and each ml of peripheral blood yields $1 \times 10^6$–$2 \times 10^6$ mononuclear cells, the latter depending obviously on the white blood cell count. Several hundred millions of mononuclear cells can usually be recovered from lymph nodes.

### 3.3.　Cell Counting

The cell count and cellular viability can be determined by hemocytometer counting and Trypan Blue dye exclusion [Freshney, 2000]. Viable cells will not take up the Trypan Blue stain.

### 3.4.　Culture Conditions

There are various types and sizes of plastic culture vessels: flasks (for example 25 cm², 80 cm², 175 cm²) or plates (with 12-, 24-, 96-wells). The number of flasks or wells used depends on the number of the primary LL cells available. As many of the malignant cells as possible should be used in attempts to establish a cell line. Although an early passage cell line may contain multiple lineages of replicating cells, it is probable that a continuous cell line arises from one cell or from a very small fraction of the total population. Thus, the greater the cell number and the more attempts made, the higher the chances will be of producing a continuous cell line.

Most LL cell lines were established by using the "direct method" of placing cells into liquid culture in a humidified incubator at 37°C and 5% $CO_2$ in air. Alternative approaches are inoculation in semi-solid media (methylcellulose, soft agar), initial heterotransplantation, and serial passage in immunodeficient mice with subsequent adaptation to in vitro culture, or culture on temporary feeder layer (e.g., on fibroblasts).

Recently, we have been successful in the establishment of various new leukemia cell lines [Drexler et al., 2001; Meyer et al., 2001; Hu et al., 2002] by using the following method.

## Protocol 13.4. Establishment of Continuous Cell Lines from Leukemia/Lymphoma

### Reagents and Materials
*Sterile*
- Culture medium
- Culture flasks, 80 cm² or 24-well plates

### Protocol
(a) Adjust the cell suspension of the original LL cells to a concentration of $2 \times 10^6$–$5 \times 10^6$/ml in culture medium.

(b) Place 5–10 ml of the cell suspension in the complete culture medium in a 80 cm² plastic culture flask. If 24-well-plates are used, add 1–2 ml cell suspension into each well. Add 100–200 µl of cell suspension into wells of 96-well flat-bottomed microtitration plates.

(c) Place the cells in a humidified incubator at 37°C and 5% $CO_2$ in air. Alternatively (but seldom an option as this requires a special type of incubator and large quantities of $N_2$), incubate the cells in a humidified 37°C incubator with 6% $CO_2$, 5% $O_2$, and 89% $N_2$.

(d) Expand the cells by exchanging half of the spent culture medium volume with fresh culture medium plus 20% FBS and 10% 5637 CM (or with appropriate concentrations of growth factors) once a week, either by centrifugation or by settling.

(e) After 4 h, some cells become adherent. These adherent cells appear to be the source of colony-stimulating factors for both normal and malignant cells. During the first two weeks, it is not necessary to remove these adherent cells from the culture unless there is a specific reason to do so, for example because of the addition of a purified growth factor to the medium in order to obtain a unique type of cell line.

(f) After 2 weeks, if the suspension cells grow very rapidly, the adherent cells can be removed simply by transferring the suspension cells into new culture vessels in order to reduce the potential for overgrowth of fibroblasts and normal lymphoblastoid cells.

(g)  During the first weeks, the neoplastic cells may appear to proliferate actively. If the medium becomes acidic quickly (yellow in the case of RPMI 1640 medium), change half of the volume of medium at 2–3 day-intervals (seldom daily). If the number of the cells increases rapidly, readjust the cell concentration weekly to at least $1 \times 10^6$/ml in fresh complete medium by dilution or subdivision into new flasks or wells of the plate. The neoplastic cells from the majority of LL patients undergo as many as four doublings in two weeks, but after 2–3 weeks most malignant cells cease proliferating. Following a lag time of 2–4 weeks ("crisis period"), a small percentage of cells from the total population may still proliferate actively and may continue to grow, forming a cell line.

(h)  If the malignant cells continue to proliferate for more than two months, there is a high possibility of generating a new LL cell line. Then, the task of characterizing the proliferating cells should be begun as soon as possible (see below). Prior to the characterization of the cells, freeze ampoules of the proliferating cells containing a minimum of $3 \times 10^6$ cells/ampoule in liquid nitrogen in order to avoid loss of the cells due to occasional contaminations or other accidents.

(i)  Use limiting dilution of the cells in 96-well-plates to generate monoclonal cell lines. However, after prolonged culture in vitro, the cell line may, effectively, become monoclonal due to the outgrowth of selected cell clones. In most cases, it is not absolutely necessary to subclone the cell line by limiting dilution. In some types of LL cell lines: for example, immature T- and precursor B-cell lines, it might be very difficult or virtually impossible to "clone" the cells.

### 3.5.  Preservation of Early Stocks

It is absolutely mandatory to freeze aliquots of the original cells and to store them in appropriate locations for later documentation, authentication, and comparisons. Once the cultured cells start to proliferate, it is wise to freeze ampoules at regular intervals during the initial period of the culture expansion (see Protocol 13.6 in Section 4).

### 3.6.  Morphology of Cells

The morphology of the primary LL cells, and later of the resulting LL cell line, is best appreciated by staining cytocentrifuge slide preparations.

## Protocol 13.5. Cytology of Leukemia/Lymphoma Cells and Cell Lines

### Reagents and Materials

*Sterile or Aseptically Prepared*
- ❑ Cell culture or primary suspension

*Non-sterile*
- ❑ Microscope slides
- ❑ Cytocentrifuge (Shandon)
- ❑ May-Grünwald and Giemsa stain
- ❑ Weise buffer (4 mM potassium di-hydrogen phosphate, 6 mM di-sodium hydrogen phosphate at pH 7.2)
- ❑ Mountant: Entellan

### Protocol

(a) Harvest cells, count and adjust the concentration to $5 \times 10^5$–$1.0 \times 10^6$ cells/ml.

(b) Spin 50 µl, 70 µl and 90 µl of cell suspension onto glass microscope slides in a cytocentrifuge at 500 rpm for 5 min and air dry at room temperature overnight (slides should not be left unstained for longer than one week).

(c) Select the best preparations under the microscope: The cells should be neither too packed nor too isolated for an appropriate morphological examination.

(d) Stain with the May-Grünwald and Giemsa solutions as follows:

(e) Fix for 5 min in 100% methanol.

(f) Transfer for 5 min to May-Grünwald stain that has been diluted 1:2 with Weise buffer, and then rinse with Weise buffer.

(g) Transfer for 20 min to Giemsa stain that has been diluted 1:10 with Weise buffer.

(h) Rinse thoroughly with Weise buffer.

(i) Let the staining dry overnight.

(j) Mount the stained cells with a coverslip by using Entellan, and examine the preparation under the microscope.

## 4. FREEZING AND STORAGE OF CELL LINES

### 4.1. Freezing of Cells

It is generally assumed that LL cell lines can be cryopreserved at $-196°C$ in liquid nitrogen for more than 10 years, if not indefinitely, without any significant changes in their biological features. The viable cell lines can be recovered at any time when needed. Prior to freezing in liquid nitrogen, the cells are suspended traditionally in the appropriate medium containing 20% FBS and 10% dimethylsulfoxide

(DMSO), which can lower the freezing point in order to protect the frozen cells from damage caused by ice crystals. Glycerol has been used as an alternative to DMSO. It appears that no single suspending medium and procedure will be ideal for processing and cryogenic storage of all cell cultures. However, the procedures described here are suitable for most LL cell lines and are compatible with prolonged preservation of viability and other characteristics of the cell lines.

### Protocol 13.6. Cryopreservation of Leukemia/Lymphoma Cell Lines

#### Reagents and Materials
*Sterile or Aseptically Prepared*
- ❑ Centrifuge tubes, 30 ml or 50 ml
- ❑ Freezing medium: freshly prepared 70% culture medium, 20% FBS, and 10% DMSO, cooled on wet ice. It is not necessary to sterilize the DMSO solution, as pure DMSO is lethal to bacteria.
- ❑ Cryovials: plastic ampoules that have been properly labeled with the name of the cell line and the date of freezing;

*Non-sterile*
- ❑ Hemocytometer
- ❑ Viability stain, for example, Trypan Blue, 0.4%

#### Protocol
(a) Harvest the cells of LL cell lines by transferring the flask contents to 30 ml or 50 ml centrifuge tubes. The cells should be harvested in their logarithmic growth phase. Freezing and storing the cells impairs the viability and quality of the cell culture prior to the freezing; at best, the status quo will be preserved, but most often will be diminished to various degrees. Primary cells can be frozen after thorough washing steps (see above Protocol 13.3, Steps (i)–(k)).

(b) Determine the total number of viable cells by counting the cells with the Trypan Blue dye exclusion method [Freshney, 2000].

(c) Centrifuge the cell suspensions at 200 $g$ for 10 min, and discard the supernate.

(d) Adjust the cells to a concentration of 5 × $10^6$–1.0 × $10^7$/ml for cell lines and 1 × $10^7$–5 × $10^7$/ml for primary material by using freshly prepared freezing solution. The freezing solution should be added quickly to the cells and then mixed thoroughly. Long-time exposure to DMSO at room temperature can trigger significant cellular changes, such as activation and so-called "induced differentiation." Keeping cells in DMSO-containing media on ice should minimize the effect of DMSO on the cells.

(e)    Distribute the cells into prelabeled cryovials with 1 ml (at least $5 \times 10^6$ cells) per ampoule. More cells per ampoule can be frozen if needed, depending on the cell type.

(f)    Seal the ampoules and make a written record of the freezing.

We are using routinely two different methods for freezing of cells: (1) Using a controlled-rate freezer (cryofreezing system) whereby a cooling rate of 1°C per minute can be achieved from room temperature to -25°C. When the temperature reaches −25°C, the cooling rate is increased to 5–10°C per minute. On reaching −100°C, the ampoules can be transferred quickly to a liquid nitrogen container. (2) If only a few ampoules are frozen, the sealed ampoules can be placed in a plastic box with an insert for ampoules (Nalgene Cryo 1°C Freezing Container); the box is half-filled with isopropanol and is stored in a −80°C freezer for at least 4 h. With this method, an approximate 1°C per minute cooling rate can also be achieved.

Δ*Safety note.* Wear cryoprotective gloves, a face-mask, and a closed lab-coat when working with liquid nitrogen. Make sure the room is well-ventilated to avoid asphyxiation from evaporated nitrogen.

### 4.2.    Storage in Liquid or Gaseous Phase

Permanent storage of ampoules with primary cells or cells from LL cell lines should be in the liquid phase of the liquid nitrogen. However, long-term storage is also possible in the vapor phase of the liquid nitrogen tank, or in a liquid nitrogen jacketed tank, which has no nitrogen in the storage space. Long-term preservation of cell lines or primary cells at −80°C (beyond one week) cannot be recommended as the cells will die under these conditions.

## 5. THAWING, EXPANSION, AND MAINTENANCE OF CELL LINES

### 5.1.    Thawing of Cells

LL cell lines, stored frozen in liquid nitrogen, must be thawed carefully in order to minimize cell loss.

### Protocol 13.7.    Thawing Cryopreserved Leukemia/ Lymphoma Cell Lines

#### Reagents and Materials

*Sterile or Aseptically Prepared*

❏    Thaw medium: 80% culture medium and 20% FBS at room temperature or 37°C

□ Universal container or centrifuge tube, ~30 ml
□ Bijou bottle or small vial for cell sample for counting.
*Non-sterile*
□ Water bath at 37°C or lidded container with 5 cm water at 37°C (water should be fresh and, preferably, sterilized)
□ Trypan Blue stain, 0.4% in PBSA
□ Hemocytometer, improved Neubauer

### Protocol

(a) Remove the frozen ampoule from liquid nitrogen, and thaw the cells rapidly in a 37°C waterbath by gently shaking the ampoule in the water.

Δ *Safety note.* An ampoule that has been stored immersed in liquid nitrogen can inspire liquid nitrogen if not properly sealed. The ampoule will then explode violently when warmed. To protect from potential explosion of ampoules, thaw in a covered container, such as a large plastic bucket with a lid. Alternatively, store in the vapor phase or in liquid nitrogen jacketed freezer. Wear gloves, a face-mask, and a closed lab-coat.

(b) It is important that the frozen cells be thawed in about 1 min. Rapid warming is necessary so that the frozen cells pass quickly through the temperature zone between − 50°C and 0°C where most cell damage is believed to occur. Slow thawing will harm the cells by formation of ice crystals in the cells, which causes hypertonicity and breakage of cellular organelles.

(c) Wipe the ampoule with a tissue pre-wetted with 70% ethanol before the vial is opened. As soon as the cell suspension is thawed, transfer the cell suspension into a centrifuge tube.

(d) Dilute the cell suspension slowly by adding 10–20 ml thaw medium to the tube, and shake the tube gently to mix. Cells frozen with DMSO are usually dehydrated. During the washing steps with medium, water will diffuse into the cells. Diluting the suspension slowly is preferable as it may reduce the loss of electrolytes, counteract extreme pH changes, and prevent denaturation of cellular proteins.

(e) Centrifuge the cells at 200 g for 10 min, and discard the supernate.

(f) Resuspend the cells again by using another 10–20 ml thaw medium, remove a sample for a cell count, centrifuge the cells again at 200 g for 10 min, and discard the supernate.

(g) During the centrifugation, determine the total cell number by hemocytometer and the percentage of viable cells by Trypan Blue exclusion.

(h) Finally, resuspend the washed cells in the desired culture medium with the recommended FBS concentration at the optimal cell density.

## 5.2. Expansion of Cells

After thawing, cells may be resuspended in complete medium at a cell concentration of $2 \times 10^5$–$1.0 \times 10^6$ cells/ml. It appears that most LL cell lines grow better at higher cell concentrations than at lower ones. Some cell lines (e.g., some precursor B-cell lines) prefer a concentration higher than $1.0 \times 10^6$/ml. Usually, the optimal concentration of a cell line for expansion must be explored empirically. If after 2–3 days of culture the cells do not grow well (perhaps due to the presence of many dead cells), it might be useful to concentrate the cells and to culture them at a higher cell density. It is recommended that the cells be resuspended first in medium containing 20% FBS; should the cells start to multiply and resume their expected growth activity, the percentage of FBS can be decreased stepwise.

Suspension cell lines may be cultured in flasks or 24-well plates. General recommendations are 10–20 ml suspension in a 80 $cm^2$ flask or 1–2 ml suspension into each well of a 24-well plate For "difficult" cell lines, it may be advantageous to suspend some cells in a flask and another aliquot in a 24-well plate (or even a 96-well microtitration plate). There are distinct differences between flask and plate regarding exposure to $CO_2$, accessibility to microscopic observation, and possibilities of manipulation.

## Protocol 13.8.  Expansion of Leukemia/Lymphoma Cell Lines

### Reagents and Materials
*Sterile or Aseptically Prepared*
❏ LL cell culture
❏ Culture medium (see section 2.1 and Table 13.2) with 10–20% FBS, as required

### Protocol
(a) Incubate the cells in a humidified 37°C incubator with 5% $CO_2$ in air. Loosen the top of the flask slightly to allow for free gaseous exchange into and out of the flask.
(b) Feed the cells by exchanging half of the culture volume with culture medium plus FBS at 2- to 3-day intervals.

(i) Before changing the medium, set the flasks upright in the incubator for at least 30 min in order to let the cells sink to the bottom of the flask.

(ii) Remove gently half of the supernatant spent medium from the flask.

(iii) Add the same volume of new complete medium to the flask.

(c) If the cells proliferate actively, the culture medium will soon change color due to a pH change caused by cellular metabolism. In this case, it is necessary to change the medium more frequently.

(d) When changing the medium, it is important to calculate the total cell number by determining the concentration as well as the viability of the cells using Trypan Blue dye exclusion.

(e) When the cell number has doubled, subdivide the cells from the original flask into a second flask by diluting the suspension 1:2 with new medium.

Careful documentation of all manipulations, macroscopic and microscopic observations, intentional and accidental changes in the cellular conditions, and data on cell concentration, viability and total cell number at different time points is mandatory.

## 5.3. Maintenance of Cell Lines

The cell lines can be maintained as long as required, and cells can be harvested for use at any time. If the cells proliferate more quickly than needed, reduce the rate of cell proliferation by decreasing the FBS concentration, changing the medium at longer intervals, or discarding a certain amount of the cells (up to 75%) during the exchange of medium. An upper limit in cell concentration in order to prevent apoptotic decline depends on the type of cell line and even on the individual cell line; for example, some cell lines with larger volume cells will not exceed $0.3 \times 10^6$ cells/ml while for instance pre B-cell lines may reach $5–7 \times 10^6$ cells/ml without problems.

There is a fundamental difference between *expansion* and *maintenance* of an LL cell line. Some cell lines will deteriorate over long-term culture under maintenance conditions. In such cases, it might then be better to freeze and rethaw the cells when needed.

After longer usage culture flasks or plates will often contain a certain amount of unused ingredients of the medium, metabolites, and cell detritus, so it is recommended that the culture vessel be changed once every 1–2 months.

**Table 13.3. Cardinal requirements for new LL cell lines.**

- Immortality of cells
- Verification of neoplasticity
- Authentication of derivation
- Scientific significance
- Characterization of cells
- Availability to other scientists

Modified after refs. [Drexler et al., 1998; Drexler and Matsuo, 1999].

## 6. CHARACTERIZATION AND PUBLICATION OF CELL LINES

### 6.1. Cardinal Requirements for New Cell Lines

We have discerned six cardinal requirements for the description and publication of new LL cell lines (Table 13.3) [Drexler et al., 1998; Drexler and Matsuo, 1999].

### 6.1.1. Immortality of Cells

To comply with the definition, a continuous cell line should have been grown in permanent uninterrupted culture for at least 6 months, even better for more than a year. This is an important aspect as on addition of growth factors, primary neoplastic cells or normal cells can sometimes be kept in culture for several months before proliferation ceases. Such cultures cannot be regarded as continuous cell lines. Furthermore, cells may live but do not proliferate in vitro (for example B-cell lymphoma and myeloma cells may survive with virtually no proliferation for up to six months). Thus, it is essential that the time period of continuous culture be indicated for permanently proliferating cells. Continuous cell lines have been defined as cultures that are apparently capable of an unlimited number of population doublings (i.e., immortal) (Table 13.4). It should be recognized further that an immortal cell is not necessarily one that is neoplastically or malignantly transformed. Clearly, a cell line established from a patient with a tumor is not necessarily a tumor cell line [Drexler, 2000].

### 6.1.2. Verification of Neoplasticity

The neoplastic nature of the cell line should be demonstrated by functional assays or by the detection of clonal cytogenetic abnormalities. Given the overwhelming preponderance of karyotypically abnormal LL cell lines (>99%), the detection of an abnormal karyotype in a cell line can be regarded as a necessary (though not sufficient) condition of neoplasticity [Drexler et al., 2000b]. Colony formation in methylcellulose or agar (clonogenic assays) is also considered an op-

**Table 13.4. Cell culture definitions.**

| | |
|---|---|
| **Primary culture:** | |
| short-term culture | lifespan of a few weeks (<10 doublings) |
| long-term culture | lifespan of a few months (maximal 50–60 doublings) |
| **Secondary culture:** | |
| Cell line | a propagated culture after the first subculture |
| **Continuous cell line:** | |
| Continuous cell line | immortalized (>150–200 doublings; >1 year continuous growth) independent of the original organism, continuously growing, individual cells |
| Continuous cell strain | a characterized cell line derived by selection of cloning |
| **Clone:** | |
| Cloned cell strain | a characterized cell line derived from a single cell of the original (parental) cell line |
| Subclone | derived from a single cell of a cloned cell strain and having divergent and/or unique features |
| **Sister Cell Line:** | |
| simultaneous | established from the same patient at the same time, but possibly from different sites or the primary sample was split into several aliquots prior to culture |
| serial/longitudial | established from the same patient, but at different time points, e.g. at diagnosis and at relapse |

Modified from ref. [Drexler et al., 2000b].

erational test of neoplasticity. Xenotransplantability of cell lines into immunodeficient mice is often mistaken as a sign of the malignant nature and allegedly indicates the immortality of the transplanted cells. Clearly, xenotransplantation cannot be used as an unequivocal marker of neoplasticity. However, in long-term culture most EBV+ B-LCLs will eventually become monoclonal and aneuploid, grow as colonies in semi-solid media, and form tumors in immunodeficient mice. Therefore, the criteria listed for neoplasticity are valid only when an EBV+ B-LCL has been excluded.

### 6.1.3. Authentication of Derivation

The cellular origin of a new cell line must be proven by authentication; that is, it must be shown that the cultured cells are indeed derived from the presumed patient's tumor and are not the result of a cross-contamination with an older cell line. We have found that about 15–20% of human LL cell lines are misidentified or cross-contaminated by the original investigator [Drexler et al., 1999; Drexler and Uphoff, 2002]. The method of choice for identity control is forensic-type DNA fingerprinting [Drexler et al., 1999]. Microsatellite analysis does not appear to be sufficient, as the loci seem to be prone to instability in certain tumor types. Immunophenotyping will not suffice either, as cell lines of the same category will often have similar if not identical immunoprofiles. The presence of unique cytogenetic marker chromosomes or molecular biological data (e.g., identical clonal gene rearrangement patterns) might also provide unequivocal evidence for the derivation of the cell line from the corresponding patient.

### 6.1.4. Scientific Significance and Characterization of Cells

With regard to novelty and scientific significance, the new cell line should carry features not yet detected in previously established cell lines. Thus, the scientific significance of a new cell line depends on the degree of its characterization. A thorough multi-parameter analysis of the cells (see below) will often unveil unique characteristics of cell lines attesting to their scientific importance.

### 6.1.5. Availability to Other Scientists

The sharing of cell lines with other scientists is of the utmost importance. Some scientific journals have adopted the policy that any readily renewable resources, including cell lines published in that journal, must be made available to all qualified investigators in the field, if not already obtainable from commercial sources. The policy stems from the long-standing scientific principle that authenticity requires reproducibility. When cell lines are proprietary and unique, suitable material transfer agreements can be drawn up between the provider and requester. By providing authenticated and unique biological material, cell line banks play a major role in this regard [Drexler et al., 2001]. Thus, authors should be encouraged to deposit their cell lines in non-profit reference cell line collections.

## 6.2. Characterization of Cell Lines

### 6.2.1. Detailed Characterization of Cellular Features

Because LL cell lines commonly grow as single or clustered cells in suspension or only loosely adherent to the flask, single cell populations can easily be prepared, and the cells can thus be characterized. Table 13.5 lists a variety of parameters useful for the description of the cells and a panel of possible tests applicable for the phenotypic and functional characterization of most cell lines. This necessary multiparametric examination of the cellular phenotype provides important information on the likely cell of origin, the variable stringency of maturation arrest, and any discrepancies in the pattern of normal gene expression. The list is not intended to cover comprehensively all possible informative parameters as with new techniques becoming available and research areas extending to new avenues, other or entirely new features might be of interest to scientists. Thus, only some of the

**Table 13.5. Characterization of LL cell lines.**

| Parameter | Details and Examples |
|---|---|
| **Most Important Data** | |
| Clinical data | Patient's data:<br>  – original disease and disease status<br>  – age, sex, race<br>  – source of material |
| *In vitro* culture | Growth kinetics, proliferative characteristics:<br>  – culture medium and culture environment<br>  – initial seeding and subcultivation routine<br>  – minimum/maximum cell density and doubling time<br>  – cell storage conditions<br>  – *in situ* growth pattern<br>  – mycoplasma and viral status |
| Immunophenotyping | Surface marker antigens (fluorescence microscopy, flow cytometry)<br>Intracytoplasmic and nuclear antigens (immunoenzymatic staining) |
| Cytogenetics | Structural and numerical abnormalities<br>Specific chromosomal rearrangements |
| **Further Characterization** | |
| Morphology | *In situ* (flask, plate) under inverted microscope<br>Light microscopy (May-Grünwald-Giemsa-staining)<br>Electron microscopy (transmission and scanning) |
| Cytochemistry | Acid phosphatase, $\alpha$-naphthyl acetate esterase, others |
| Genotyping | Southern blot analysis of T-cell receptor and immunoglobulin heavy and light chain gene rearrangements<br>Northern analysis of expression of receptor transcripts |
| Cytokines | Production of cytokines<br>Expression of cytokine receptors<br>Response to cytokines, dependency on cytokines |
| Functional aspects/ specific features | Phagocytosis<br>Antigen presentation<br>Immunoglobulin production/secretion<br>(Hemo)globin synthesis<br>Capacity for (spontaneous or induced) differentiation<br>Positivity for EBV or HTLV-I or other viruses<br>Heterotransplantability into mice or other animals<br>Colony formation in agar/methylcelluose - clonogenicity<br>Production/secretion of specific proteins<br>NK cytotoxic activity<br>Oncogene expression<br>Transcription factor expression<br>Unique point mutations |
| Date of analysis | Age of cell line at time of analysis<br>Possible changes in the specific marker profile during prolonged culture |

Adapted from refs. [Drexler et al., 1998; Drexler and Matsuo, 1999]. Additional abbreviations: HTLV, human T-cell leukemia virus; NK, natural killer.

features of the phenotypic profiles of cell lines, which are most often studied, are highlighted here. It is also important to indicate when in the life of a cell line individual data were generated and also whether alterations in the phenotypic features of the cells might occur during prolonged culture.

### 6.2.2. Core Data

Immunophenotypic analysis and cytogenetic karyotyping currently appear to be the most important and informative examinations (Table 13.5). Although the scope and extent of the analytical characterization of LL cell lines is certainly variable, a core data set is obligatory and essential for the identification, description, and culture of a cell line. These data include the clinical background and cell culture description of the cell line. Clearly, the origin of an established cell line must be documented sufficiently. It is a cardinal principle to obtain the informed consent of the donor of the malignant cell line and to uphold the anonymity of the donor. It is important to check for mycoplasma contamination and to indicate whether the cells are EBV$^+$. We have found that about 25–30% of LL cell lines are mycoplasma-contaminated [Drexler and Uphoff, 2002].

### 6.2.3. Sister Cell Lines and Subclones

Subclones of any given cell line and sister cell lines must be distinguished and properly presented (see Table 13.4). Serial or longitudinal sister cell lines (e.g., established at diagnosis and relapse of the same patient) may provide unique opportunities to study the molecular mechanisms involved in disease progression and transformation. Of particular interest also are pairs of cell lines consisting of one cell line with diseased cells and one cell line with normal, albeit EBV-transformed cells from the same patient (which additionally provide reliable data allowing for subsequent authentication).

## 6.3. Classification of Cell Lines

### 6.3.1. Distinction of Malignant from Normal Hematopoietic Cell Lines

Normal hematopoietic cells survive in vitro for days or weeks only, even in the presence of physiological or pharmacological stimulators. However, normal hematopoietic cells can be immortalized by certain viruses, the most prominent are EBV and human T-cell leukemia virus (HTLV)-I. A more detailed discussion of this topic and the distinction between normal and malignant hematopoietic cell lines has been provided elsewhere [Drexler, 2000; Drexler et al., 2000b].

### 6.3.2. Classification of LL Cell Lines

There are a number of possibilities for classifying LL cell lines: according to the diagnosis of the patient or according to immunophenotypes, specific cytogenetic aberrations, functional characteristics, or other salient features of the cultured cells. The most often used (and clearly most practical) classification is based on the physiological spectrum of the normal hematopoietic cell lineages [Drexler and Minowada, 2000]. The primary distinction separates lymphoid from myeloid cells (see Table 13.6). Within the lymphoid and myeloid categories, B-cell, T-cell, and NK-cells, on the one hand, and myelocytic, monocytic, erythrocytic and megakaryocytic cells, on the other hand, are discerned.

Undoubtedly, the most useful technique for assigning a cell line to one of the major cell lineages is immunophenotyping. The more extensive and complete the immunoprofile, the more precise is the classification of the cell lineage derivation and of the status of arrested differentiation along this cell axis. Immunomarker expression is also the basis for an even finer subclassification of B- and T-cell lines (Table 13.6). Other techniques may add highly valuable information in cases of uncertain cell lineage assignment. The original diagnosis of the patient is also of importance for the cell lineage assignment.

## 7. BIOLOGICAL SAFETY

Biosafety practices must be followed rigorously during work with human blood or other tissues, including primary LL cells, established LL cell lines, and pathogenic and infectious agents. Some cell lines are

**Table 13.6. Classification of LL cell lines.**

| Main Cell Type | Physiological Cell Lineage | Type and Subtype of Cell Line |
|---|---|---|
| **Lymphoid** | B-cell | Precursor B-cell line |
| | | Mature B-cell line |
| | | Plasma cell line |
| | T-cell | Immature T-cell line |
| | | Mature T-cell line |
| | NK cell | NK cell line |
| **Myeloid** | Myelocytic | Myelocytic cell line |
| | Monocytic | Monocytic cell line |
| | Erythrocytic | Erythrocytic cell line |
| | Megakaryocytic | Megakaryocytic cell line |

For practical purposes, it may be preferable to use the term "erythrocytic-megakaryocytic cell line" and to combine such cell lines under this heading; this may be also true for myelocytic and monocytic cell lines which may be combined under "myelomonocytic" cell lines. Adapted from refs. [Drexler 2000; Drexler et al., 2000b].

virally-infected: for example, EBV and human herpesvirus-8 (HHV-8) are assigned to biological safety risk category 2; HTLV-I/-II and human immunodeficiency virus (HIV) fall into risk category 3. Fresh primary material may contain hepatitis viruses B or C (both in risk category 3).

A standard code of practice, including safety and legal considerations, has been published by the UKCCCR (United Kingdom Coordinating Committee on Cancer Research) Guidelines for the Use of Cell lines in Cancer Research [UKCCCR, 2000].

## 8. COMMENTARIES

### 8.1. General Considerations

Although malignant cells enjoy in vivo a selective growth advantage over normal hematopoietic cells, in vitro LL cells are so difficult to grow and to maintain that attempts to establish cell lines meet much more often with failure than with success. Although currently there is no one single cell culture system, which assures consistent establishment of LL cell lines, several methods for immortalizing neoplastic cells have been developed. The technique of seeding cells in suspension cultures as described in this chapter is certainly the most often used. Other methods recommended by several researchers have their advantages and might meet with success in some attempts.

Growth of LL cells in soft agar or methylcellulose offers the advantage that the colonies formed are discrete and can be removed easily from the supporting medium for further culture in other environments. Thus, a cell line might be established by passaging single colonies.

Some lymphocytic leukemia cells can be immortalized by using transforming viruses. EBV can promote growth of malignant B-cell lines from some patients with mature B-cell malignancies. But the EBV can also transform normal B-lymphocytes. HTLV-I allows the growth of malignant T-cells by inducing the IL-2 receptor. Considering the fact that EBV and HTLV can also transform normal cells, it is necessary to ascertain the malignant origin of the established cell lines by means of karyotype and molecular genetic analysis.

The growth of malignant and normal hematopoietic cells in vitro and in vivo is the result of complex interactions between growth factors and their respective receptors. The addition of some growth factors and cytokines into the culture medium can support the proliferation of the neoplastic cells and induce the formation of cell lines.

As purified or recombinant factors are expensive, the CM of some malignant human or murine cell lines can be used instead. Such cell lines are, for instance, 5637 (an adherent cell line from a patient with bladder carcinoma), Mo-T (HTLV-II transformed T-leukemia cell line), and WEHI-3 (mouse monocytic cell line) [Lange et al., 1987; Quentmeier et al., 1997; Drexler et al., 2001]. These cell lines generate several growth factors that are secreted into the culture supernatant. The supernatant from cultures of these tumor cell lines can be stored at $-20°C$ for several months before use. These CM should be used at a final concentration of 10–20% (v/v). CM from PHA-stimulated lymphocytes is also an ideal and inexpensive source of these molecules.

A number of completely synthetic media, including several media designed specifically for unique types of LL cells, have been used by researchers for establishment and maintenance of cell lines in suspension cultures (listed in Table 13.2). It appears that no single medium is well suited for the growth of all types of LL cells. Should one medium fail to support cell growth, it might become necessary to try another.

### 8.2. Culture Environment

A minimal amount of oxygen is essential for the growth of most types of LL cells in suspension cultures. The majority of cell lines grows well when they are incubated in a humidified $37°C$ incubator with 5% $CO_2$ in air. However, the partial pressure of oxygen ($pO_2$) in normal body fluids is significantly less than that of air. Studies have shown that the $pO_2$ in human bone marrow is 2–5%. This is considerably lower than the $pO_2$ (15–20%) existing in the typical cell culture incubator maintained at 5% $CO_2$ in air. It has been reported that growth of cultured cells could be improved by reducing the percentage of oxygen in the gaseous phase to between 1–10%. Growing LL cells under low oxygen conditions of 6% $CO_2$, 5% $O_2$, and 89% $N_2$ may be a useful method for the establishment of LL cell lines [Smith et al., 1984, 1989; Zhang et al., 1993].

### 8.3. Common Cell Culture Problems

In the attempts to establish LL cell lines, the overgrowth of fibroblasts and normal EBV[+] lymphoblastoid cells is the most common problem. Should the nutrients in the medium become exhausted too quickly, the adherent cells should be removed by passaging the suspension cells into new flasks containing fresh medium. The overgrowth of EBV[+] B-lymphocytes can become visible as early as 2 weeks after seeding of the new culture. The EBV[+] B-cells look small and have irregular contours with some short villi. They proliferate preferentially in big floating clusters or colonies. The colonies can be picked out with a Pasteur pipette, and the EBV genome should be detected as soon as possible.

It is always necessary to freeze aliquots of the fresh primary cells in liquid nitrogen before culture. If a LL cell line should subsequently become established, the original cells can be used as a control for characterization of the established cell line. In case of failure to immortalize a cell line, the frozen primary back-up cells can be used for further attempts.

Maintenance of cell lines requires careful attention. Every cell line appears to have its optimal growth environment; parameters such as culture medium, cell concentration, nutritional supplements, and pH all play a major role. If cell growth becomes suboptimal or if cells inexplicably die during culture, some of the following problems should be considered: suitability of the culture medium or the growth supplements for this particular cell line; proper functioning of the incubator at the appropriate temperature, humidity and $CO_2$ levels; and selection of an adequate cell concentration. Some cell lines clearly grow better in 24-well plates than in culture flasks.

Contamination with mycoplasma, bacteria, fungi, and viruses and cross-contamination with other "foreign" cells are the most common problems encountered in the maintenance of LL cell lines [Drexler et al., 2001; Drexler and Uphoff, 2002]. Therefore, analyses must be undertaken at regular intervals to ensure a contamination-free environment for cell growth. All solutions and utensils coming into contact with cells must be sterilized before use; sterile techniques and good laboratory practices must be followed strictly. Although antibiotics can be added to the culture medium to prevent bacterial infection, they do not usually inhibit virus, fungus, or mycoplasma infection. Therefore, antibiotics such as penicillin and streptomycin are not necessary if care is taken regarding cell culture techniques. Because contamination can cause the loss of valuable cell lines, it is important to cryopreserve a sufficient amount of cell material from each cell line. In order to prevent cellular contamination and misidentification, it is mandatory to use a separate bottle of medium for each cell line. Furthermore, cell culturists should not deal with and feed more than one cell line at the same time.

## 8.4. Time Considerations

Generation of a LL cell line may require 2–6 months. However, a cell culture should not be considered a continuous cell line until the cells have been passaged and expanded for at least half a year, or better for one year (see Table 13.4). Expansion of the cell line takes 2–3 weeks. Depending on the parameters analyzed, 2–4 months might be needed for a thorough characterization of the established LL cell line (Table 13.5).

# SOURCES OF MATERIALS

| Item | Supplier | Catalog Number |
| --- | --- | --- |
| Cell line 5637 | DSMZ | ACC 35 |
| Cell line M-07e | DSMZ | ACC 104 |
| Centrifuge (bench-top) | Kendro | Megafuge 1.0 |
| Centrifuge tubes (15 ml, plastic, Falcon Becton Dickinson) | Omnilab | 352 095 |
| Centrifuge tubes (30 ml, plastic, Universals containers) | Greiner | 201 172 |
| Centrifuge tubes (50 ml, plastic) | Greiner | 722 7261 |
| Cryofreezing container (Cryo Box) | Nalge | 510 000 01 |
| Cryofreezing system (controlled rate freezer) | Messer | Planer Kryo 10 |
| Cryovials (1.8 ml, plastic) | Nunc | 368 632 |
| Cytocentrifuge | Shandon | Cytospin 3 |
| Dimethylsulfoxide | Merck | 1.02931 |
| Entellan | Merck | 1.07961 |
| Fetal bovine serum (FBS) | PAA | A15-002 |
| Ficoll-Hypaque | Pharmacia | 17-1440-03 |
| Filters (0.2 μm, Minisart/Sartorius) | Omnilab | 178 23K |
| Flasks (80 cm$^2$, plastic) | Nunc | 147 589 |
| Giemsa solution | Merck | 1.09204 |
| Granulocyte-macrophage colony-stimulating factor (GM-CSF) | Roche | 1 087 762 |
| Hematocytometer (Neubauer improved) | Omnilab | 9.161.078 |
| Incubator | Forma Scientific | Steri-Cult 200 |
| Interleukin-3 (IL-3) | Roche | 1 381 547 |
| Iscove's modified Dulbecco's medium (IMDM) | Invitrogen | 21980-032 |
| L-glutamine | Invitrogen | 25030-024 |
| May-Grünwald solution | Merck | 1.01424 |
| McCoy's 5A medium | Invitrogen | 26600-023 |
| 2-Mercaptoethanol | Merck | 15 433 |
| Minimum essential medium-alpha (MEM-α) | Invitrogen | 22571-020 |
| Multiwell plates (24-well, plastic) | Nunc | 143 982 |
| Multiwell plates (96-well, plastic, flat-bottomed) | Nunc | 167 008 |
| Nitrogen freezers | Taylor Wharton | |
| Penicillin–streptomycin | Invitrogen | 15070-022 |
| RPMI 1640 medium | Invitrogen | 21875-034 |
| Stem cell factor (SCF) | Roche | 1 485 130 |
| Trypan blue solution | Sigma | T 8154 |
| Trypsin/EDTA | Invitrogen | 25300-054 |
| Weise buffer | Merck | 1.09879 |

Indicated here are only those products that we use and our source of suppliers. Most items (reagents, materials, and equipment) of equal quality are also available from many other suppliers, which, however, cannot be catalogued here.

# REFERENCES

Drexler, H.G. (1998) Review of alterations of the cyclin-dependent kinase inhibitor INK4 family genes p15, p16, p18 and p19 in human leukemia-lymphoma cells. *Leukemia* 12:845–859.

Drexler, H.G., Matsuo, Y., Minowada, J. (1998) Proposals for the characterization and description of new human leukemia-lymphoma cell lines. *Human Cell* 11:51–60.

Drexler, H.G., Matsuo, Y. (1999) Guidelines for the characterization and publication of human malignant hematopoietic cell lines. *Leukemia* 13:835–842.

Drexler, H.G., Dirks, W.G., MacLeod, R.A.F. (1999) False human hematopoietic cell lines: Cross-contaminations and misinterpretations. *Leukemia* 13:1601–1607.

Drexler, H.G. (2000) *The Leukemia-Lymphoma Cell Line FactsBook*. San Diego: Academic Press.

Drexler, H.G., and Matsuo, Y. (2000) Malignant hematopoietic cell lines: In vitro models for the study of multiple myeloma and plasma cell leukemia. *Leukemia Res.* 24:681–703.

Drexler, H.G., and Minowada, J. (2000) Human leukemia-lymphoma cell lines: Historical perspective, state of the art and future prospects. In Masters, J.R.W., Palsson, B.O. (eds) *Human Cell Culture, Vol. III Cancer Cell Lines Part 3: Leukemias and Lymphomas*. Kluwer Academic Publishers: Dordrecht, The Netherlands, pp 1–18.

Drexler, H.G., and Uphoff, C.C. (2000) Contamination of Cell Cultures, Mycoplasma. In Spier, E., Griffiths, B., Scragg, A.H. (eds) *The Encyclopedia of Cell Technology*. Wiley: New York, pp 609–627.

Drexler, H.G., Fombonne, S., Matsuo, Y., Hu, Z.B., Hamaguchi, H., Uphoff, C.C. (2000a) p53 alterations in human leukemia-lymphoma cell lines: In vitro arifact or prerequisite for cell immortalization? *Leukemia* 14:198–206.

Drexler, H.G., Matsuo, Y., MacLeod, R.A.F. (2000b) Continuous hematopoietic cell lines as model systems for leukemia-lymphoma research. *Leukemia Res.* 24:881–911.

Drexler, H.G., and Uphoff, C.C. (2002) Mix-ups and mycoplasma: The enemies within. *Leukemia Res.* 26: 329-333.Drexler, H.G., et al. (eds) (2001) *DSMZ Catalogue of Human and Animal Cell Lines, 8th Edition*, DSMZ: Braunschweig, Germany.

Freshney, R.I. (2000) *Culture of Animal Cells*, 4th ed. John Wiley & Sons: New York, NY, pp 309–312; 331.

Hu, Z.B., MacLeod, R.A.F., Meyer, C., Quentmeier, H., Drexler, H.G. (2002) New acute myeloid leukemia-derived cell line: MUTZ-8 with 5q-. *Leukemia* 16: 1556–1561.

Kees, U.R., Ranford, P.R., Hatzis, M. (1996) Deletions of the p16 gene in pediatric leukemia and corresponding cell lines. *Oncogene* 12:2235–2239.

Lange, B., Valtieri, M., Santoli, D., Caracciolo, D., Mavilio, F., Gemperlein, I., Griffin, C., Emanual, B., Finan, J., Nowell, P., Rovera, G. (1987) Growth factopr requirements of childhood acute leukemia: Establishment of GM-CSF-dependent cell lines. *Blood* 70:192–199.

Lange, B. et al. (1992) Pediatric leukemia/lymphoma with t(8;14)(q24;q11). *Leukemia* 6: 613-618.

Matsuo, Y., and Drexler, H.G. (1998) Establishment and characterization of human B cell precursor-leukemia cell lines. *Leukemia Res.* 22:567–579.

Meyer, C., MacLeod, R.A.F., Quentmeier, H., K. Janssen, J.W.G., Coignet, L.J., Dyer, M.J.S., Drexler, H.G. (2001) Establishment of the B-cell precursor-acute lymphoblastic leukemia cell line MUTZ-5 carrying a /12;13) translocation. *Leukemia* 15:1471–1474.

O'Connor, R., Cesano, A., Lange, B., Finan, J., Nowell, P.C., Clark, S.C., Raimondi, S. C., Rovera, G., Santoli, D. (1991) Growth factor requirements of childhood T-lymphoblastic leukemia: Correlation between presence of chromosomal abnormalities and ability to grow permanently in vitro. *Blood* 77:1534–1545.

Quentmeier, H., Zaborski, M., Drexler, H.G. (1997) The human bladder carcinoma cell line 5637 constitutively secretes functional cytokines. *Leukemia Res.* 21:343–350.

Schneider, U., Schwenk, H.U., Bornkamm, G. (1977) Characterization of EBV-genome negative "Null" and "T" cell lines derived from children with acute lymphoblastic leukemia and leukemic transformed non-Hodgkin lymphoma. *Int. J. Cancer* 19:621–626.

Smith, S.D., Shitsky, M., Cohen, P.S., Warnke, R., Link, M.P., Glader, B.E. (1984) Monoclonal antibody and enzymatic profiles of human malignant T-lymphoid cells and derived cell lines. *Cancer Res.* 44:5657–5660.

Smith, S.D., et al. (1989) Long-term growth of malignant thymocytes in vitro. *Blood* 73:2182–2187.

UKCCCR (2000). UKCCCR guidelines for the use of cell lines in cancer research. *Br. J. Cancer* 82:1495–1509.

Zhang, L.Q., Downie, P.A., Goodell, W.R., McCabe, N.R., LeBeau, M.M., Morgan, R., Sklar, J., Raimondi, S.C., Miley, D., Goldberg, A., Lu, M.M., Montag, A., Smith, S.D. (1993) Establishment of cell lines from B-cell precursor acute lymphoblastic leukemia. *Leukemia* 7:1865–1874.

# 14

# In Vitro Culture of Malignant Brain Tumors

John L. Darling

*Division of Biomedical Science, School of Applied Sciences, University of Wolverhampton, United Kingdom*
*Corresponding author: John L. Darling, Professor of Biomedical Science, Division of Biomedical Science, School of Applied Sciences, University of Wolverhampton, Wulfruna Street, Wolverhampton WV1 1SB United Kingdom. Phone: 01902 321155. Fax: 01902 322714.*
*E-mail: J.Darling@wlv.ac.uk*

*Culture of Human Tumor Cells*, Edited by Roswitha Pfragner and R. Ian Freshney.
ISBN 0-471-43853-7   Copyright © 2004 Wiley-Liss, Inc.

# I. INTRODUCTION

## 1.1. Incidence, Histology, and Prognosis of Malignant Brain Tumors

Within the next 12 months about 6,000 people in the United Kingdom will be diagnosed with a primary brain tumor. The brain is the 10th most common site for the development of cancer in men and about the 12th most common site in women. Brain tumors are the second leading cause of cancer death in children under age 15 and in young adults up to age 34. They are the second fastest-growing cause of cancer death among those over 65, and no behavioral change has been shown to reduce risk. The prognosis of many types of malignant brain tumor is very poor [Berger et al., 1996].

The most common primary brain tumors are derived from supporting glial cells in the brain [Kleihues et al., 1993; Kleihues and Cavenee, 1997]. These tumors, known collectively as glioma, comprise astrocytoma, derived from astrocytes, oligodendroglioma, derived from oligodendrocytes, and ependymoma, derived from ependymal cells.

Low-grade astrocytomas usually express glial filaments composed of the astrocyte specific protein, glial fibrillary acidic protein (GFAP). Anaplastic astrocytomas show less evidence of astrocytic differentiation and display additional sinister features, including nuclear pleomorphism, mitotic activity, and vascular endothelial proliferation, but necrosis is uncommon. Glioblastoma multiforme, the most common glioma, is a highly malignant neoplasm, featuring areas of necrosis bordered by areas of glial cells displaying characteristic pseudopallisading and extensive endothelial proliferation. Both within individual tumors and among different glioblastomas there is often marked heterogeneity in histological appearance. The histological appearance and grading criteria have been codified by using either the WHO grading system [Kleihues and Cavenee, 1997] or the St. Anne/Mayo Clinic scheme [Daumas-Duport et al., 1988] (Table 14.1). The prognosis of these tumors is related to histological grade. Patients with low-grade tumors typically have a 5-year survival rate of 30–50%, whereas anaplastic astrocytoma or glioblastoma multiforme produces virtually no 5-year survivors, and the majority of patients die within 1–2 years of diagnosis.

Oligodendrogliomas comprise about 5% of brain tumors. Although most are well differentiated with little evidence of mitotic activity, about one-third display more aggressive tendencies with increased evidence of mitotic activity, nuclear pleomorphism, and vascular proliferation. The prognosis of patients with well-differentiated oligodendroglioma is similar to those with low-grade astrocytoma. However, it

**Table 14.1. Classification of astrocytoma. A four point scale is commonly applied to the classification of astrocytic brain tumors.**

| WHO designation | WHO grade | St Anne/Mayo Scheme grade | St Anne/Mayo Scheme histological criteria |
|---|---|---|---|
| Pilocytic astrocytoma | I | Grade 1 | No criteria |
| Low grade diffuse astrocytoma | II | Grade 2 | One criterion (nuclear atypia) |
| Anaplastic astrocytoma | III | Grade 3 | Two criteria (nuclear atypia and mitotic activity) |
| Glioblastoma multiforme | IV | Grade 4 | Three + criteria (as above, plus endothelial proliferation and/or necrosis) |

has been recognized that patients with malignant oligodendroglioma usually respond to combination chemotherapy with procarbazine, lomustine (1-(2-chloroethyl)-3- cyclohexyl-1-nitrosourea, CCNU), and vincristine (the PCV protocol) and consequently have a more favorable prognosis than patients with malignant astrocytoma [Cairncross et al., 1998]. This feature appears to be associated with deletions of all, or part, of chromosome 1p and 19q [Cairncross et al., 1998].

Tumors derived from ependymal cells that line the cerebral ventricles and spinal column are rather rare and usually occur in children. They often express GFAP, and cells have a tendency to form rosettes or perivascular pseudorosettes. Anaplastic ependymoma appears to be composed of cells that retain some ependymal differentiation, but also display nuclear atypia, evidence of increased mitotic activity, necrosis, and endothelial cell proliferation.

In children, although low-grade astrocytoma is common, the so-called primitive neuroectodermal tumors (PNET) are the most common malignant tumors of the brain. These usually occur in the cerebellum and display a capacity for divergent differentiation that may be neuronal, astrocytic, ependymal, muscular, or melanotic in nature. Diagnostically, these are now classified into specific entities based on their apparent tissue differentiation; for example, ependymoblastoma and rhabdomyosarcoma. The most common variant is medulloblastoma, which is composed of cells derived from the external granular layer or embryonic cells in the medullary velum. They contain small, round, or oval cells with hyperchromatic nuclei and scanty cytoplasm. The mitotic rate is high, and cells are arranged in sheets, parallel rows, or in Homer-Wright rosettes. Despite its histological anaplasia, medulloblastoma is relatively sensitive to radiotherapy, which produces cures in between 60% and 70% of children.

## 1.2. Cell Culture of Malignant Brain Tumors

It has been recognized for many years that virtually all high-grade astrocytomas can be grown as short-term cultures for between 5 and 7 passages and approximately half give rise to established cell lines [Ponten and Macintyre, 1968; Bigner et al., 1981a; Bigner et al., 1981b; Jacobsen et al., 1987; Westphal et al., 1988; Westphal and Meissner, 1998; Darling, 2000]. Numerous reports have been published, which extensively review the in vitro characteristics of cell lines established from malignant astrocytoma [See Darling, 1991; 2000]. It has also become clear that these short-term cultures display biological features that are consistent with both their neoplastic and astrocytic nature. They are capable of growth on confluent monolayers of normal glia [MacDonald et al., 1985], they produce factors capable of inducing angiogenesis on chick chorioallantoic membrane [Frame et al., 1984], they produce plasminogen activator [Frame et al., 1984], are less sensitive to growth inhibition at high cells density [Westermark, et al. 1973] and some, at least, are capable of producing tumors when implanted subcutaneously or intracranially in nude mice [Bigner et al. 1981]. These short-term cultures are usually aneuploid and often with distinct characteristic karyotypic abnormalities [Bigner et al., 1981b; Rey et al., 1983; Rey, et al. 1987; Rey et al, 1989; Rey et al. 1992; Bello et al., 1994a–d].

The growth and cellular behavior of cultures derived from malignant astrocytoma appear to be influenced by the expression of a wide range of growth factors and their cognate receptors, indicating that there are complex patterns of autocrine and paracrine growth stimulatory pathways in these tumors. One growth factor that seems to be of particular importance is platelet-derived growth factor (PDGF). There is clear experimental evidence that malignant astrocytoma, both in situ [Maxwell et al., 1990; Hermanson et al., 1992] and in vitro [Nister et al., 1991] over-produce PDGF A and B chain and their respective receptors, suggesting that autocrine and paracrine growth stimulatory pathways are important in the pathogenesis of these tumors and that these relationships are maintained in vitro.

Although there is little doubt that these cultures display unequivocal neoplastic features, evidence of astrocytic differentiation has been more difficult to substantiate in vitro. Normal astrocytes characteristically express GFAP as their major intermediate filament protein [Eng et al., 1971], but cells derived from malignant astrocytoma usually do not express this antigen after prolonged culture in vitro. Westphal and colleagues [Westphal et al., 1990] have demonstrated that, although GFAP-positive cells are present in significant numbers in about 50% of primary cultures derived from malignant astrocytoma, by the time these cultures had reached passage level 8, GFAP-positive cells constituted less than 1% of cells in only about 10% of the cultures. The cell type that had come to predominate in these cultures was GFAP-negative but expressed cell surface fibronectin. However, similar cell lines exhibited other astrocytic features, including high-affinity GABA uptake that was inducible by steroids and glutamine synthetase activity [Frame et al., 1984].

In contrast to the malignant astrocytomas, cell lines derived from oligodendrogliomas appear to be very rare. One cell line, KG-1 [Miyake, 1979], has been described derived from a mixed oligoastrocytoma. This culture remained viable for at least 23 passages and was aneuploid, with chromosome numbers ranging from 78 to 139 with a modal number of 102.

There are no reported established cell lines derived from ependymoma, although it has been possible to establish these tumors as xenografts [Horowitz, 1987]. In the past, it has proved difficult to establish cell lines from medulloblastoma [Jacobsen et al., 1985], although it has become clearer that many of these lines, when they do establish, have little propensity to adhere to the growth substratum [Friedman et al., 1983; Friedman et al. 1985; Friedman et al. 1988], suggesting that different strategies need to be used to develop cell lines from these tumors. More recently, using combinations of high glucose concentration and human umbilical cord serum under conditions favoring non-anchorage-dependent cell growth has proved more successful in establishing cell lines from these tumors [Pietsch et al., 1994].

## 2. PREPARATION OF MEDIA AND REAGENTS

### 2.1. Collection and Handling of Biopsy Material

#### 2.1.1. Biopsy Collection Medium

Prepare an antibiotic mix as follows and store frozen at $-20°C$ in aliquots comprising:

(i)   10 ml penicillin (5000 U/ml) and streptomycin (5 mg/ml) solution mixture
(ii)   5 ml kanamycin solution (2.5 mg/ml)
(iii)   5 ml amphotericin B solution (62.5 µg/ml)

For use, remove individual aliquots of antibiotic from the freezer, thaw, and add to a 500 ml bottle of Ham's F10 medium buffered with 20 mM $N$-(2-hydroxyethyl)piperazine-$N$-ethanesulfonic acid (HEPES) producing final concentrations of penicillin, 200 units/ml; streptomycin, 200 µg/ml; kanamycin, 100 µg/ml; and amphotericin B, 2.5 µg/ml. This is then divided into 15 ml aliquots in 30 ml plastic Universal containers and stored at 4°C. Under these conditions, these preparations are stable for at least 2 months.

### 2.1.2. Tumor Biopsies

Samples should be collected into biopsy collection medium immediately after removal to prevent drying out and the possibility of microbial infection. Although samples taken directly from the operating theater are rarely contaminated with microorganisms, samples that are supplied indirectly by neuropathology departments after samples have been taken for diagnostic purposes or samples taken during postmortem are more often contaminated.

We have asked neurosurgeons to provide a minimum of five pieces of tumor tissue (each approximately $2.5 \times 2.5 \times 2.5$ mm) in 15 ml of biopsy collection medium. This provides ample material for culture, as well as DNA extraction for molecular genetic purposes. If samples are not to be processed immediately, they can be stored at 4°C. Under these conditions, the tissue is well oxygenated and viability is maintained for long periods of time. The success rates for samples that have been taken 24–48 h before processing is no different than for samples prepared immediately after removal.

## 2.2. Culture Media and Serum

A wide range of culture media have been used to produce primary cultures and maintain established cell lines from malignant brain tumors. These range from the simple defined media, such as Eagle's Minimal Essential Medium, to more complex formations, such as Dulbecco's modified Eagle's medium (DMEM) to even richer media, such as Ham's F10 and F12 medium [for review, see Darling, 1991]. Fetal bovine serum (FBS) appears to be the most effective mitogen for cells derived from malignant astrocytoma, probably because of the high concentrations of PDGF. In our hands, newborn calf serum has not proven to be capable of supporting the growth of human brain tumor cells. We have no preferred suppliers for FBS, and we batch test from several suppliers usually on an annual basis. Batch testing is mandatory, as there is often considerable variation in the ability of batches of FBS to supplement the growth of astrocytoma cells in vitro. It is unnecessary to heat-inactivate serum before use. Pietsch and colleagues have reported the establishment of medulloblastoma cell lines by using DMEM supplemented with high concentrations of glucose and 10%

v/v human umbilical cord serum, which had been preselected for the enhancement of growth of other medulloblastoma cell lines [Pietsch et al., 1994]. On delivery, stocks of serum should be thawed, aliquoted in 55 ml lots suitable for addition to 500 ml of culture medium, and stored frozen at –20°C in order to avoid repeated freezing and thawing.

### 2.2.1. Complete Growth Medium

Ham's F10 medium, buffered with 20 mM HEPES and supplemented with 10 % v/v FCS, 100 units/ml penicillin, and 100 µg/ml streptomycin.

## 2.3. Enzymes

### 2.3.1. Collagenase

Collagenase, Type IA, is preferred. Check data sheet for specific activity of each batch and dissolve in Hanks' balanced salts solution (HBSS) to give a final concentration of 2000 units/ml. It may be necessary to warm in a 37°C water bath to dissolve all the enzyme powder. If there is any insoluble matter, this should be removed by centrifugation at 3000 rpm for 10–15 min. The supernate is removed and passed through a 0.45 µm filter, then a 0.22 µm filter by using a 50 ml syringe. The enzyme solution is then aliquoted in 1 ml lots in plastic bijou containers and stored frozen at –20°C. For use, aliquots are thawed and diluted with complete growth medium to give a final working concentration of 200 units/ml.

### 2.3.2. Trypsin

0.25% trypsin in HBSS with 0.5 g/l sodium bicarbonate and antibiotics, but without calcium, magnesium, and Phenol Red, (Gibco trypsin 0.25%). To avoid repeated freeze/thawing on delivery, trypsin is aliquotted into 10 and 20 ml lots that are stored at –20°C until use.

## 2.4. Other Reagents

### 2.4.1. Poly-L-lysine Solution (PLL)

PLL 13.1 µg/ml in distilled water.

### 2.4.2. Hoechst 33258

Prepare a stock solution of Hoechst dye (bisbenzimide Hoechst No. 33258) by adding 5 mg of Hoechst 33258 dye to 100 ml of HBSS without Phenol Red and stirring for 30 min at room temperature until all the solute is dissolved. The solution can be stored at 4°C in a glass bottle wrapped in aluminum foil. For use, 0.5 ml of stock solution is added to 100 ml HBSS without Phenol Red, filtered through a 0.45 μm syringe filter, and stored in the dark at 4°C.

### 2.4.3.  Acidified Ethanol

1 ml hydrochloric acid in 99 ml of 70% v/v ethanol in distilled water.

### 2.4.4.  Propidium Iodide

Dissolve 40 μg/ml in HBSS or in glycerol.

### 2.4.5.  PBSA

Dulbecco's phosphate buffered saline solution A (lacking $Ca^{2+}$ and $Mg^{2+}$).

## 3.  METHODS FOR PRODUCING SHORT-TERM CULTURES

### 3.1.  Monolayer Cell Cultures Produced from Collagenase Disaggregated Material

This technique, which has been found to be satisfactory for the majority of biopsy samples derived from human brain tumors, is a modification of the one described by Freshney and colleagues [Freshney, 1972; Guner et al., 1977]. The major modification is the reduction of the length of enzyme exposure from 24 to 4 h. It has the advantage that collagenase can be used in the presence of FBS, unlike trypsin, thereby enhancing cell viability. Crude preparations of collagenase are to be preferred to highly purified preparations of the enzyme. This is, presumably, because of the nonspecific proteases present in crude preparations that may contribute significantly to the disaggregating properties of these preparations.

### Protocol 14.1.  Collagenase Disaggregation and Primary Culture of Brain Tumors

*Reagents and Materials*
*Sterile*
- ❑  Plastic Petri dishes, 5 cm diameter
- ❑  Biopsy collection medium (see Section 2.1.1)
- ❑  Scalpel blade holders and scalpel blades (number 11.)

□  Collagenase solution (see Section 2.3.1)
□  Complete growth medium (see Section 2.2.1)
□  Cell culture flasks (25 cm$^2$ growth area)

***Protocol***

(a)  Transfer tumor biopsies to a 5-cm-diameter plastic Petri dish, together with some fresh biopsy collection medium and, using forceps and a sterile scalpel, remove obvious non-tumor material.

(b)  Transfer the tumor fragments to a fresh Petri dish with medium, and chop into pieces approximately 1 mm × 1 mm by using crossed scalpels. These are sterilized before use by flaming in ethanol [Freshney, 2000]. It is very important to change the blades between samples to ensure that the blades are sharp and that the tumor fragments are cleanly cut and not mangled. This produces a larger number of viable cells from a given weight of tumor.

(c)  To remove necrotic non-viable material, transfer the tissue fragments to a 30 ml universal container and add 10–15 ml of biopsy collection medium. The larger pieces of tissue are allowed to settle, and the supernatant is discarded. Repeat this process twice.

(d)  After the final wash, remove the majority of the supernate and resuspend the tissue fragments in complete growth medium.

(e)  Incubate the tissue in collagenase solution for 4 h at 37°C. Monitor disaggregation periodically by using an inverted microscope until the tissue is reduced to small cell aggregates.

(f)  Pipette the preparation several times to promote further disaggregation. Transfer the tissue suspension to a universal container, and centrifuge at 250 g for 5 min and discard the enzyme solution.

(g)  Resuspend the cells in complete growth medium, and transfer to a 25 cm$^2$ cell culture flask and incubate at 37°C overnight.

(h)  The next day discard the medium and non-adherent material. Refeed the culture with fresh complete growth medium and return to the incubator. This process removes dead cells that may autolyze and releases non-specific proteases that interfere with tumor cell adhesion.

(i)  Examine cultures twice a week and re-feed as required (see below). As buffering in this system is achieved by HEPES, flasks can be incubated at 37°C with their caps tightly sealed in a non-humidified incubator or hot room. This markedly reduces microbial contamination and provides superior buffering capacity than bicarbonate/$CO_2$.

Although there is some variation in the time taken for primary cultures to become confluent, most are capable of being passaged within 2–4 weeks of initial plating.

### 3.2. Other Disaggregation Methods

Other workers have used alternative, often more aggressive enzyme preparations containing trypsin or enzyme cocktails or mechanical methods to successfully produce short-term cell lines and established cell lines from human brain tumors as reviewed by [Darling, 1991].

### 3.3. Explant Cultures from Small Surgical Biopsies

Explanted biopsy material has been used successfully to grow short-term cultures from human brain tumors. For these to be successful, it is essential that the tumor fragments be anchored firmly to the growth substratum. We have developed a method for small biopsy samples taken during stereotactic neurosurgical operations.

## Protocol 14.2. Primary Explant Culture of Brain Tumors

### Reagents and Materials

*Sterile*
- ❑ Silicone grease (sterilized in a hot-air oven)
- ❑ Glass coverslips (9 × 22 mm, No. 1 thickness, sterilized in a hot-air oven).
- ❑ Plastic cell culture flasks, 25 cm$^2$
- ❑ Forceps
- ❑ Plastic pipette, 1 ml
- ❑ Biopsy collection medium (see Section 2.1.1)
- ❑ Complete growth medium (see Section 2.2.1)

### Protocol

(a) Wash tumor biopsies in biopsy collection medium.

(b) Using a sterile 1 ml plastic pipette, place two small spots of silicone grease about 1 cm apart on the base of a 25 cm$^2$ cell culture flask.

(c) Place a single tumor fragment in a drop of biopsy collection medium between the spots of silicone grease by using a pair of sterile forceps.

(d) Place a sterile glass coverslip over the fragment and spots of silicone grease, and gently press it down by using sterile forceps until the tumor biopsy is held firmly down against the plastic surface and the coverslip is firmly anchored by the grease. It is essential that the tumor fragment is held down with some of the biopsy under the coverslip and some outside.

(e) Add 5 ml of complete growth medium slowly down the side of the flask in order not to dislodge the explant.

(f) Incubate at 37°C for 24–48 h, and then add a further volume of complete growth medium to bring the total volume to 10 ml per flask.

(g) Examine the flask regularly for signs of cell proliferation; once this has occurred, gently remove the coverslip and the remains of the tumor biopsy with a pair of sterile forceps and refeed the culture with complete growth medium.

(h) When a large area of cells has developed the cultures can be passaged in the normal manner into larger culture flasks.

## 3.4. Other Explant Methods

Cell adhesion may be accomplished in other ways, including allowing tumor fragments to dry slightly to the growth surface of a plastic flask. Adherence is effected by incubating the flask for 15–20 min with the flask growth surface orientated normally, or by inverting the flask and placing it in a well-humidified incubator at 37°C for 30 min before the addition of a small amount of culture medium. Once there are signs of cell proliferation, the volume of culture medium can be increased progressively over a period of 1–2 weeks, and the remains of the biopsy sample can be removed.

## 3.5. Re-feeding, Subculture, and Contamination Testing

### 3.5.1. Re-feeding Cultures

Primary cultures should be re-fed the day after plating, following enzymatic disaggregation, to remove any dead cells and residual enzyme. Thereafter, cultures should be re-fed every 2–3 days. Cultures that are slow growing need feeding less often. To feed the cultures, remove the old medium by using an aspiration system and add fresh complete growth medium by using an appropriate-sized plastic pipette. Always use plastic pipettes for transferring cell culture medium. Never pour medium into or out of culture flasks, as this contributes to cell line cross contamination, through the generation of aerosols, and to the spread of microbial contamination.

### 3.5.2. Passaging Cultures

### Protocol 14.3. Subculture of Brain Tumor Derived Primary Cultures

*Reagents and Materials*

*Sterile*
❑ HBSS
❑ Trypsin solution (thawed aliquot from Section 2.3.2)
❑ Complete growth medium (see Section 2.2.1)

### Protocol

(a)  Remove spent medium with an aspiration device, and wash the monolayer three times with 10 ml of HBSS.

(b)  Rinse the monolayer with 3 ml of trypsin solution, and incubate the flask for between 5 and 10 min at 37°C.

(c)  At periodic intervals, examine the monolayer and, as soon as the cells begin to detach, add 7 ml of complete growth medium to each flask to inactivate the trypsin.

(d)  Pipette the medium to resuspend the cells and break down any large cell aggregates.

(e)  Transfer the cell suspension to a sterile universal, and centrifuge for 5 min at 250 $g$.

(f)  Remove the supernate and resuspend the cells in 10 ml of fresh complete growth medium.

(g)  Count the cells and use for either freezing, further passage, or experiments.

It is our practice to passage the total contents of our primary cultures from 25 cm$^2$ flasks into a 75 cm$^2$ flask (in effect, a 1 to 3 split). If larger cell numbers are required, a further passage of the contents of the 75 cm$^2$ flask into one or two 175 cm$^2$ flasks is performed (Fig. 14.1).

Once the primary culture has been passaged successfully, we do not find that it is necessary to add penicillin and streptomycin to the culture medium. This has a marked advantage because the routine use of antibiotics may mask low-level microbial contamination, sloppy aseptic technique, or the development of antibiotic resistant strains of bacteria within the laboratory. It is also unnecessary to switch cultures to antibioticfree medium for a period of time before routine Mycoplasma testing.

### 3.5.3.  Procedures to Avoid Cell Line Cross-Contamination

Over the past thirty years, biomedical scientists have produced numerous cell lines from a wide variety of human cancers that have high plating efficiencies and short doubling times. This has led to a large number of cases where it is apparent that some lines have become contaminated with cells from another. The use of single-wrapped pipettes, together with enforcing a rule that once a pipette has been used it must be immediately discarded, markedly reduces the possibility of cell line cross-contamination. However, as it does not reduce the possibility to zero, and we have adopted the following addition safeguards:

(i)  Each cell line is fed from its own bottle of medium. When initiating short-term cultures, it is usually sufficient to aliquot 50–100 ml

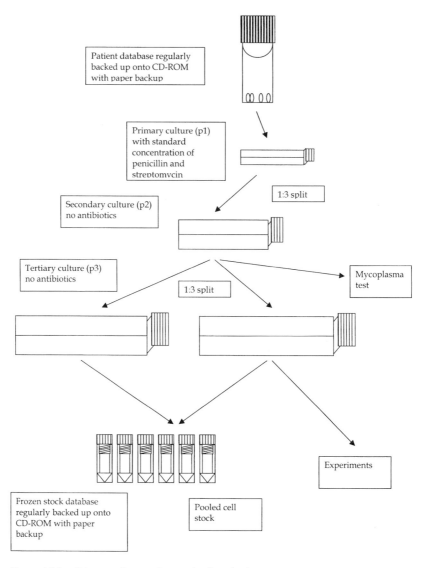

**Figure 14.1.** Primary culture and expansion by subcultures.

of complete growth medium into an appropriate sterile container. Single 500 ml bottles are allocated to cell lines that are being used for more extensive series of experiments. All media bottles must be labeled with the cell line to which it is allocated and the name of the individual research worker.

(ii) All ancillary bottles containing, for example, HBSS and other reagents such as trypsin, must also be allocated to individual cultures.

(iii) Any established cell lines must be attended to after any manipulation of short term lines.

(iv) Only one culture and its allocated bottles of culture medium and other reagents should be in the cell culture hood at any one time.

(v) Record keeping and labeling of flasks and freezing vials must be meticulous, as cell line cross-contamination or misidentification may result from misreading labels on vessels or records cards or computer databases.

### 3.5.4. Mycoplasma Screening

It is essential that all cultures taken from human tumors be examined to ensure that there is no evidence of Mycoplasma infection before frozen stocks are prepared and that regular rechecking takes place thereafter on all cultures. In general cell culture laboratories, this can be performed most easily by using DNA staining.

## Protocol 14.4. Mycoplasma Screening of Brain Tumor Derived Cultures

### *Reagents and Materials*

*Sterile*
- ❑ Poly-L-lysine solution (PLL, 13.1 μg/ml in distilled water)
- ❑ HBSS
- ❑ Coverslips (13 mm, round, glass coverslips, sterilized by dry heat)
- ❑ Forceps
- ❑ Multi-well plates, 24-well
- ❑ Micropipette with sterile tips
- ❑ Mylar plastic plate sealers

*Non-sterile*
- ❑ Methanol
- ❑ Hoechst 33258 (see Section 2.4.2)
- ❑ Glycerol/PBSA solution
- ❑ Clear nail varnish
- ❑ Fluorescence microscope equipped with an LP 440 nm barrier filter and a 330/380 nm excitation filter

### *Protocol*
(a) Place a sterile coverslip into each well of a 24-well plate aseptically and add about 0.5 ml of PLL.

(b) Leave for 45 min, aspirate the PLL solution, and leave to dry for a few minutes in a cell culture cabinet, with the lid of the plate slightly ajar to aid evaporation.

(c) Prepare a cell suspension as described in Protocol 14.3.

(d) Add about 50 μl of cell suspension into each well by using a Gilson pipette with a sterile tip.

(e) Add 1 ml of complete growth medium to each well.

(f) Seal with a self-adhesive Mylar plate sealer.

(g) After 3–4 days incubation, or when the culture is 60% confluent, remove the coverslip and place cells uppermost on a humidified staining tray.

(h) Wash the coverslip gently with HBSS.

(i) Fix the cells for 10 min with 100% methanol cooled to −20°C.

(j) Wash again with HBSS.

(k) Flood each coverslip with a solution of Hoechst stain for 30 min at room temperature.

(l) Wash with HBSS three times and twice with distilled water.

(m) Mount the coverslip in a solution of glycerol/PBSA with the cells downwards.

(n) Seal the coverslip with clear nail varnish.

(o) The stained preparations should be examined with a fluorescence microscope under a 100× oil immersion objective. The nuclei of the cells will be stained brightly. However, the presence of stained material between cells or over the cytoplasm of cells, in the form of discrete dots or strings of dots indicates the presence of Mycoplasma contamination.


### 3.5.5. Cell Freezing

**Protocol 14.5. Cryopreservation of Brain Tumor Derived Cultures**

*Reagents and Materials*
*Sterile*
❑ 10% v/v solution of dimethysulphoxide (DMSO) in neat FBS. This should be prepared immediately before use and not stored.
❑ Complete growth medium (see Section 2.2.1)
❑ Cryovials: Plastic freezing vials and colored plastic inserts for the caps. To simplify the retrieval of individual vials, we label the color-coded plastic inserts that fit within the cap of each vial with the same cell line information as on the side of the vial.


*Protocol*
(a) Examine cultures, selecting those in mid- to late-exponential phase.

(b) Prepare a cell suspension as described above (Protocol 14.3), and count the cells.

(c) Centrifuge at 100 *g* for 3 min and discard the supernate.

(d) Resuspend the cell pellet in DMSO/serum solution at a density of $1 \times 10^6$ cells per ml, and transfer to cryovials in 1 ml aliquots.

(e) Tighten the vial caps finger-tight, and immediately place in a polystyrene box with approximately 2-cm-thick walls.

(f) Place the box in a $-70°C$ freezer for at least 8 h, and then transfer the vials to the liquid phase of a liquid nitrogen refrigerator.

ΔSafety note. Wear cryoprotective gloves, a face-mask, and a closed lab-coat, when working with liquid nitrogen. Make sure the room is well ventilated to avoid asphyxiation from evaporated nitrogen.

(g) To revive cells, remove the vial from the liquid nitrogen and drop immediately into a covered plastic beaker containing water at 37°C, to avoid the possibility of injury should any liquid nitrogen trapped within an incompletely sealed vial cause an explosion.

(h) Once the cell suspension has thawed, swab to outside of the vial with 70% ethanol.

(i) Transfer the cell suspension to a 75-cm² flask, dilute slowly with 10 ml complete growth medium, and incubate at 37°C overnight.

(j) The next day, remove the culture medium and re-feed the cells with complete growth medium to remove any residual DMSO.

## 4. DETERMINING THE ANTIGENIC PHENOTYPE OF BRAIN TUMOR CULTURES

In the following section, three generic protocols are provided to visualize three classes of antigens useful in the characterization of brain tumor cultures, namely intracytoplasmic antigens, such as intermediate filaments, nuclear antigens, such as p53 protein, and cell surface antigens, such as growth factor receptors.

### 4.1. Staining Cytoplasmic Antigens

### Protocol 14.6. Immunostaining of Cytoplasmic Antigens in Brain Tumor Derived Cultures

**Reagents and Materials**

*Sterile*
❑ HBSS or phosphate buffered saline (PBSA)
❑ Acidified ethanol (see Section 2.4.3.)
❑ Chamber slides, 8-well

*Non-sterile*

❑ Antibodies: Anti-GFAP, anti-vimentin and other intermediate filament protein antibodies (see Sources of Materials at the end of this chapter.)

❑ Propidium iodide in HBSS (see Section 2.4.4)

❑ Citifluor aqueous mountant

❑ Fluoresence microscope

**Protocol**

(a) Prepare a cell suspension as described in Protocol 14.3.

(b) Seed into chamber slides at a density of 5000 cells in 0.5 ml of complete growth medium per well.

(c) Incubate at 37°C for 2–4 d, until 60–85% confluence is achieved.

(d) Wash the cells with HBSS or PBSA three times, and fix in acidified 70% ethanol for 10 min.

(e) Wash three more times with HBSS, and apply the primary antibody (Table 14.2) for 2 h at room temperature. Keep slides in a humidified environment to prevent evaporation.

(f) Wash another three times with HBSS or PBSA, and apply the species-specific biotinylated secondary antibody (Table 14.2) for 30 min.

(g) Wash the cells another three times, and add the fluorochrome biotin-streptavidin complex (Table 14.2), and incubate for 15 min. Wash another three times in HBSS or PBSA.

(h) Briefly immerse coverslip or slides in propidium iodide in HBSS for 1 to 2 s between the final washes to give the nuclei a red counterstain.

(i) Mount the cells in Citifluor, and examine with an fluorescence microscope.

**Table 14.2.  Type and working dilution of intermediate filament protein antibodies.**

| Antibody | Antigen | Source | Working Dilution | Supplier |
|----------|---------|--------|------------------|----------|
| Primary | Glial fibrillary acidic protein (GFAP) | mouse (mono) | 1:10 | Amersham RPN 1106 |
| | Neurofilament protein (NF) (160kD) | mouse (mono) | 1:10 | Amersham RPN 1104 |
| | Vimentin (VIM) | mouse (mono) | 1:10 | Amersham RPN 1102 |
| | Cytokeratin (CYT) | mouse (mono) | 1:10 | Amersham RPN 1100 |
| | Desmin (DES) | mouse (mono) | 1:10 | Amersham RPN 1101 |
| Secondary | Anti-mouse Ig | sheep (mono) | 1:50 | Amersham RPN 1001 |
| | Anti-rabbit Ig | donkey (poly) | 1:50 | Amersham RPN 1004 |

## 4.2. Staining Nuclear Antigens

## Protocol 14.7. Immunostaining of Nuclear Antigens in Brain Tumor Derived Cultures

### Reagents and Materials

*Sterile*
- ❏ Chamber slides, 8-well
- ❏ Complete growth medium

*Non-sterile*
- ❏ Acetone/methanol solution (1:1)
- ❏ PBSA solution (see Section 2.5.5)
- ❏ PBSA with 0.1% Tween 20
- ❏ Three mouse monoclonal antibodies against p53 protein can be used, D0-1 and PAb421, which recognize both wild-type and mutant forms of the protein, and PAb240, which recognizes only the mutant form. Antibodies are in the form of tissue culture supernatant and are used undiluted. All incubations are performed at room temperature.
- ❏ Biotinylated sheep anti-mouse Ig at 1:50 dilution in 5% FBS in PBSA
- ❏ Streptavidin fluorescein conjugate, 1:50 dilution in PBSA
- ❏ Citifluor mountant

### Protocol

(a) Prepare a cell suspension as described in P14.3.
(b) Seed cells into eight-well glass chamber slides at 5000 cells in 0.5 ml of complete growth medium per well. Incubate for 3–4 days at 37°C or until approximately 50% confluent.
(c) Fix the cells in ice-cold acetone/methanol for 2 min, and wash in PBSA.
(d) Incubate cells with 100 μl of monoclonal antibody for 2 h.
(e) Remove the primary antibody, and wash cells once in PBSA with 0.1% Tween 20 and twice in PBSA alone.
(f) Incubate with 100 μl biotinylated second antibody for 2 h.
(g) Remove the antibody, and wash the cells once in PBSA with 0.1% Tween 20 and twice in PBSA alone.
(h) Incubate the cells with the steptavidin conjugate for 40 min.
(i) Remove the conjugate, and wash the cells once in PBSA with 0.1% Tween 20 and twice in PBSA alone.
(j) Air-dry the slides, and mount in Citifluor.

### Protocol 14.8.   Immunostaining of Cell Surface Antigens in Brain Tumor Derived Cultures

*Reagents and Materials*

*Sterile*

❏ Cells on coverslips (see Protocol 14.4) or 8-well chamber slides (see Protocol 14.6, 14.7)

*Non-sterile*

❏ Mouse monoclonal antibodies against platelet-derived growth factor (PDGF) alpha receptor subunit (Ab-3) or insulin-like growth factor receptor type I (ILGF-I receptor, Ab-1). Dilute each antibody to 10 µg/ml in PBSA containing 0.1% albumin.

❏ PBSA

❏ Biotinylated anti-mouse IgG at a working strength of 5 µg/ml

❏ Solution of FITC-avidin, 2 µg/ml

❏ Propidium iodide in glycerol (see section 2.5.4)

*Protocol*

(a)   Grow cells on coverslips or in multi-chamber slides, as described in Protocol 14.4, 14.6, 14.7, until the cells are about 50% confluent.

(b)   Remove the culture medium and wash three times in PBSA, removing as much of the last wash as possible.

(c)   Add 100 µl of diluted primary antibody to each well, and allow to disperse completely across the cell monolayer.

(d)   Incubate in a humidified atmosphere at room temperature for 1 h.

(e)   Aspirate the antibody solution, and wash cells three times in PBSA.

(f)   Add 100 µl biotinylated second antibody to each well.

(g)   Incubate for 45 min at room temperature.

(h)   Aspirate the antibody solution, and wash cells three times in PBSA.

(i)   Add 100 µl of FITC-avidin to each well.

(j)   Incubate for 15 min at room temperature in the dark.

(k)   Aspirate and wash cells five times in PBSA.

(l)   Mount in a solution of propidium iodide in glycerol to counterstain the nuclei.

## 5.   SAFETY

The safety precautions that are necessary for culturing human brain tumors are broadly similar for other human cancers, with the added caveat that human brain tissue is being cultured. Samples from patients who have or who are suspected to have hepatitis, HIV, or tuberculosis should not be processed. Laboratory workers who use human tissue should be encouraged to obtain immunization against hepatitis B. All samples and cultures derived from human brain tumors must be handled in a Class II biological safety cabinet. Laboratory coats should always be worn, and these are reserved for use only in the cell culture laboratory. Latex gloves should always be worn when handing either cultures or samples. These should be changed regularly during the working day to ensure they do not tear.

All waste must be discarded into plastic waste boxes that are discarded after incineration by an authorized contractor. Sharps (needles and scalpel blades) must be discarded into appropriate containers. Scalpel blades should only be attached and detached from handles by using appropriate safety aids.

Blunt hypodermic needles should be used for aspirations of small volumes of liquid. Sharp hypodermic needles are dangerous with not only the risk of producing a needle stick injury; when they are used on a vacuum line to aspirate tissue culture supernatants from microtitration plates or multichamber slides, they occasionally "catch" in the plastic of the plate, which results in the plate being jerked out of the hand of the experimenter and causing the contents to spill and contaminate other wells on the plate, spreading potentially toxic or hazardous chemicals, such as cytotoxic drugs or radioactive isotopes within the cell culture hood.

Although in the past cell culture medium was aspirated routinely from flasks by using unplugged glass Pasteur pipettes attached to a vacuum line, for safety reasons these pipettes are now being replaced with single-wrapped, unplugged polystyrene aspiration pipettes, which are available in a number of different lengths for use in different sizes of flasks.

## ACKNOWLEDGMENTS

The author would like to thank the Samantha Dickson Research Trust, Charlie's Challenge, the United Kingdom Brain Tumour Society (UKBTS), Cancer Research Campaign, the Brain Research Trust, and the Medical Research Council who currently support or have supported work in his laboratory. The assistance of Tracey Collins, Dr. Sally Ashmore, Karen Bevan, Suzanne Clark, and Dr. Grazyna Lewandowicz in the development of these protocols over the past 10 years is gratefully acknowledged.

## SOURCES OF MATERIALS

| Item | Supplier |
|------|----------|
| Amphotericin B | ICN |
| Albumin (BSA) | Sigma |
| Aspiration pipettes, unplugged | Corning Costar |
| Bijou containers | Bibby-Sterilin |
| Biotinylated anti-mouse IgG | Vector Laboratories |
| Biotinylated anti-mouse IgG | Vector Laboratories |
| Chamber Slides | Nalge Nunc |
| Collagenase, Type IA | Sigma |
| Cryovials: Cryocolour color-coded | Nalge Nunc |
| DMSO | Amersham Pharmacia |
| FITC-avidin | Sigma |
| Flasks, plastic, cell culture | Vector Laboratories |
| GFAP monoclonal antibody | Nalge Nunc |
| Glass coverslips | Chance Propper |
| Hanks' balanced salts solution (HBSS) | Oncogene Science |
| ILGF-I receptor antibody, Ab-1 | Gibco |
| Intermediate filament protein antibodies (see Table 14.2) | Amersham Pharmacia |
| Kanamycin | ICN |
| Micropipettes | Gilson |
| Monoclonal antibodies | Amersham Pharmacia |
| Mountant | Citifluor |
| Multichamber slides (Lab-Tek Chamber Slides) | Nalge Nunc |
| Multi-well plates | Nalge Nunc |
| Mylar plastic plate sealers | ICN; Gibco |
| p53 antibodies: D0-1, PAb421, PAb240. | Oncogene Science |
| PBSA | Sigma |
| PDGF alpha receptor subunit antibody, Ab-3 | Oncogene Science |
| Penicillin/streptomycin | Gibco |
| Petri dishes | Nalge Nunc |
| Polyclonal antibodies | Dako |
| Poly-L-lysine | Sigma |
| Scalpel handles and blades | Swan Morton |
| Silicone grease; Edwards High Vacuum grease | Girovac |
| Tween 20 | Sigma |
| Universal containers | Nalge Nunc |
| Vitmentin monoclonal antibody | Amersham Pharmaci |

## REFERENCES

Bello M.J., de Campos J.M., Kusak M.E., Vaquero J., Sarasa J.L., Pestana A., Rey J.A. (1994a) Molecular analysis of genomic abnormalities in human gliomas. *Cancer Genet. Cytogenet.* 73(2):122–129.

Bello M.J., de Campos J.M., Kusak M.E., Vaquero J., Sarasa J.L., Petana A., Rey J.A. (1994b) Ascertainment of chromosome 7 gains in malignant gliomas by cytogenetic and RFLP analyses. *Cancer Genet. Cytogenet.* 72(1):55–58.

Bello M.J., de Campos J.M., Vaquero J., Kusak M.E., Sarasa J.L., Pestana A., Rey J.A. (1994c) Molecular and cytogenetic analysis of chromosome 9 deletions in 75 malignant gliomas. *Genes Chrom. Cancer* 9(1):33–41.

Bello M.J., Vaquero J., de Campos J.M., Kusak M.E., Sarasa J.L., Saez-Castresana J., Pestana A., Rey J.A. (1994d) Molecular analysis of chromosome 1 abnormalities in human gliomas reveals frequent loss of 1p in oligodendroglial tumors. *Int. J. Cancer* 57(2):172–175.

Berger, M., et al. (1996) Primary central nervous system tumors of the supratentorial compartment. In Levin, V. (ed) Cancer in the Nervous System. Churchill Livingstone: New York, NY; 57–126.

Bigner D.D., Bigner S.H., Ponten J., Westermark B., Mahaley M.S., Ruoslahti E., Herschman H., Eng L.F., Wikstrand C.J. (1981a) Heterogeneity of genotypic and phenotypic characteristics of fifteen permanent cell lines derived from human gliomas. *J. Neuropathol. Exp. Neurol.* 40(3):201–229.

Bigner S.H., Bullard D.E., Pegram C.N., Wikstrand C.J., Bigner D.D. (1981b) Relationship of in vitro morphologic and growth characteristics of established human glioma-derived cell lines to their tumorigenicity in athymic nude mice. *J. Neuropathol. Exp. Neurol.* 40(4):390–409.

Cairncross J.G., Ueki K., Zlatescu M.C., Lisle D.K., Finkelstein D.M., Hammond R.R., Silver J.S., Stark P.C., Macdonald D.R., Ino Y., Ramsay D.A., Louis D.N. (1998) Specific genetic predictors of chemotherapeutic response and survival in patients with anaplastic oligodendrogliomas. *J. Natl. Cancer Inst.* 90(19):1473–1479.

Darling, J.L. (1991) Brain. In Masters J.R.W.[D9]. (ed) Human Cancer In Primary Culture. Kluwer Academic Publishers Dordrecht, The Netherlands; 231–251.

Darling, J.L. (2000) Neuronal and glial tumours in vitro: An overview. In Doyle, A., and Griffiths, J.B. (eds) Cell and Tissue Culture for Medical Research. Chichester, John Wiley and Sons Ltd, 306–320. [D11]

Daumas-Duport[D12], C., Scheichauer, B., O'Fallon. J., and Kelly, P. (1988) Grading of astrocytomas. A simple and reproducible method. *Cancer* 15; 62(10):2152–2165.

Frame M.C., Freshney R.I., Vaughan P.F., Graham D.I., Shaw R. (1984). Interrelationship between differentiation and malignancy-associated properties in glioma. *Br. J. Cancer* 49(3):269–280.

Freshney, R.I. (1972) Tumour cells disaggregated in collagenase. *Lancet* 2(7775):488–489.

Freshney, R.I. (2000). Culture of Animal Cells, a Manual of Basic Technique, New York, John Wiley & Sons, p 72.

Friedman H.S., Bigner S.H., McComb R.D., Schold S.C. Jr., Pasternak J.F., Groothuis D.R., Bigner D.D. (1983) A model for human medulloblastoma. Growth, morphology, and chromosomal analysis in vitro and in athymic mice. *J. Neuropathol. Exp. Neurol.* 42(5): 485–503.

Friedman H.S., Burger P.C., Bigner S.H., Trojanowski J.Q., Brodeur G.M., He X.M., Wikstrand C.J., Kurtzberg J., Berens M.E., Halperin E.C., et al. (1988) Phenotypic and genotypic analysis of a human medulloblastoma cell line and transplantable xenograft (D341 Med) demonstrating amplification of c-myc. *Am. J. Pathol.* 130(3):472–484.

Friedman H.S., Burger P.C., Bigner S.H., Trojanowski J.Q., Wikstrand C.J., Halperin E.C., Bigner D.D. (1985) Establishment and characterization of the human medulloblastoma cell line and transplantable xenograft D283 Med. *J. Neuropathol. Exp. Neurol.* 44(6):592–605.

Guner M., Freshney R.I., Morgan D., Freshney M.G., Thomas D.G., Graham D.I. (1977) Effects of dexamethasone and betamethasone on in vitro cultures from human astrocytoma. *Br. J. Cancer* 35(4):439–447.

Hermanson M., Funa K., Hartman M., Claesson-Welsh L., Heldin C.H., Westermark B., Nister M. (1992) Platelet-derived growth factor and its receptors in human glioma tissue: expression of messenger RNA and protein suggests the presence of autocrine and paracrine loops. *Cancer Res.* 52(11):3213–3219.

Horowitz, M.E., Parham, D.M., Douglass, E.C., Kun, L.E., Houghton, J.A., Houghton, P.J. (1987) Development and characterization of human ependymoma xenograft HxB. *Cancer Res.* 47:499–504.

Jacobsen P.F., Jenkyn D.J., Papadimitriou J.M. (1985) Establishment of a human medulloblastoma cell line and its heterotransplantation into nude mice. *J. Neuropathol. Exp. Neurol.* 44(5):472–485.

Jacobsen P.F., Jenkyn D.J., Papadimitriou J.M. (1987) Four permanent cell lines established from human malignant gliomas: three exhibiting striated muscle differentiation. *J. Neuropathol. Exp. Neurol.* 46(4):431–450.

Kleihues P., Burger P.C., Scheithauer B.W. (1993) The new WHO classification of brain tumours. *Brain Pathol.* 3(3):255–268.

MacDonald C.M., Freshney R.I., Hart E., Graham D.I. (1985) Selective control of human glioma cell proliferation by specific cell interaction. *Exp. Cell Biol.* 53(3):130–137.

Maxwell M., Naber S.P., Wolfe H.J., Galanopoulos T., Hedley-Whyte E.T., Black P.M., Antoniades H.N. (1990) Coexpression of platelet-derived growth factor (PDGF) and PDGF-receptor genes by primary human astrocytomas may contribute to their development and maintenance. *J. Clin. Invest.* 86(1):131–140.

Nister M., Claesson-Welsh L., Eriksson A., Heldin C.H., Westermark B. (1991) Differential expression of platelet-derived growth factor receptors in human malignant glioma cell lines. *J. Biol. Chem.* 266(25):16755–16763.

Pietsch T., Scharmann T., Fonatsch C., Schmidt D., Ockler R., Freihoff D., Albrecht S., Wiestler O.D., Zeltzer P., Riehm H. (1994) Characterization of five new cell lines derived from human primitive neuroectodermal tumors of the central nervous system. *Cancer Res.* 54(12):3278–3287.

Ponten, J., and Macintyre, E.H. (1968) Long term culture of normal and neoplastic human glia. *Acta Pathol. Microbiol. Scand.* 74(4):465–486.

Rey J.A., Bello M.J., de Campos J.M., Benitez J., Ayuso M.C., Valcarcel E. (1983) Chromosome studies in two human brain tumors. *Cancer Genet. Cytogenet.* 10(2):159–165.

Rey J.A., Bello M.J., de Campos J.M., Kusak M.E., Ramos C., Benitez J. (1987) Chromosomal patterns in human malignant astrocytomas. *Cancer Genet. Cytogenet.* 29(2):201–221.

Rey, J.A., et al. (1989) Cytogenetic follow-up from direct preparation to advanced in vitro passages of a human malignant glioma. *Cancer Genet. Cytogenet.* 41(2):175–183.

Rey J.A., Pestana A., Bello M.J. (1992) Cytogenetics and molecular genetics of nervous system tumors. *Oncol. Res.* 4(8–9):321–331.

Westermark B., Ponten J., Hugosson R. (1973) Determinants for the establishment of permanent tissue culture lines from human gliomas. *Acta. Pathol. Microbiol. Scand.* [A] 81(6):791–805.

Westphal, M., and Meissner, H. (1998) Establishing human glioma-derived cell lines. *Methods Cell Biol.* 57:147–165.

Westphal, M., et al. (1988) Glioma biology in vitro: goals and concepts. *Acta Neurochir. Suppl.* 43:107–113.

Westphal M., Nausch H., Herrmann H.D. (1990) Antigenic staining patterns of human glioma cultures: primary cultures, long-term cultures and cell lines. *J. Neurocytol.* 19(4):466–477.

# 15

## Culture of Human Neuroendocrine Tumor Cells

Roswitha Pfragner[1], Annemarie Behmel[2], Elisabeth Ingolic[3] and Gerhard H. Wirnsberger[4]

[1] *Department of Pathophysiology and* [2] *Department of Medical Biology and Human Genetics, Medical University Graz, Graz, Austria;* [3] *Research Institute for Electron Microscopy, Technical University Graz, Austria; and* [4] *Department of Internal Medicine; Medical University Graz, Graz, Austria. Email: roswitha.pfragner@uni-graz.at*

*Culture of Human Tumor Cells*, Edited by Roswitha Pfragner and R. Ian Freshney.
ISBN 0-471-43853-7   Copyright © 2004 Wiley-Liss, Inc.

## I. BACKGROUND

Neuroendocrine tumors are rare tumors that mainly occur sporadically, but also as dominantly inherited cancer syndromes, e.g. as the multiple endocrine neoplasia syndromes type 1 (MEN1), type 2 (FMTC, MEN2), and others.

**Multiple endocrine neoplasia type 1 (MEN1)** is characterized by the combined occurrence of tumors of the parathyroid gland, the endocrine pancreas, the pituitary gland and the adrenal cortex. In addition, carcinoids in the lung, stomach, and thymus may be included, as well as various non-endocrine tumors, including lipomas and facial angiofibromas. The MEN1-gene is located on chromosome 11q13 and has been identified as a tumor-suppressor gene encoding *menin*. Somatic and germline mutations show similar spreads across the open reading frame [Chandrasekharappa et al., 1997]. Most mutations lead to truncation of the encoded *menin* and its inactivation. Germline mutation screening is recommended to limit clinical screening to gene carriers. Unfortunately, there is no effective malignancy prevention for MEN1.

**Multiple endocrine neoplasia type 2 (MEN2)** presents in three distinct phenotypes: familial medullary thyroid carcinoma only **(FMTC)**; multiple endocrine neoplasia type 2A **(MEN2A)**, in which MTC is associated with pheochromocytoma and primary hyperparathyroidism (PHPT); multiple endocrine neoplasia type 2B **(MEN2B)**, in which MTC is associated with pheochromocytoma, PHPT, mucosal neurinomas, and a marphanoid phenotype. In MEN2A, a cutaneous lichen amyloidosis can be present [Nunziata et al., 1989]. Sporadic and MEN2-associated tumors are based on somatic and germline *RET* gene mutations. *RET* is located on chromosome 10q11 and encodes a receptor tyrosine kinase [Mulligan et al., 1993]. The true dominant *RET* mutations lead to activation of the signal transduction cascade and show a nonrandom distribution across the gene with strong genotype-phenotype correlations. Screening for *RET* discloses nearly 100% of the familial cases. Early prophylactic thyroidectomy in gene carriers prevents the development of MTCs, that originate from the calcitonin-producing C-cells of the thyroid and are resistant to chemotherapy and radiation [Brandi et al., 2001 Ponder, 1999; Mulligan et al., 1994; Eng et al., 1996; Eng 2000; Hansford and Mulligan, 2000.]

Due to generally small samples and poor initial proliferation in vitro, only a few cell lines from neuroendocrine tumors have been established. Continuous cell lines from rodents are MTC-M [Tischler et al., 1976; Tischler et al., 1986]; rMTC-23, 44-2-C [Zeytin and Delellis, 1987; Zeytinoglu et al., 1980]; and CA77 [Muszynski, 1983]. Since 1981, the only human MTC-cell line available was TT, established by S. Leong [Leong et al., 1981].

We aimed to develop further cell lines from human neuroendocrine tumors, representing sporadic and hereditary forms, different tumor stages, growth dynamics, endocrine differentiation, and ages of patients at surgery, to use as models for studies of biological and genetic characterization of the primary tumor, progressing, and metastatic tumors, for anticancer drug testing and for future therapies.

The following protocols focus on the handling of MTC, but can be applied equally to the other MEN-related tumors [Pfragner et al., 1990, 1992, 1993, 1996, 1998; Sigl et al., 1999].

## 2. REAGENTS AND MATERIALS

### 2.1. Transport Medium

(i)  Ham's F12 without serum
(ii)  100 U penicillin/ml + 100 μg streptomycin/ml
(iii)  2.5 μg/ml amphothericin B (Fungizone)

Prepare 10 ml aliquots in sterile plastic universal containers and store at 4°C for up to 3 months.

### 2.2. Collagenase-Medium for Separating Tumor Cells from Fibroblasts

(i)  Dissolve 0.1 g collagenase type IV in 180 ml growth medium (Ham's F12 with glutamine, without FBS)
(ii)  Sterilize by filtration
(iii)  Add 20 ml FBS
(iv)  Apply for 24 to 48 h.

### 2.3. Coating of Culture Vessels with Poly-D-Lysine

(i)  Dilute 20 mg poly-D-lysine in 40 ml Ham's F12 (without serum).
(ii)  Store 1.5 ml aliquots at −20°C.
(iii)  Dilute to final concentration of 5 μg poly-D-lysine/ml Ham's F12 (without serum)
(iv)  Add appropriate amount to culture surface and spread evenly
(v)  Incubate at room temperature for 1 h.
(vi)  Aspirate remaining material and store at 4°C until use.

### 2.4. Growth Medium

(i)  Ham's F12 (liquid medium with 10 mg/l Phenol Red, with 1.176 g/l NaHCO$_3$, with stable L-glutamine)
(ii)  10% FBS
(iii)  100 U penicillin/ml + 100 μg streptomycin/ml, if necessary
(iv)  2.5 μg/ml amphothericin B (fungizone), if necessary

## 2.5.  Growth Medium for Cytogenetic Analyses

(i)   Ham's F12 (liquid medium with 10 mg/l Phenol Red, with 1.176 g/l NaHCO₃, with stable glutamine, Glutamax)

(ii)   20% FBS

(iii)   Apply 24 h before chromosome preparation

(iv)   100 U penicillin/ml + 100 μ/ml streptomycin, if necessary

(v)   2.5 μg/ml ampothericin B (fungizone), if necessary

## 2.6.  Serum-Free Medium (modified after Brower et al., 1986)

(i)   Ham's F12 (liquid medium with 10 mg/ml Phenol Red, with 1.176 g/l NaHCO₃, with stable glutamine)

(ii)   Insulin 20 μg/ml

(iii)   Transferrin 10 μg/ml

(iv)   Hydrocortisone 50 nM

(v)   Sodium selenite 25 nM

(vi)   EGF 10 ng/ml

(vii)   Sodium pyruvate 0.5 nM

(viii)   Triiodothyronine 0.1 nM

(ix)   100 U penicillin/ml + 100 μg/ml streptomycin, if necessary

Use in combination with coated culture vessels, e.g. poly-D-lysine.

# 3.  TUMOR COLLECTION AND FREEZING

Collaboration with an interested surgeon is the prerequisite for obtaining tumor tissue. Otherwise, all tissue will be fixed with formalin for histological diagnosis. The surgeon should provide such information on the patient, as age, sex, primary tumor or metastases, and whether the tumor is sporadic or hereditary, and family history over the past three generations. With hereditary forms, the information on further tumors in other family members will be of pivotal importance for genetic studies.

## 3.1.  Collection of Tumor Samples

As clinics and cell culture laboratory frequently are not in the same town, it is useful to have warning of a sample by telephone, especially before weekends. Samples should not be in transit for more than 24 h (use overnight mail); otherwise the tissue becomes necrotic. Note: Carriers of *RET* gene mutations undergo pre-symptomatic thyroidectomy, and tumors are excised at an early stage. Only small tissue samples may be available; but, a minimum size of 5 mm³ should usually suffice for chromosome analyses.

## Protocol 15.1.  Collecting Neuroendocrine Tumors

### Reagents and Materials
*Sterile*
- ❏ Specimen tubes (25 ml) with wide-neck and leak-proof caps; for example, universal containers, labeled with patient's data, containing 10 ml transport medium
- ❏ Transport medium: Ham's F12 without serum, with 100 U/ml penicillin, 100 μg/ml streptomycin, stored at 4°C in the operating room refrigerator.

*Non-sterile*
- ❏ Patient forms should contain patient's data and the contact address and telephone number of cell culture laboratory
- ❏ Autoclavable bags
- ❏ Well-insulated transport box (e.g., styrofoam)

### Protocol
(a)  Put an unfixed sample of tumor tissue into the transport medium, close and wrap the specimen tube in an autoclavable bag.

(b)  Send by express mail to cell culture laboratory. No cooling is required.


### 3.2.  Cryopreservation of Biopsies

It is useful to preserve pieces of the larger biopsies by freezing. It may be necessary to repeat unsatisfactorily growing cultures or the tissue will be used for other studies and confirmation of origin of any cell line that arises.


## Protocol 15.2.  Freezing Neuroendocrine Tumor Biopsies

### Reagents and Materials
*Sterile*
- ❏ Growth medium: Ham's F12, supplemented with 10% FBS, 100 U penicillin/ml, 100 μg streptomycin/ml
- ❏ Freezing medium: Growth medium containing 10% dimethylsulfoxide (DMSO), or ORIGEN™ freezing medium
- ❏ Hanks' BSS (HBSS)
- ❏ Concentrated antibiotic solution: 1000 U penicillin and 1000 μg streptomycin/ml in HBSS
- ❏ Biopsy
- ❏ Petri dishes, 9 cm, for dissection
- ❏ Scalpels, scissors, curved forceps
- ❏ Cryotubes, 1.8 ml

### Protocol
(a) Soak the fresh tumor tissue in concentrated antibiotic solution at room temperature for 20 min.
(b) Remove fat and visible connective tissue.
(c) Transfer the tissue to HBSS and cut pieces of approximately 3 to 4 mm diameter.
(d) Pipette 1 ml freezing medium into each cryotube, and add four to five pieces of tissue to each tube.
(e) Freeze slowly at 1°C/min, and store in liquid nitrogen.
(f) When required, thaw tube rapidly in 37°C water.
(g) Place the tissues into fresh nutrient mixture, dissociate mechanically, and set up primary cultures as described in Protocol 15.3.

## 4. PRIMARY CULTURE

For very small samples, the primary explant culture method is most suitable. For larger samples, we found that the best way to prepare cell suspensions is by mechanical dissociation. Enzymatic digestion is not recommended, as it has proven to be destructive to the delicate tumor cells, and it enhanced fibroblast growth.

### Protocol 15.3. Primary Explant Culture of Neuroendocrine Tumors

**Reagents and Materials**
*Sterile*
❑ Concentrated antibiotic solution: 1000 U penicillin/ml, 1000 μg streptomycin/ml in HBSS
❑ Growth medium: Ham's F12, supplemented with 10% FBS, 100 U penicillin/ml, 100 μg streptomycin/ml
❑ Petri dishes, non-tissue culture grade, 9 cm × 1 cm
❑ Petri dishes, tissue culture grade 3.5 × 1.0 cm
❑ Glass coverslips
❑ Scalpels, scissors, curved forceps
❑ Pipettes with wide tips (e.g., old glass pipettes with broken tips, ground or heated to remove sharp edges)

### Protocol
(a) On arrival, remove fat and visible connective tissue.
(b) Transfer the biopsy from the collection medium to the concentrated antibiotic solution for 20 min at room temperature.
(c) Cut the biopsy to 1 mm$^3$ size fragments.

(d) Transfer by pipette one or more fragments to the center of a Petri dish (3.5 cm diameter), and remove most of the medium.

(e) Cover the fragments with a glass coverslip.

(f) Gently add 1 ml growth medium on top of coverslip, close the dish, and place it into the incubator.

(g) Incubate without agitation.

(h) Feed once a week by partial medium exchange.

(i) Subculture the outgrowing cells by conventional trypsinization at 37°C.


## Protocol 15.4. Primary Culture of Neuroendocrine Tumors by Mechanical Disaggregation

### Reagents and Materials
*Sterile*
- ❑ Ham's F12, supplemented with 10% FBS
- ❑ Complete growth medium: Ham's F12, 10% FBS with 100 U/ml penicillin, 100 $\mu$g/ml streptomycin
- ❑ Panserin 401$^{TM}$ serum-free medium
  *Note: Use nutrient mixture Ham's F12, 10% FBS for primary cultures and early passages. When established as continuous cell lines, they may also be grown in serum-free medium Panserin$^{TM}$.*
- ❑ Red blood cell lysing buffer: 8.29 g $NH_4Cl$, 10 g $KHCO_3$, 0.0371 g ethylene diamine tetraacetate (EDTA) in 1 L $H_2O$
- ❑ Concentrated antibiotic solution: 1000 U penicillin/ml, 1000 $\mu$g streptomycin/ml HBSS. Amphotericin B: 0.25 mg/ml; final concentration 2.5 $\mu$g/ml
- ❑ Collagenase IV, 0.05% in Ham's F12 medium with 10% FBS
- ❑ Trypsin (0.25% in PBSA) or trypsin-EDTA (0.5g/l and 0.2g/l EDTA (4Na) in PBSA
- ❑ Cell culture flasks for suspension culture (Sarstedt green color code)
- ❑ Cell+ culture flasks (Sarstedt yellow color code: positively charged surface for delicate cells)
- ❑ Scissors, curved forceps
- ❑ Pipettes with wide bore tip
- ❑ Six-well dishes or 12.5 $cm^2$ flasks, coated with poly-D-Lysine
- ❑ Six-well dishes or 12.5 $cm^2$ flasks coated with poly-D-Lysine and Laminin

*Non-sterile*
- ❑ Trypan Blue
- ❑ Hemocytometer

*Protocol*

(a)  On arrival, remove fat and visible connective tissue.

(b)  Transfer the biopsy from the collection medium to a concentrated antibiotic solution for 20 min at room temperature.

(c)  Mince the tissue to pieces <1 mm$^3$ then centrifuge the suspension at 300 $g$ for 5 min.

(d)  Discard the supernate and resuspend with red blood cell lysis buffer for 10 min (this step can be omitted in tissues with minor amounts of erythrocytes).

(e)  Centrifuge at 300 $g$ for 5 min.

(f)  Discard the supernate and resuspend with complete growth medium.

(g)  Count the number of viable cells using Trypan Blue exclusion.
*Note:* Precise cell counting will not be practicable at this stage due to undissociated aggregates of cells. However, the resuspension in medium with an estimated cell number will be sufficient for setting up cultures.

(h)  Set up cultures in flasks or multiwell dishes with as many different coatings as the number of cells will allow, e.g. poly-D-lysine or poly-D-lysine with laminin.

(i)  Incubate undisturbed for 3 to 4 days in a separate incubator, which is designated only for primary cultures. Check for contaminations.

(j)  If absolutely necessary, elimination of bacterial contaminations may be attempted with penicillin (1000 U, streptomycin 1000 µg/ml HBSS) for 20 min at 37°C.

(k)  It is not recommended to attempt to eliminate fungi and yeasts. However, if the culture is irreplaceable, use amphotericin B (2.5 µg/ml medium) but not Nystatin, which would damage the cells.

(l)  If the pH drops, add a small amount of complete growth medium as long as the level of medium in the vessel is not increased unduly. At this stage of cultivation the cell number is usually low, a complete medium change would deprive the cultures of valuable autocrine or paracrine growth factors.

(m)  Continue by partial replacement of medium: Withdraw half medium containing floating cells, centrifuge at 300 $g$ for 5 min, and return the pelleted cells with fresh medium to the culture vessel. Repeat this once or twice a week for several weeks.

(n)  Observe the cultures; it will take a few days for attachment of the cells. Finally monolayers will form containing mixed populations of epithelioid cells, fibroblasts, and others. Within these monolayers, multicellular spheroids will arise. Initially adherent, they will tower up and detach spontaneously.

(o)  Isolate the detached cells by pipetting.

(p)  Transfer the isolated cells into vessels with special surface for suspension culture (Sarstedt green-capped flasks).

(q)  Trypsinize and transfer the adherent cells to vessels with a special surface for adherent growth (Cell+™ Sarstedt yellow-capped flasks).

*Note:* Most probably the suspension cells represent the MTC cells; while the adherent cells are probably fibroblasts, but one can never be sure. Do not discard any fraction before characterization, e.g., by immunocytochemical analysis for calcitonin. We established one continuous cell line from the anchorage-dependent tumor cells (OEE-III). Another reason for further culture of both adherent and nonadherent cells is the detachment of additional tumor cells, even after many weeks.

## 5.  LONG-TERM CULTURES AND CONTINUOUS CELL LINES

When cell cultures are subcultured and maintained for extended periods, cross contamination becomes a serious potential risk. To avoid cross contamination, it is essential to prepare an individual medium bottle for each cell line. Write the code of the individual culture on the flask. Do not open simultaneously the medium bottles and/or culture vessels of different cell lines. Cell lines derived from tumors of several different patients must be handled one after the other under the sterile hood. Precautions against cross-contamination are crucial, especially in regard to cytogenetic studies.

A proportion of short term cultures will give rise to continuous cell lines (i.e., continuous growth for >100 doublings or >1 year). Generally, the continuous MTC cell lines grow as non-adherent, multicellular aggregates. Before subculture, dissociate the aggregates by gentle pipetting, and count by hemocytometer or electronic particle counter.

### Protocol 15.5.  Subculture of Non-Adherent Cultures

***Reagents and Materials***

*Sterile*

❑  Centrifuge tubes

❑  Ham's F12, 10% FBS, omit antibiotics if possible

❑  Serum-free medium (Brower et al., 1986) can be used from higher passages on, with slower growth

❑  Cell culture flasks

*Protocol*

(a) Disperse and count the cells by using a hemocytometer or an electronic particle counter.

(b) Centrifuge the cell suspension at 300 *g* for 5 min.

(c) Discard the supernate.

(d) Add medium, and resuspend gently by pipetting.

(e) Dilute the suspension to 3–5 × $10^5$ cells/ml and seed culture flasks.


## Protocol 15.6.  Elimination of Fibroblasts from Neuroendocrine Cultures

### *Reagents and Materials*

❑ Complete growth medium: Ham's F12 with 10% FBS

❑ Collagenase, 0.05% in growth medium Ham's F12 with 10% FBS

❑ Two different sorts of vessels (for adherent and non adherent cells, Sarstedt)

### *Protocol*

(a) Treat monolayers with collagenase-medium for 24 h. If no effect is apparent, continue for another 24 h.

(b) When some cells begin to detach, separate the floating cells from the adherent ones, and centrifuge at 300 *g* for 5 min.

(c) Discard supernate, and re-suspend the pellet in complete growth medium.

(d) Repeat the collagenase-treatment after few weeks, as soon as fibroblastic overgrowth reappears.


## 6.  CRYOPRESERVATION OF CULTURED CELLS

## Protocol 15.7.  Cryopreservation of Cultured Neuroendocrine Cells

### *Reagents and Materials*

*Sterile*

❑ Cell line in log-phase

❑ Freeze medium: Ham's F12, 10% FBS, without antibiotics, with added 10% DMSO or use commercial freeze medium ORIGEN™

❑ Cryotubes 1.8 ml

*Non-sterile*

❑ Hemocytometer or electronic cell counter

❑ Trypan Blue

❑ Insulated box (styrofoam) or controlled-rate freezer

### Protocol

(a)  Count the cells and centrifuge at 300 *g* for 5 min.

(b)  Resuspend in freeze medium at 1 × 10⁶ cells/ml.

(c)  Pipette 1 ml cell suspension into each cryotube.

(d)  Freeze at 1°C/min to <50°C, in insulated box or by using a controlled-rate freezer.

(e)  Transfer to liquid nitrogen freezer.

Δ*Safety note.* Wear cryoprotective gloves, a face-mask, and a closed lab-coat, when working with liquid nitrogen. Make sure the room is well ventilated to avoid asphyxiation from evaporated nitrogen.

(f)  Thaw rapidly in 37°C waterbath, centrifuge at 300 *g* for 5 min, decant the freeze medium, and resuspend at 3–5 × 10⁵ cells/ml in growth medium.

Δ*Safety note.* An ampoule which has been stored in liquid nitrogen can inspire liquid nitrogen if not properly sealed. The ampoule will then explode violently when warmed. To protect from potential explosion of ampoules, thaw in a covered container, such as a large plastic bucket with a lid, or store in the vapor phase or a liquid nitrogen jacketed freezer. Wear gloves, a face-mask, and a closed lab-coat.

## 7.  CHARACTERIZATION

Various methods are available for characterization of cell lines. We primarily examine our cell lines by immunocytochemistry, electron microscopy and cytogenetic analyses. These methods were modified for neuroendocrine tumor cells. Further protocols on characterization and authentication have been described [Freshney, 2000].

### 7.1.  Immunocytochemistry

### Protocol 15.8.  Immunocytochemistry of Cultured Neuroendocrine Cells

#### *Reagents and Materials*
*Non-sterile*
Store all reagents at 4°C

❏  Sörensen's phosphate buffer , pH 7.4

❏  0.05 % glutaraldehyde buffered with Sörensen's

❏  Storage solution: dissolve 42 g sucrose and 0.35 g $MgCl_2 \cdot 6\,H_2O$ in 250 ml glycerol, and bring this to a total volume of 500 ml with Sörensen's buffer.

❏  Veronal acetate buffer (pH 9.2)

- Fast Red TR working solution: Dissolve 10 mg naphthol-AS-MX-phosphate in 0.5 ml N,N-dimethylformamide and 50 ml veronal acetate buffer, and add 50 mg Fast Red TR salt and 10 mg levamisol. Use the solution after filtration and adjustment of the pH to 9.2 to 9.8 with 2 N NaOH (stable for about 2 h).
- Biotinylated IgG (1:20 diluted in Sörensen's buffer)
- Bovine serum albumin (BSA)
- Polyclonal secondary antibody (rabbit anti-mouse pan-Ig)
- APAAP complex
- Silanized slides: Treat glass slides with 100% chloroform for 30 min, 100% ethanol for 30 min, 100% acetone for 30 min, and 3-aminopropyltriethoxysilane for 5 min, then dip in 100% acetone, dip in distilled $H_2O$, and air dry.
- Press Flexiperm chambers™ (8-well) onto the silanized slides.

## *Protocol*

*(1) Preparation of samples*

(a) Pipette into each well 0.2 ml cell suspension, which should not be too dense; for example, $1-2 \times 10^5$ cells/ml.

(b) Fix the undried preparations immediately in 0.05% glutaraldehyde for 5 min at room temperature.

(c) Store in Coplin jars filled with the precooled storage solution at $-20°C$ (for up to 1 year).

*(2) Labeled Avidin-Biotin (LAB) Technique*

(a) Wash the cell preparations gently and briefly in Sörensen's buffer. (If necessary, block endogenous peroxidase by bathing in 3% $H_2O_2$, diluted in $dH_2O$ for 5 min at room temperature, then rinse in Sörensen's buffer for 1 to 2 min.)

(b) Incubate the primary antibody (see Table 15.1) containing 1% BSA in a humid chamber for 60 min at room temperature.

(c) After two 2-min rinses, incubate in Sörensen's buffer with a bio-

Table 15.1. **Primary antibodies for immunocytochemical characterization of MTC cells.**

| Primary Antibody | Dilution | Source |
|---|---|---|
| ra-calcitonin | 1:500 | INCstar, United States |
| ra-calcitonin gene related peptide (CGRP) | 1:200 | Milab, Sweden |
| m-bombesin (gastrin-releasing peptide, GRP) | 1:200 | Roche, Germany |
| m-neuron specific enolase (γ-subtype) | 1:10 | Imogenetics, Belgium |
| m-chromogranin (clone PHE-5) | neat | Enzo, USA |
| m-chromogranin A and related peptides (clone LK2H10) | 1:100 | Roche, Germany |
| m-somatostatin (SRIF) | 1:20 | Novo Biolabs, Denmark |
| ra-serotonin (5-HT) | 1:1ßßß | Bioscience, Switzerland |

m = mouse, ra = rabbit, *with slight modifications

tinylated IgG, diluted 1:20 in Sörensen's buffer, for 30 min at room temperature, followed by two 2-min rinses in Sörensen's buffer.

(d)  Incubate ExtrAvidin® diluted 1:20 in Sörensen's buffer for 30 min at room temperature.

(e)  Visualize the reaction product by application of 0.05% 3,3-diaminobenzidine-tetrahydrochloride (w/v) and 0.01% $H_2O_2$ (v/v) at room temperature; the DAB staining product obtained is a brown non-soluble precipitate.

(f)  After a short wash in tap water, dehydrate in graded alcohols and mount with Eukitt®.

*(3)  Alkaline Phosphatase Anti-alkaline Phosphatase (APAAP) Technique*

(a)  After incubation with the primary antibody, wash briefly in Sörensen's buffer.

(b)  Use a polyclonal secondary antibody (rabbit anti-mouse pan-Ig) in a dilution of 1:30 in Sörensen's buffer for 30 min in a humid chamber at room temperature. For polyclonal primary antibodies generated in rabbits, use additionally an affinity-purified mouse anti-rabbit IgG as a linking antibody (diluted 1:50).

(c)  Before adding the APAAP complex, wash the cell preparations twice briefly in Sörensen's buffer.

(d)  After incubation of the APAAP complex (diluted 1:200 in Sörensen's buffer), perform the color reaction with Fast Red TR working solution, resulting in an alcohol-soluble, distinct red-staining product.

(d)  After washing the stained cell preparation in tap water, counterstain with Mayer's hematoxylin for 30 s and mount a coverslip with a water based mounting medium.

*Note:* Controls of the specificity of the immunocytochemical procedure always include the application of non-immune sera instead of primary antibodies and omission of essential steps in the staining procedure. As a positive control for the specificity of the primary antibodies, adequate cell preparations or tissue sections should be added in each experiment [Wirnsberger et al., 1992].

## 7.2.  Electron Microscopy

A general feature of neuroendocrine cells and their tumor counterparts is the presence of typical secretory granules in the cytoplasm; for example, in medullary thyroid carcinomas the parafollicular C-cells contain dense core granules that easily distinguish them from follicular cells. Note that each MTC has individual amounts of granules. It is therefore important to study the ultrastructure of each original tumor tissue and later to compare with the ultrastructure of the cultured cells. During cultivation, the number of granules will decrease

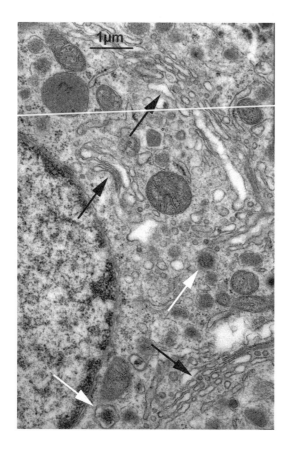

**Figure 15.1.** Transmission electron microscopy of HST-cell line, neuroendocrine granules (white arrows). Note the enlarged Golgi-apparatus (black arrows).

due to the grade of dedifferentiation. Cells with initially few granules will later have even fewer granules or will loose them completely. Whereas, in cells with many granules initially, a moderate number of granules will remain during cultivation time. In the tissue of origin, C-cells show a moderate Golgi apparatus; however, in cultured MTC-cells, the Golgi apparatus is enlarged (Figs. 15.1, 15.2).

### 7.2.1. Transmission Electron Microscopy (TEM)

### Protocol 15.9. Transmission Electron Microscopy of Neuroendocrine Cultures

#### Reagents and Materials

*Sterile*
- Transwell™ polycarbonate filter inserts in 24-well plates (Costar)

*Non-sterile*
- 0.1 M Cacodylate buffer, pH 7.2
- Glutaraldehyde, 3% in 0.1 M cacodylate buffer, pH 7.2, on ice (Plano)
- Osmium tetroxide, 1% in 0.1 M cacodylate buffer, pH 7.2

**Figure 15.2.** Tumor formation in nude mouse, TEM of SINJ-continuous cell line–mouse transplant. Increased number of neuroendocrine granules.

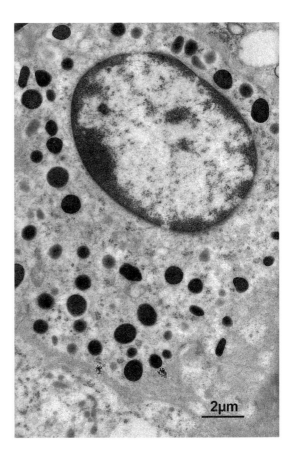

2µm

❏ Staining solution: Uranyl acetate, 1.0% in deionized water lead citrate (mix 1.33 g lead citrate and 1.76 g sodium citrate in 30 ml deionized water, add 8.0 ml N NaOH and dilute to 50 ml with deionizerd water (pH 12.0)
❏ Graded ethanol concentrations from 30% to absolute in stages of 20%
❏ Propylene oxide (omit this step for filters)
❏ Agar
❏ Razor blade
❏ 75 mesh Cu/Pd grids coated with Pioloform
❏ Epon 812 epoxy-embedding medium
❏ Ultramicrotome Leica UCT
❏ Glass and diamond knives
❏ TEM EM 420 Philips

### Protocol

*(1)  Monolayers in situ*

(a)  Grow monolayers on Transwell™ polycarbonate membranes

(Corning), and fix by immerging the whole culture on its support in 3% glutaraldehyde in 0.1 cacodylate buffer.

(b) Rinse the monolayer in cacodylate buffer.

(c) Fix at room temperature in osmium tetroxide (1% in cacodylate buffer) for 2 h.

(d) Dehydrate with alcohol. (Do not use propylene oxide, which will dissolve filters!)

(e) Cut filter into stripes, and embed in flat molds in Epon 812 (Merck).

(2) *Agar Method for Suspension Cultures [Modified after Ryter and Kellenberger, 1958]*

(a) Prepare a 2% solution of agar by dissolving the agar in boiling distilled water.

(b) Pour the solution into a test tube while it is still molten

(c) Place the tube in a water bath at 45°C. At this temperature the agar remains liquid.

(d) Place the cell suspension in a centrifuge tube, add the fixative and mix with the cells by gently shaking the tube.

(e) Centrifuge at 400 g in Eppendorf tubes. Try to obtain a small and tightly packed pellet.

(f) Remove supernatant and overlay pellet with glutaraldehyde solution. Fix at room temperature for ~1 hour or on ice, or, if necessary, overnight at room temperature.

(g) Wash three times for 30 min in cold 0.1 M cacodylate buffer, with at least one overnight submersion.

(h) Postfix with 1% OsO4 in cacodylate buffer, pH 7.0 for 2 h at room temperature or on ice.

(i) Rinse the pellet with several changes of water.

(j) Remove the water and transfer a small drop of agar to the centrifuge tube with a warm pipette and mix the agar with the pellet gently to suspend the cells.

(k) Transfer the mix with the pipette to a cool glass microscope slide.

(l) After the agar has set, cut the solidified agar containing the cells into small cubes of about 1 mm$^3$, using a sharp single-edged razor blade.

(m) Transfer the agar cubes to a glass vial and rinse with water. Then dehydrate specimen by 10–15 min incubations in alcohol:water (or acetone:water) mixtures of 30, 50, 70%, and at least two 100% alcohol rinses and at last dehydrate with propylene oxide (propylene oxide (1,2-epoxy propane) is used routinely between ethanol dehydration and infiltration with epoxy embedding media. EPON 812 [Luft, 1961].

(n) Cut sections and stain in 1% uranyl acetate and lead citrate for 20 min [Reynolds 1963]

### 7.2.2. Scanning Electron Microscopy (SEM) of Monolayers and Suspensions

The scanning electron microscope provides a three-dimensional, high-resolution image of cells surface. This is important for studying effects of test substances on cell membranes of tumor cells.

## Protocol 15.10. Scanning Electron Microscopy of Neuroendocrine Cultures

### Reagents and Materials
*Sterile*
- ❑ Transwell™ poycarbonate filter inserts in 24-well plates

*Non-sterile*
- ❑ Paper filter, 9 cm diameter
- ❑ Glutaraldehyde 3% in 0.1 M cacodylate buffer, pH 7.2, +4°C
- ❑ Osmiumtetroxide 1% in 0-1 M cacodylate buffer

### Protocol
(a) Grow adherent cells in Transwell™-polycarbonate filter inserts for 24 h.

(b) Remove filter from plate, and place shortly on a pad of paper filter until approximately 90% of the supernatant medium has trickled through. Note: the filter well must not run dry.

(c) Pipette the fixation fluid (glutaraldehyde in cacodylate buffer) carefully towards the inner rim of filter well, and let approximately 90% trickle through.

(d) Carefully immerse the filter well in glutaraldehyde, and leave overnight/4°C.

(e) Postfix with osmium tetroxide (see Protocol 15.11.)

The following procedures of critical point drying, coating with Au-Pd, follows routine methods [Anderson, 1951]. All cells to be illustrated were studied in a Leitz Scanning electron microscope AMR 1000 operating at accelerating voltages ranging from 5 to 30 KV. Images were recorded photographically on black-and-white, negative film TMY 120 Kodak TMAX 400 ASA.

## 7.3. Cytogenetic Analyses

Cytogenetic analyses of neuroendocrine tumors need experienced tumor cytogeneticists that can modify their protocol depending on the microscopic observations during the preparation of the chromosomes.

Because of small sample size and poor growth of neuroendocrine tumors, direct preparation of chromosomes was successful in metas-

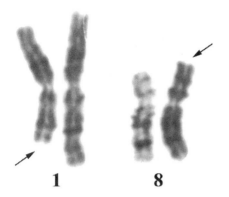

1  8

**Figure 15.3.** EDR, short-term culture, consistent chromosome aberration: 46, XX, t(1;8)(q25 or q32.1;p11.1 or p21.1) (arrows).

tases only. Usually first harvests for chromosome preparation were performed after 3 days (EDR, Fig. 15.3) to 8 weeks in culture. Long-term cultures and continuous cell lines were analyzed repeatedly and showed stable marker chromosomes (SINJ). In one cell line with three consistent marker chromosomes (MTC-SK, Fig. 15.4a) a new clonal marker developed only after 4 years in culture (Fig. 15.4b).

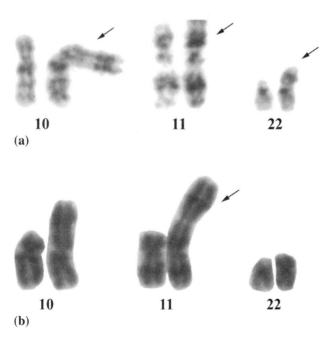

10   11   22

(a)

10   11   22

(b)

**Figure 15.4.** (a) MTC-SK, continuous cell line. (a) Passage 50. Consistent chromosome aberration: 46,XX,10p+,11p+,22p+ = 46,XX,−10,−11,−22,+der(10)t(3;10)(q21;p15), +der(11)t(11;11)(p15;q24), + der(22)t(17;22)(q21;p13) (arrows). (b) Passage 130. Di/tetra/octoploid, consistent chromosome aberration as in (a), new clonal marker: t(1q;11p+)(q12;pter) (arrow).

**Protocol 15.11. Chromosomal Analyses of Adherent and Suspension Neuroendocrine Cells**

*Reagents and Materials*

*Sterile*

❑  Ham's F12 with 20% FBS

❑  Colchicine 50 μM, dilute to 0.5 μM for use

*Non-sterile*

❑  Hypotonic solution 0.056 M KCl, 8.5 mM sodium citrate

❑  Fixative: 1 part w/v glacial acetic acid to 3 parts anhydrous methanol, freshly prepared, and kept on ice

❑  SSC and Giemsa stain (4%) for GAG banding

*Protocol*

(a)  Add 2 ml of cell-suspension to each slide-flask, and incubate at 37°C in a 5% $CO_2$-enriched atmosphere. Control cell growth by inverted microscope.

(b)  Add colchicine at final concentrations of 0.01 to 0.04 ng/ml medium 15 min to 12 h prior to harvesting.

(c)  Prepare adherent and floating cells separately; transfer the supernatant medium containing floating cells from the slide flask to a centrifuge tube; centrifuge at 250 g.

(d)  Expose slides with adherent cells and tubes to hypotonic treatment with 0.05 M to 0.075 M KCl from 15 min (continuous cell lines, suspension cultures) to 25 min (adherent cells), control cell swelling by inverted microscope.

(e)  Wash three times in 3:1 methanol:acetic acid, resuspend floating cells in 3:1 fixative; drop onto wet slides and air dry.

(f)  Prepare adherent cells in situ, air dry the slides; note the morphology and cell type of the cells surrounding the metaphases (distinguish tumor cells from stromal cells) by using a phase contrast microscope.

(g)  Age the slides 3 days at room temperature or overnight at 37°C.

(h)  Stain the slides by using banding and molecular cytogenetic methods.

### 7.4  Tumorigenicity Testing of MTC Cell Lines

There are different reasons for injection of MTC cells into nude mice: one is the testing of malignancy in vivo for the characterization of cell lines; another is to create transplantable tumors for in vivo experiments; and finally there is a possibility to re-differentiate a cell line with diminished properties. The take rate of MTC-grafts differs for each individual cell line. Usually 3 to 5 mice out of 10 will develop tumors, but the take rate will increase in the following mouse-passages (Fig. 15.6).

## Protocol 15.12. Tumorigenicity of Neuroendocrine Cells in Nude Mice

### Reagents and Materials
- ❑ 10 Athymic nude mice nu/nu-BALB/c for testing one cell line
- ❑ 75% Ethanol for disinfection of mouse skin
- ❑ Syringe, 1 ml
- ❑ MTC cells in log phase

### Protocol
(a) Harvest tumor cells in log phase.

(b) Count the cells, and calculate the cell concentration.

(c) Centrifuge at 300 g for 5 min, and resuspend in Ham's F12 without FBS.

(d) For each mouse prepare $3 \times 10^7$ MTC cells in 0.5 ml.

(e) Wipe the skin of mouse with ethanol, and inject the cell suspension subcutaneously into the flank.

(f) Watch the mice for 6 months for tumor growth.

(g) Use arising tumors for further transplantations, and characterize the tumor tissue by immunocytochemistry and electron microscopy, or other methods.

## 8. RESULTS AND APPLICATIONS

### 8.1. Establishment of Cell Lines

For almost two decades, only one human MTC cell line has been available, i.e. TT [Leong et al., 1981]. The rare occurrence of MTC 30 MTCs/year in Austria), the slow growth, and the usually very small size of the samples are some reasons for the lack of continuous cell lines.

The aim of our work was therefore to establish further MTC cell lines in order to provide new experimental models to be used in testing of prospective anticancer drugs or antiproliferative drugs. We collected tumor tissue samples from Austrian hospitals and set up cultures of 60 MTCs by the methods described above. Continuous cell lines could be established from nine cases (MTC-SK, SINJ [Pfragner et al., 1990, 1993], and GRS-IV, GRS-V, OEE-III, BOJO, RARE, SHER-I, GSJO, HOKA-I [personal observations]. The other 51 grew as cell lines with limited lifespans. Other neuroendocrine tumors, which we established as continuous cell lines by the above methods, were KRJ-I, a human carcinoid of the small intestine [Pfragner, 1996], and KNA, a human pheochromocytoma [Pfragner et al., 1998].

**Table 15.2.** Neuroendocrine characteristics of seven MTC-cell lines (Immunocytochemistry).

| CELL LINE | CT | CGRP | GRP | SRiF | 5-HT | NSE | PHE5 | LK2H10 | ER | Pgr |
|---|---|---|---|---|---|---|---|---|---|---|
| BOJO | ++ | ++ | n.d. | n.d. | n.d. | +++ | ++ | n.d. | n.d. | n.d. |
| GRS-IV | ++ | ++ | ++ | − | − | ++ | ++ | +++ | + | + |
| GRS-V | + | + | ++ | − | − | + | ++ | ++ | ++ | ++ |
| MTC-SK | ++ | +++ | +++ | − | − | +++ | +++ | n.d. | +/++ | +++ |
| SINJ | +/++ | +/++ | ++ | − | − | ++ | ++ | +++ | + | + |
| OEE-III | ++ | n.d. | n.d. | (+) | n.d. | +++ | +++ | n.d. | ++ | n.d. |
| RARE | n.d. | ++ | + | n.d. | n.d. | +++ | n.d. | ++ | (+) | n.d. |

ra-CT = calcitonin, CGRP = ra-calcitonin gene-related peptide, m-GRP, bombesin, m-SRIF = somatistatin, ra-5-HT = serotonin, m-NSE = neuron-specific enolase; PHE5 = m-chromogranin clone PHE5; LK2H10 = m-chromogranin A and related peptides clone LK2H10, ER = m-estrogen receptor, Pgr = m-progesterone receptor, tumor = tumorigenicty, m = nouse, r = rat, + = weak, ++ = moderate, +++ = strong, n.d. = not done

## 8.2. Characterization

The continuous cell lines were characterized by immunocytochemistry, in situ hybridization, electron microscopy, and chromosome analyses. They differed in several characteristics, such as age of patients, stage of tumors, and origin in sporadic or familial cases. They exhibited characteristic features of MTC, such as neuroendocrine granules and positive immunoreactivity to antibodies to calcitonin (CT), calcitonin gene-related peptide (CGRP) and bombesin (GRP) (Table 15.2 and Figs. 15.5, 15.6). In situ hybridization confirmed our findings (Fig. 15.7). SEM of MTC-SK showed numerous microvilli (Fig. 15.8).

Our cell lines have a variety of applications. Chromosome analyses were performed not only for the genetic characterization of sporadic and hereditary neuroendocrine tumors, but also for studies of tumor progression studying primary tumors, tumor recurrences, and metas-

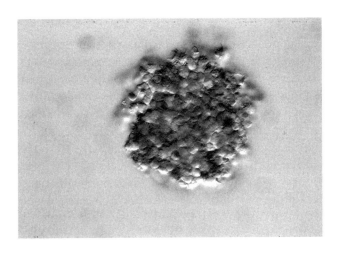

**Figure 15.5.** Strongly calcitonin-positive multicellular aggregate of the MTC-SK cell line; LAB technique, no counterstain, ×300.

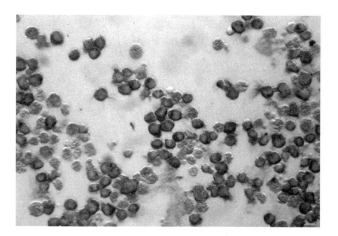

**Figure 15.6.** Immunocytochemical detection of CGRP in a cell suspension of the MTC-SK cell line; LAB technique, no countertstain, ×250.

tases of one and the same patient [OEE-I, -II, -III, -IV, -V cell lines and GRS-I, -II, -III, -IV cell lines; unpublished observations]. Cytogenetic findings played an important role in follow-up of continuous cell lines, which in part lost their endocrine characteristics, but not their marker chromosomes), in tumors generated in nude mice (clonal 11q- in the original tumor, consistent 11q- in the nude mouse transplant; SINJ). Early molecular cytogenetic analyses in MTC-SK refined cytogenetic data of a 3;10 translocation [Telenius et al., 1992] and later of the two 11p+ and 22p+ marker chromosomes as 11;11 and 17;22 translocation [Charlotte Benn, personal communication]. In all tumors, the breakpoints found in structural aberrations and the loss or gain of chromosomal material concerned tumor-relevant genes, but not all of the aberrations occurred repeatedly (trisomy 7, 11q-, loss of X- or Y-chromosome ao). Further molecular studies will elucidate the pathways of tumor development and progression once the basal

**Figure 15.7.** MTC-SK cell line, in situ hybridization with $^{32}$P-labeled RNA probe complementary to calcitonin. The specific autoradiographic signal is concentrated over the tumor cells compared with the background, revealing a significantly lower signal. Dark field illumination, ×200.

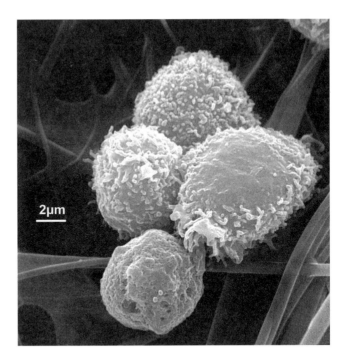

**Figure 15.8.** Scanning electron microscopy of the MTC-SK continuous cell line; multicellular aggregate with numerous microvilli.

mutation in the relevant genes (*RET, MEN1* ao) has occurred. *RET*-mutation studies were performed in MTC-SK, SINJ, GRS-IV [Eng et al., 1995], which proved the presence of a sporadic tumor in the *RET* negative patients. Since 1996, genomic DNA of all patients with MTC and other neuroendocrine tumors is screened for *RET* or MEN1-mutations, respectively [Weinhäusel et al., 2001].

## 8.3. Anticancer Drug Screening

To modulate the growth kinetics and viability, we investigated the effects of factors known for their anticancer activity in several other tumors. Little was known however, of their effects in MTC. The MTC cell lines were treated with different concentrations of methotrexate—an antagonist to folinic acid [Pfragner et al., 1990], interferon α-2B—as biological modulating substance; rapamycin, a powerful immuno-suppressant and antiproliferative agent; and four vitamin D3 analogs (1-alpha-dihydroxyvitamin D3, CB 1093, KH 1060, EB 1089). Growth kinetics and viability were examined by using the Casy-1 cell counter and analyzer and the MTT-based cytotoxicity assay. In each cell line, methotrexate, rapamycin, and interferon α-2B caused significant, dose-dependent growth inhibition, but no effect was seen after treatment with vitamin D analogs. Flow cytometry revealed individual responses for each cell line, obviously associated with individual characteristics of the tumor of origin. Unfortunately, no normal human C-cell lines are available to act as control cells. Control cell lines are usually grown from any normal human cells; we used human skin fibroblasts.

## 8.4. Apoptosis

The survival factor *bcl-2* is frequently found to be deregulated in human cancers. Bcl-2 family proteins are critical regulators of apoptotic cell death. An overexpression of the antiapoptotic gene bcl-2 leads to extended viability of cells by blocking apoptotic cell death. MTCs are well known for the phenomenon of bcl-2-based drug resistance. We appllied Western blotting with chemoluminescent detection to our MTC cell lines and found the anti-apoptotic gene *bcl-2* to be expressed in each cell line. To modulate apoptotic rates, we investigated the effects of different concentrations of interferon-α-2B, rapamycin, and four vitamin D3 analogs [Tschernatsch et al., Girsch et al., unpublished observations] and found significant up-regulation of *bcl-2*-expression. We investigated ten agents from plants of the genera *Stemona* (Stemonaceae), *Aglaia* (Meliaceae) and *Artemisia* (Asteraceae) for their effects on proliferation and apoptotic rates. Extracts have been used in traditional Chinese medicine, however no experience on their effects on MTC has been reported so far. Apoptosis was studied by measuring of caspase-3 activity and *bcl-2* expression. A strong antiproliferative effect was recognized in each *Aglaia* species and with Artemisia, whereas an enhancement of apoptosis was provoked particularly by *Stemona* species [Rinner et al, 2003]. The activity of the novel plant extracts offers a new approach towards successful chemotherapy of MTC.

## 8.5. Therapy

A new approach to dendritic cell-based immunotherapy is being developed with MTC cell lines [Bachleitner-Hofmann et al., 2002]. MTC cell cultures were established from patients with progressive MTC. The originally low cell numbers were multiplied in vitro. Dendritic cells (DCs) from the same patients were generated from peripheral blood monocytes by using GM-CSF and IL-4 (immature DCs) or GM-CSF, IL-4, and TNFα (mature DCs). MTC-cells were lysed and DCs were pulsed with these autologous MTC-cell lysates. Thereafter, pulsed DCs were co-cultured with naive autologous T-cells. Finally the capacity of the T-cells to lyse autologous tumor cells was tested by using a standard 4-h europium release cytotoxicity assay. Only mature tumor-lysate-pulsed DCs could elicit an HLA class I-restricted T cell response against autologous MTC cells, whereas immature tumor lysate-pulsed DCs did not stimulate significant antitumor activity. The maturation step of DCs is of importance for immunotherapy protocols. The final aim of our in vitro findings is an application for future therapy of patients with residual or distant MTC.

## 9. AVAILABLE CELL LINES IN CELL BANKS

### 9.1. American Type Culture Collection (ATCC)

**TT:** medullary thyroid carcinoma, human, female, [Leong S., et al., 1981]

**6-23 (Clone 6):** medullary thyroid carcinoma, Rattus norvegicus [Zeytinoglu, F.N., et al., 1980 and Gagel R.F., et al., 1980]

**MTC-M:** medullary thyroid carcinoma, Mus musculus (Balb/c), [Tischler, A.S., et al., 1986]

### 9.2. European Collection of Cell Cultures (ECACC)

**TT:** medullar thyroid carcinoma, human, female [Leong S., et al., 1981],

**MTC-SK:** medullary thyroid carcinoma, human, female [Pfragner R., et al., 1990]

**SINJ:** medullary thyroid carcinoma, human, male [Pfragner, R., et al., 1993]

**KNA:** pheochromocytoma, human, female [Pfragner, R., et al., 1998]

**KRJ-I:** carcinoid of small intestine, human, male [Pfragner, R. et al., 1996]

## ACKNOWLEDGMENTS

This chapter is dedicated by R. P. to her husband Julius Pfragner and sons Stefan, Matthias, and Michael for their continuous support. The authors thank Veronika Siegl for excellent technical assistance. This work was supported in part by funds of Österreichische Nationalbank (7984) and Austrian Cancer Aid/ Styria (EF 02/2003).

## SOURCES OF MATERIALS
(see also Table 15.1)

| Material | Catalog No. | Supplier |
| --- | --- | --- |
| 3-Aminopropyltriethoxysilane | 09324 | Fluka |
| Agar 100 Kit (Epon 812) | R 1031 | Plano |
| Amphotericin B 0.25 mg/ml (Fungizone) | P 11-001 | PAA |
| APAAP complex | D 0651 | Dako Diagnostics |
| Autoclavable bags | E706.1 | Roth |
| BioCoat Cellware™ | | BD Biosciences |
| Biotinylated IgG | | Sigma |
| Bovine serum albumin (BSA) | A 9906 | Sigma |

| Material | Catalog No. | Supplier |
|---|---|---|
| Cacodylate buffer (Dimethylarsinsäure Natriumsäure) | R 1031 | Plano |
| Cell Counter + Analyzer, Casy™ | Model TTC | Schärfe System |
| Cell culture dishes: 60/15 mm | 83 1801.003 | Sarstedt |
| Cell culture flasks: 12.5 cm² (to be coated with collagen IV or poly-D-lysine) | 353018 | BD Falcon |
| Cell culture flasks | 83 181001 | Sarstedt |
| with green color code (for suspension culture) | 83 1810500 | |
| with yellow color code (C+ ™ for delicate cells) | 83 1810300 | |
| Cell lines | | American Type Culture Collection (ATCC) 10801 University Boulevard, Manassas, Virginia 20110-2209 |
| Cell lines | | European Collection of Cell Cultures (ECACC) CAMR, Porton Down, Salisbury, Wilts SP4 ÜJG, England |
| Collagen IV | 354245 | BD Biocoat |
| Collagenase V | C 5138 | Sigma |
| Cryotubes, 1.8 ml | 375418 | Nunc |
| 3,3-Diaminobenzidine-tetrahydro-chloride | D1,238-4 | Sigma |
| DMSO | D2650 | Sigma |
| Epidermal growth factor | E 9644 | Sigma |
| Epon 812 (Agar 100 Kit) | R 1031 | Merck |
| Eukitt® | 03989 | Fluka |
| ExtrAvidin (incl. Biotinylated IgG) Universal LSAB® 2 Kit | K 0675 | Dako Diagnostics |
| Fast red TR salt | 21317.01 | SERVA |
| Fetal bovine serum (FBS) | A15-043 | PAA |
| FlexiPERM slide™ (8 chambers) | 50 03 20 39 | Kendro |
| Formamide | 104008 | Merck |
| Giemsa | 1.09204.100 | Merck |
| Glutamax | 35050-038 | Gibco |
| Glutaraldehyde | R 1011 | Plano |
| Ham's F12 (powder) with L-Glutamin, without NaHCO₃ | 5 081-01 | Biochrom |
| Hydrocortison | H-6909 | Sigma |
| Insulin | I 0516 | Sigma |
| Lead II nitrate | 107398 | Merck |
| Levamisol | 27750.02 | SERVA |
| Mounting medium, water based (Gel/Mount®) | MO1 | Biomeda |
| Mouse anti-rabbit Immunoglobulins | M 0737 | Dako Diagnostics |
| Multiwell plates with | | BD Biosciences |
| collagen IV, | 35 4428 | |
| poly D-lysine, | 35 4413 | |
| poly-D-lysine + laminin, | 35 4595 | |

## SOURCES OF MATERIALS *(Continued)*

| Material | Catalog No. | Supplier |
|---|---|---|
| NaHCO$_3$ (7.5%) | L 1713 | Biochrom |
| Naphthol-AS-MX-phosphate | 30002 01 | Serva |
| ORIGEN™ DMSO freeze medium | 21 0002 | IGEN™ International, Inc |
| Osmium tetroxide | R 1015 | Plano |
| Panserin™ 401 serum free medium | P04-710401 | CoaChrom |
| Paraformaldehyde | 104005 | Merck |
| Paper filter | | Sigma-Aldrich |
| Penicillin streptomycin 100× (10.000 U/ml Penicillin G, 10 mg/ml Streptiomycin, 0.9% NaCl) | P11-010 | PAA |
| Petri dishes, 3.5 × 1.0 cm | 83.1800.003 | Sarstedt |
| Pioloform | 63148-65-2 | SPI Supplies or Wacker |
| Poly-D-Lysine | 354210 | BD Biocoat |
| RPMI 1640 | FG 1213 | Biochrom |
| Rabbit anti-mouse immunoglobulins | Z 0109 | Dako Diagnostics |
| Slide-flasks | 177453 (glass) 170920 (polystyrene) | Nunc |
| Sodium pyruvate | S 8636 | Sigma |
| Sodium selenite | S 9133 | Sigma |
| Streptavidin-FITC | F 0422 | Dako Diagnostics |
| Transferrin | T 1428 | Sigma |
| Transmission electron microscope | EM 420 | Philips |
| Transwell cell culture chamber inserts, polycarbonate | 3415 | Corning Costar |
| Triiodothyronine | T 5516 | Sigma |
| Trypan blue | S11-004 | PAA |
| Trypsin (1:250) | L11-001 | PAA |
| Trypsin-EDTA (1:250) | L11-003 | PAA |
| Ultramicrotome | UCT | Leica |
| Uranyl acetate | R 1260 A | Plano |

## REFERENCES

Anderson, T.F. (1951) Techniques for the preservation of three-dimensional structure in preparing specimens for the electron microscope. *Trans.N.Y. Acad. Sci. Ser. II* 13:130–134.

Bachleitner-Hofmann, T., Stift, A., Friedl, J., Pfragner, R., Radelbauer, K., Dubsky, P., Schüller, G., Benkö, T., Niederle, B., Brostjan, C., Jakesz, R., Gnant, M. (2002) Stimulation of autologous antitumor T-cell responses against medullary thyroid carcinoma using tumor lysate-pulsed dendritic cells. *J. Clin. Endocr. Metab.* 87(3):1098–1104.

Brandi, M.L., Gagel, R.F., Angeli, A., Bilezikian, J.P., Beck-Peccoz, P., Bordi, C., Conte-Devolx, B., Falchetti, A., Gheri, R.G., Libroia, A., Lips, C.J., Lombardi, G., Mannelli, M., Pacini, F., Ponder, B.A., Raue, F., Skohseid, B., Tamnurrano, G., Thakker, R.V., Thompson, N.W., Tomassetti, P., Tonelli, F., Wells, S.A. Jr., Marx, S.J. (2001) Guidelines for diagnosis and therapy of MEN type 1 and type 2.J. Clin. Endocrinol. Metab. 86(12):5658–5671.

Brower, M., Carney, D.N., Oie, H.K., Gazdar, A.F., Minna, J.D. (1986) Growth of cell lines and clinical specimens opf human non-small cell lung cancer in a serum-free defined medium. Cancer Res. 46:798–806.

Chandrasekharappa, S.C., Guru, S.C., Manickam, P., Olufemi, S.E., Collins, F.S., Emmert-Buck, M.R., Debelanko, L.V., Zhuang, Z., Lubensky, I.A., Liotta, L.A., Crabtree, J.S., Wang, Y., Roe, B.A., Weisemann, J.M., Boguski, M.S., Agarwal, S.K., Kester, M.B., Kim, Y.S., Heppner, C., Dong, Q., Spiegel, A.M., Burns, A.L., Marx, S.J. et al. Positional cloning of the gene for multiple endocrine neoplasia type 1. *Science* 276:404–407.

Eng, C. (2000) Multiple endocrine neoplasia type 2 and the practice of molecular medicine. Rev. Endocr Metab Disord 1(4):283–290.

Eng, C., and Mulligan, L.M. (1991) Mutations of the RET proto-oncogene in the multiple endocrine neoplasia type 2 syndromes, related sporadic tumors, and Hirschsprung disease. *Hum. Mutat.* 9:97–109.

Eng, C., Mulligan, L.M., Smith, D.P., Healey, C.S., Frilling, A., Raue, F., Neumann, H.P., Pfragner, R., Behmel, A., Lorenzo, M.J., Timothy, J., Stonehouse, T.J., Ponder, M.A., Ponder, B.A.J. (1995) Mutation of the RET protooncogene in sporadic medullary thyroid carcinoma. *Genes Chrom. Cancer* 12:209–212.

Eng, C., Clayton, D., Schuffenecker, I., Lenoir, G. Cote, G., Gagel, R.F., van Amstel, H.K., Lips, C.J., Nishisho, I., Takai, S.I., Marsh, D.J., Robinson, B.G., Frank-Raue, K., Xue, F., Noll, W.W., Romei, C., Pacini, F., Fink, M., Niederle, B., Zedenius, J., Nordenskjold, M., Komminoth, P., Hendy, G.N., Mulligan, L.M. et al. (1996) The relationship between specific RET proto-oncogene mutations and disease phenotype in multiple endocrine neoplasia type 2. International RET Mutation Consortium Analysis. *JAMA* 276:1575–1579.

Freshney, R.I. (2000) Culture of Animal Cells. A Manual of Basic Technique. Wiley-Liss, Inc., 4th ed., pp 229–257.

Gagel, R.F., Zeytinoglu, F.N., Voelkel, E.F., Tashjian, A.H., Jr. (1980) Establishment of a calcitonin-producing rat medullary thyroid carcinoma cell line. II. Secretory studies of the tumor and cells in culture. Endocrinology 107:509–516.

Hansford, J.R., and Mulligan, L.M. (2000) Multiple endocrine neoplasia type 2 and RET: from neoplasia to neurogenesis. *J. Med. Genet.* 37:817–827.

Höfler, H., Pütz, B., Ruhri, C., Wirnsberger, G., Klimpfinger, M., Smolle, J. (1987) Simultaneous localization of calcitonin mRNA and peptide in a medullary thyroid carcinoma. *Virchows Arch* 54:144–151.

Leong, S.S., Horoszewicz, J.S., Shimaoka, K., Friedman, M., Kawinski, E., Song, M.J., Chu, T.M., Baylin, S., Mirand, E.A. (1981) A new cell line for study of human medullary thyroid carcinoma. In Andreoli, M., Monaco, H., Robbins, J. (eds) *Advances in Thyroid Neoplasia*. Field Educational Italia: Rome, pp 95–108.

Luft, J.H. (1961) Improvements in epoxy resin embedding methods. *J. Biophys. Biochem. Cytol.* 9:409.

Mathew, C.G., Chin, K.S., Easton, D.F., Thorpe, K., Carter, C., Liou, G.I., Fong, S.L., Bridges, C.D.B., Haak, H., Nieuwenhuijzen Kruseman, A.C., Telenius, H., Telenius Berg, M., Ponder, B.A.J. (1987) A linked genetic marker for multiple endocrine neoplasia type 2A on chromosome 10. *Nature (London)* 328:527–528.

Mercer, E.H., Birbeck, and Birbeck, M.S.C. (1961) *Electron Microscopy Handbook for Biologists*. Blackwell Scientific Publications: Oxford, UK, p. 104.

Mulligan, L.M., Kwok, J.B., Healey, C.S., Elsdon, M.J., Eng, C., Gardner, E., Love, D.R., Mole, S.E., Moorem, J-K, Pap, L., Ponder, M.A., Telenius, H., Tunnacliffe, A., Ponder, B.A.J. (1993) Germline mutations of the *RET* proto-oncogene in multiple endocrine neoplasia syndrome type 2A. *Nature (London)* 363:458–460.

Muszynski, M.B., Birnbaum, R.S., and Roos, B.A. (1983) Glucucorticoids stimulate the production of prepro calcitonin-derived serotonin peptides by rat medullary thyroid casrcinoma cell line. *J. Biol. Chem.* 258(19):11678–11683.

Nunziata, V., Giannatasio, R., di Giovanni, G., D'Armiento, M.R., Mancini, M. (1989) Hereditary localized pruritus in affected members of kindred with multiple endocrine neoplasia (Sipple's syndrome). *Clin. Endocrinol.* 30:57–63.

Pfragner, R., Höfler, H., Behmel, A., Ingolic, E., Walser, V. (1990) Establishment and characterization of continuous cell line MTC-SK deived from a human medullary thyroid carcinoma. *Cancer Res.* 50:4160–4166.

Pfragner, R., Behmel, A., Langsteger, W., Florian, W., Lanner, N. (1991) Preliminary report on the influence of methotrexate on a continuous cell line from a medullary thyroid carcinoma. In: Modern Aspects of Tumor Diagnosis and Treatment. Eber, O. (ed) Blackwell Special Projects, Vienna, pp. 108–112.

Pfragner, R., Wirnsberger, G., Behmel, A., Niederle, B., Längle, F., Roka, R., Mandl, A., Pürstner, P., Auner, J., Tatzber, F. (1992) Biologic and cytogenetic characterization of three human medullary tharoid carcinomas in culture. *Henry Ford Hosp. Med.J.* 40(3–4):299–302.

Pfragner, R., Wirnsberger, G., Behmel, A., Wolf, G., Passath, A., Ingolic, E., Adamiker, D., Schauenstein, K. (1993) New continuous cell line from human medullary thyroid carcinoma SINJ. Phenotypic analysis and in vivo carcinogenesis. *Int.J. Oncol.* 2:831–836.

Pfragner, R., Wirnsberger, G., Niederle, B., Behmel, A., Rinner, I., Mandl, A., Wawrina, F., Luo, J.S., Adamiker, D., Höger, H., Ingolic, E., Schauenstein, K. (1996) Establishment of a continuous cell line from a human carcinoid of the small intestine (KRJ-I): characterization and effects of 5-azacytidine on proliferation. *Int.J. Oncol.* 8:513–520.

Pfragner, R., Behmel, A., Smith, D.P., Ponder, B.A.J., Rinner, I., Porta, S., Henn, T., Niederle, B. (1998) First continuous human pheochromocytoma cell line: KNA. Biological, cytogenetic and molecular characterization of KNA cells. *J. Neurocytol.* 27:175–186.

Pourani, J., Kaserer, K., and Pfragner, R. (2002) Cytogenetic and molecular analyses of multiple endocrine neoplasias of the MEN1 syndrome. *Int.J. Oncol.* 20:971–976.

Reynolds, E.S. (1963) Liver parenchymal cell injury. 3. The nature of calcium-associated electron-opaque masses in rat liver mitochondria following poisining with carbon tetrachloride. *J.Cell Biol.* 17:208–212.

Rinner, B., Siegl, V., Pürstner, P., Efferth, T., Brem, B., Greger, H., Pfagner, R. (2003) Activity of novel plant extracts against medullary thyroid carcinoma. *Anticancer Res.* (in press).

Sabatini, D.D., Bensch, K., and Barnett, R.J. (1963) Cytochemistry and electron microscopy. The preservation of cellular ultrastructure and enzymatic activity by aldehyde fixatation. *J. Cell Biol.* 17:19.

Sigl, E., Behmel, A., Henn, T., Wirnsberger, G., Weinhäusel, A., Kaserer, K., Niederle, B., Pfragner, R. (1999) Cytogenetic and CGH studies of four neuroendocrine tumors and tumor-derived cell lines of a patient with multiple endocrine neoplasia type 1. *Intl.J. Oncol.* 15:41–51.

Telenius, H., Pelmear, A.H., Tunnacliffe, A., Carter, N.P., Behmel, A., Ferguson-Smith, M.A., Nordenskjöld, M., Pfragner, R. Ponder, B.A.J. (1992) Cytogenetic analysis by chromosome painting using DOP-PCR amplified flow-sorted chromosomes. *Genes Chrom. Cancer* 4:257–263.

Tischler, A.S., Dichter, M.A., Biales, B., De Lellis, R.A., Wolfe, H. (1976) Neural properties of cultured human endocrine tumor cells of proposed neural crest origin. *Science* 192:902–904.

Tischler, A.S., Lee, Y.C., Costopoulos, D., Nunnemacher, G., DeLellis, R.A., Van Zwieten, M.J., Wolfe, H.J., Bloom, S.R. (1986) Establishment of a continuous somatostatin-producing line of medullary thyroid carcinoma cells from BALB/c mice. *J. Endocrinol.* 110:309–313.

Weinhäusel, A., Fink, M., Haas, O., Niederle, B. (2001) Clinical management of patients with MEN syndromes. *Top. Endocrin.* 18:13–15.

Wirnsberger, G.H., Becker, H., Ziervogel, F., Höfler, H. (1992) Diagnostic immunohistochemistry of neuroblastic tumors. *Am. J. Surg. Pathol.* 16:4957.

Zeytinoglu, F.N., De Lellis, R.A., Gagel, R.F., Wolfe, H.J., Tashjian, A.H. Jr. (1980) Establishment of a calcitonin-producing rat medullary thyroid carcinoma cell line.I. Morphological studies of the tumor and cells in culture. *Endocrinology* 107:509–515.

Zeytinoglu, F.N., Gagel, R.F., Tashjian, A.H., Hammer, R.A., Leeman, S.E. (1980) Characterization of neurotensin production by a line of rat medullary thyroid carcinoma cells. *Proc. Natl. Acad. Sci. USA* 77:3741–3745.

Zeytin, F.N., and De Lellis, R. (1987) The neuropeptide-synthesizing rat 44-2C cell line: regulation of peptide synthesis, secretion, $3',5'$cycle adenosine monophosphate efflux, and adenylate cyclase activation. *Endocrinology* 121:352–360.

# Suppliers List

*Note:* The international dialing code for the USA is +1 and is not given with every entry. Other codes are given; the + indicates the local prefix used, usually 00.

**Aldrich Chemical Co.**
1001 W. St. Paul Ave., Milwaukee, WI 53233, USA
*Phone:* 414 273 3850; 800 558 9160
*Fax:* 414 273 4979
*Web site:* www.sigma-aldrich.com

**American Type Culture Collection, Inc. (see ATCC)**

**Amersham Pharmacia Biotech UK Limited**
Amersham Place, Little Chalfont, Buckinghamshire HP7 9NA, U.K.
*Phone:* 0800 515 313
*Fax:* 0800 616 927
*E-mail:* cust.servde@eu.apbiotech.com
*Web site:* www.amershambiosciences.com

**ATCC (American Type Culture Collection)**
10801 University Boulevard, Manassas, VA 20110-2209, USA
*Phone:* 703 365 2700
*Fax:* 703 365 2701
*E-mail:* tech@atcc.org
*Web site:* www.atcc.org

**BD Biosciences**
Am Concordpark E1/7, A-2320 Schwechat, Austria
*Phone:* +43 (0)1 706 36 60-20
*Fax:* +43 (0)1 706 36 60-11
*Web site:* www.bdbiosciences.com

**BD Biosciences**
1 Becton Drive, Franklin Lakes, NJ 07417-1883, USA
*Phone:* 201 847 6800; 888 237 2762
*E-mail:* mail@bdl.com
*Web site:* www.bdbiosciences.com/labware; www.difco.com/difco/

**BD Biosciences**
Between Towns Road, Cowley, Oxford, OX4 3LY, UK
*Phone:* +44 (0)1865 748844
*Fax:* +44 (0)1865 781627
*Web site:* www.bdbiosciences.com/labware

**BD Biosciences**
10974 Torreyana Road, San Diego, CA 92121, USA
*Phone:* 877 232 8995
*Fax:* 858 812 8888
*Web site:* www.bdbiosciences.com

**Beckman Coulter Corporation**
P.O. Box 169015, Miami, FL 33116-9015, USA
*Phone:* 305 380 2530; 800 327 6531
*Fax:* 305 380 3699
*Web site:* www.BeckmanCoulter.com

**Beckman Coulter, Inc.**
Diagnostics Division, 11800 SW 147th Ave., M/S 42-C03, P.O. Box 169015, Miami, FL 33116-9015, USA
*Phone:* 800 526 7695
*Web site:* www.beckmancoulter.com

**Becton Dickinson (see BD Biosciences)**

**Bibby Sterilin Ltd.**
Tilling Drive, Stone, Staffordshire ST15 0SA, U.K.
*Phone:* +44 (0)1785 812121
*Fax:* +44 (0)1785 815066
*E-mail:* bsl@bibby-sterilin.com
*Web site:* www.bibby-sterilin.co.uk

**Biochrom KG**
Leonorenstrasse 2-6m D-q12247 Berlin, Germany
*Phone:* +49 307799060
*Fax:* +49 307710012
*E-mail:* biochrom@tonline.de
*Web site:* www.biochchrom.de

**BioDesign Inc.**
P.O. Box 1050, Carmel, NY 10512, USA
*Phone:* 845 454 6610
*Fax:* 845 454 6077
*E-mail:* service@biodesignofny.com
*Web site:* biodesignofny.com

**Biomeda**
P.O. BOX 8045, Foster City, CA 94404, USA
or 21343 Cabot Blvd., Hayward, CA 94545, USA
*Phone:* 800 341 8787
*Fax:* 510 783 2299
*E-mail:* support@biomeda.com
*Web site:* www.biomeda.com/

**Bio-Rad Laboratories**
Life Science Group Div., 2000 Alfred Nobel Dr., Hercules, CA 94547, USA
*Phone:* 510 741 1000; 800 4BIORAD
*Fax:* 800 879 2289
*Web site:* www.bio-rad.com

**Biosource International**
542 Flynn Road, Camerillo, CA 93012, USA
*Phone:* 800 242 9697
*E-mail:* sales.desk@biosource.com
*Web site:* www.biosource.com

**BioWhittaker, Inc.**
8830 Biggs Ford Road, P.O. Box 127, Walkersville, MD 21793-0127, USA
*Phone:* 301 898 7025; 800 638 8174; 800 654 4452
*Fax:* 301 845 8338
*E-mail:* sales@biowhittaker.com
*Web site:* www.biowhittaker.com

**Boehringer Mannheim Corp (now Roche Molecular Biochemicals)**
9115 Hague Rd, P.O. Box 50414, Indianapolis, IN 46250-0414, USA
*Phone:* 800 262 1640
*Fax:* 317 521 7317
*E-mail:* mannheim.biocheminfo@roche.com
*Web site:* www.biochem.roche.com

**CABI Bioscience Switzerland Center**
1 Rue des Grillons, 2800 Delemont, Switzerland
*Phone:* +41 (0)32 421 4870
*Fax:* +41 (0)32 421 4871
*E-mail:* swiss.centre@cabi-bioscience.ch.com
*Web site:* www.cabi-bioscience.org

**Calbiochem-Novabiochem Corp**
10394 Pacific Ctr. Ct., San Diego, CA 92121, USA
*Phone:* 619 450 9600; 800 854 3417
*Fax:* 619 453 3552
*Web site:* www.calbiochem.com

**Calbiochem-Novabiochem Corporation**
P.O. Box 12087, La Jolla, CA 92039-2087, USA
*Phone:* 800 628 8470
*Web site:* www.calbiochem.com

**Cambrex (see BioWhittaker)**

**Carl Roth GmbH & Co**
Schoemperlenstr. 1-5, D-76161 Karlsruhe, Germany
*Phone:* +49 (0)721 56 06 0
*Fax:* +49 (0)721/ 56 06 149
*E-Mail:* info@CarlRoth.de
*Web site:* www.Carl-Roth.de

**Cellgro (see Mediatech)**

**Chance Propper Ltd.**
P.O. Box 53, Spon Lane South, Smethwick, West Midlands B66 1NZ, UK
*Phone:* +44 (0)121 553 5551
*Fax:* +44 (0)121 525 0139

**Chemicon International**
28820 Single Oak Drive, Temecula, CA 92590, USA
*Phone:* 800 437 7500; 909 676 8080
*Fax:* 800 437 7592; 909 676 9209
*Web site:* www.chemicon.com

**CELOX Laboratories**
1311 Helmo Avenue, St. Paul, MA 55128, USA
*Phone:* 612 730 1500
*Fax:* 612 730 8900
*Web site:* www.celox.com

**Citifluor Ltd.**
18 Enfield Cloisters, Fanshaw Street, London N1 6LD, UK
*Phone:* 020 7739 6561
*Fax:* 020 7729 2936
*E-mail:* enquiries@citifluor.co.uk; sales@citifluor.co.uk.
*Web site:* www.citifluor.co.uk

**Clonetics Technical Sales Department**
P.O. Box 127, 0245 Brown Deer Road, San Diego, CA 92121, USA
*Phone:* 800 852 5663
*Fax:* 858 824 0826
*Web site:* www.clonetics.com

**CN Biosciences (UK) Ltd.**
Boulevard Industrial Park, Padge Road, Beeston, Nottingham NG9 2JR, UK
*Phone:* 0800 622935; +44 (0)115 943 0840
*Fax:* +44 (0)115 943 0951
*E-mail:* customer.service@cnuk.co.uk.
*Web site:* http://www.cnuk.co.uk

**Coachrom Diagnostica**
Leo Mathauser Gasse 71, A-1230 Vienna, Austria
*Phone:* +43 (0)1699 97 97 0
*Fax:* +43 (0)1699 18 97
*E-mail:* office@coachrom.com
*Web site:* www.coachrom.com/

**Collaborative Biomedical Products** (*see also* **BD Biosciences**)
Two Oak Park, Bedford, MA 01370, USA
*Phone:* 800 343 2035
*Fax:* 617 275 0043
*E-mail:* bedfordcustomerservice@bd.com
*Web site:* www.bdbiosciences.com

**Collagen Corporation**
2500 Faber Place, Palo Alto, CA 94303, USA
*Phone:* 415 856 0200
*Fax:* 415 856 2238
*Web site:* www.collagen.com/

**Corning Inc.**
P.O. Box 5000, Corning, NY 14830, USA
*Phone:* 607 974 7740; 800 222 7740; 800 492 1110
*Fax:* 607 974 0345; 978 635 2476
*E-mail:* labware@corning.com
*Web site:* www.corninglabware.com

**Corning Costar Corporation**
Am Kümmerling 21-25, D-55294 Bodenheim, Germany
*Phone:* +49(0)6135 9315 0
*Fax:* +49(0)6135 5148

**DAKO Ltd.**
Denmark House, Angel Drove, Ely, Cambridgeshire, CB7 4ET, England, UK
*Phone:* +44 (0)1353 669911
*Fax:* +44 (0)1353 668989
*E-mail:* techsupport@dako.co.uk
*Web site:* www.dako.com/

**DAKO Corporation**
6392, Via Real, Carpinteria, CA 93013, USA
*Phone:* 800 400 DAKO
*Fax:* 805 684 5935
*E-mail:* litreq@dakousa.com

**Digene Corporation**
1201 Clopper Road, Gaithersburg, MD 20878, USA
*Phone:* 800 344 3631; 301 944 7000
*Fax:* 301 944 7121
*E-mail:* inforeq@digene.com
*Web site:* www.digene.com/

**DSMZ – German Collection of Microorganisms and Cell Cultures**
Department of Human and Animal Cell Cultures, Mascheroder Weg 1B, D-38124 Braunschweig, Germany
*Phone:* +49 531 2616 161
*Fax:* +49 531 2616 150

*E-mail:* mutz@dsmz.de
*Web site:* www.dsmz.de

**Eastman Kodak, Co.**
Scientific Imaging Systems Div, 343 State St, Rochester, NY 14652-4115, USA
*Phone:* 203 786 5600; 800 225 5352
*Fax:* 203 786 5694
*Web site:* www.kodak.com

**ECACC European Collection of Cell Cultures**
CAMR, Salisbury, Wiltshire SP4-JG, UK
*Tel:* +44(0)1980 612512
*Fax:* +44(0)1980 611315
*E-mail:* ecacc@camr.org
*Web site:* ecacc.org.uk

**Eli Lilly & Company**
Lilly Corporate Center, Indianapolis, IN 46285, USA
*Phone:* 317 276 2000; 888 885 4559
*Web site:* www.LillyMedical.com

**Enzo Life Sciences, Inc.**
60 Executive Blvd., Farmingdale, NY 11735, USA
*Phone:* 631 694 7070; 800 221 7705
*Fax:* 631 694 7501
*Web site:* www.enzo.com

**Eppendorf Scientific, Inc.**
6524 Seybold Road, Madison, WI 53719, USA
*Phone:* 608 276 9855; 800 421 9988
*Fax:* 608 276 9866
*E-mail:* custservice@eppendorf-usa.com
*Web site:* www.eppendorf.com

**Fine Science Tools Inc. (FST)**
373-G Vintage Park Drive, Foster City, CA 94404, USA
*Phone:* 800 521 2109
*Fax:* 800 523 2109; 650 349 3729
*E-mail:* info@finescience.com

**Fisher Scientific U.K.**
Bishops Meadow Road, Loughborough, Leicestershire LE11 5RG, UK
*Phone:* +44 (0)1509 231166
*Fax:* +44 (0)1509 231893
*E-mail:* sales@fisher.co.uk
*Web site:* www.catalogue.fisher.co.uk

**Fisher Scientific Corp.**
2000 Park Ln, Pittsburgh, PA 15275, USA
*Phone:* 412 490 8300; 800 766 7000
*Web site:* www.fishersci.com

**Fluka (see Sigma)**

**Forma Scientific, Inc. (see also Thermo Life Sciences)**
Millcreek Road, P.O. Box 649, Marietta, OH 45750, USA
*Phone:* 614 373 4763; 800 848 3080
*Fax:* 614 374 1817
*E-mail:* dbergene@forma.com
*Web site:* www.forma.com

**Gemini Bio-Products**
1 Harter Ave. Suite B, Woodland, CA 95776, USA
*Phone:* 800 543 6464

**GIBCO (see Invitrogen)**

**Girovac**
Units 1 & 2, Douglas Bader Close, Folgate Road, North Walsham, North Norfolk, NR28 0TZ, UK
*Phone:* +44 (0)1692 403008
*Fax:* +44 (0)1692 404611

**Globepharm – Surrey Diagnostics**
*Phone:* +44 (0)1483 259 739
*Fax:* +44 (0)1483 259 059
*E-mail:* jayne@globepharm.co.uk
*Web site:* www.globepharm.co.uk

**Greiner Bio-One**
Maybachstrasse 2, D-72636 Frickenhausen, Germany
*Phone:* +49 7022 948-0
*Fax:* +49 7022 948-580
*E-mail:* info@greinerbiooneinc.com
*Web site:* www.greinerbiooneinc.com

**Harlan Sera-Lab**
Dodgeford Lane, Loughborough, Leicestershire, LE12 9TE, UK
*Phone:* +44 (0)1530 222 123
*Fax:* +44 (0)1530 224 970
*Web site:* www.harlanseralab.com

**Heraeus (now Kendro Laboratory Products)**
D-63450 Hanau, Germany
*Phone:* +43(0)1805 536 376
*Fax:* +43(0)1805 112114
*E-mail:* info@kendro.de
*Web site:* www.sorvall.com or www.kendro.de

**Hyclone**
1725 South Hyclone Road, Logan, Utah 84321-6212, USA
*Phone:* 800 492 5663
*Fax:* 800 533 9450
*Web site:* www.hyclone.com

**ICN Biomedicals Ltd.**
Cedarwood, Chineham Business Park, Crockford Lane, Basingstoke, Hants RG24 8WG, UK
*Phone:* 0800 282 474
*Fax:* 0800 614735
*Web site:* www.icnbiomed.com

**ICN Biochemicals, Inc.**
3300 Hyland Ave., Costa Mesa, CA 92626, USA
*Phone:* 714 545 0100; 800 854 0530
*Fax:* 714 557 4872; 800 334 6999
*Web site:* www.icnbiomed.com

**IGEN Europe Inc.**
Oxford Bio Business Center Littlemore Park, Littlemore, Oxford OX4 2SS, UK
*Phone:* +44 (0)7000 443 683
*Fax:* +44 (0)7000 443 632
*E-mail:* m-series@igen.com
*Web site:* www.igen.com/

**Innogenetics**
Canadastraat 21, 2070 Zwijndrecht, Belgium
*Phone:* +32 (0)3 252 3711
*Fax:* +32 (0)3 252 3799

**INCSTAR Corp.**
P.O. Box 285, 1990 Industrial Boulevard, Stillwater, MN 55082-0285, USA
*Phone:* 612 439 9710; 800 328 1482
*Fax:* 612 779 7847
*E-mail:* corpinfo@incstar.com
*Web site:* www.incstar.com/

**Invitrogen**
9800 Medical Center Dr, P.O. Box 6482, Rockville, MD 20849-6482, USA
*Phone:* 800 828 6686
*Fax:* 800 331 2286
*Web site:* www.invitrogen.com

**Invitrogen**
8400 Helgerman Ct., P.O. Box 6009, Gaithersburg, MD 20884, USA
*Phone:* 301 840 8000; 800 828 6686
*Fax:* 301 258 8238; 800 331 2286
*Web site:* www.invitrogen.com

**Invitrogen**
3 Fountain Drive, Inchinnan Business Park, Paisley, PA4 9RF, Scotland, UK
*Phone:* +44 (0)141 814 6100
*Fax:* +44 (0)141 814 6287
*Web site:* www.invitrogen.com

**Invitrogen (Tokyo)**
*Phone:* +81 (0)3 3633 8241
*Fax:* +81 (0)3 3663 8242
*Web site:* www.invitrogen.com

**Iwaki Glass (Tokyo)**
*Phone:* +81 3 5645 2751
*Web site:* www.igc.co.jp/zigyo/rika

**Jackson ImmunoResearch Laboratories, Inc.**
872 West Baltimore Pike, P.O. Box 9, West Grove, PA 19390, USA
*Phone:* 610 869 4024; 800 367 5296
*Fax:* 610 869 0171
*E-mail:* cuserjaxn@aol.com
*Web site:* www.jacksonimmuno.com

**JRH (USA)**
13804 West 107th St., P.O. Box 14848, Lenexa, KS 66215, USA
*Phone:* 800 255 6032
*Fax:* 913 469 5584
*E-mail:* info@jrhbio.com
*Web site:* www.jrhbio.com/about_jrh/index.html

**Kendro (see Heraeus)**

**Kindler O. Mikroskopische Gläser**
Ziegelhofstrasse 214, D-79110 Freiburg, Germany
*Phone:* +49 (0)761 81077
*Fax:* +49 (0)761 892535

**Life Technologies (see Invitrogen)**

**Life Technologies GmbH**
Haus 223, 5090 Lofer, Austria
*Phone:* 0800 20 1087
*Fax:* 0800 21 6317

**List Biologic Laboratories., Inc.**
Campbell, CA, USA

*Phone:* 800 726 3213
*Fax:* 408 866 6364
*E-mail:* info@listlabs.com

**Mediatech Inc.**
13884 Park Center Road, Herndon, VA 20171, USA
*Phone:* 1 800 CELLGRO
*Web site:* www.cellgro.com

**Merck GmbH**
Zimbagasse, 51147 Vienna, Austria
*Phone:* +43 (0)1 576 00 0
*Fax:* +43 (0)1 577 3370
*E-mail:* merck-wien@merck.at
*Web site:* www.merck.at/flabor.htm

**Merck, Inc.**
P.O. Box 2000, RY7-220, Rahway, NJ 07065, USA
*Phone:* 908 594 4600; 800 672 6372
*Fax:* 908 388 9778
*Web site:* www.merck.com

**Merck Ltd.**
Merck House, Poole, Dorset BH15 1TD, UK
*Phone:* +44 (0)1202 669700
*Fax:* +44 (0)1202 665599
*E-mail:* info@merck-ltd.co.uk
*Web site:* www.merckeurolab.ltd.uk

**Merck Eurolab (see also VWR International)**
*E-mail:* info@merckeurolab.de
*Web site:* www.merckeurolab.de

**Messer Cryotherm GmbH**
Euteneuen 4, D 57548 Kirchen/ Sieg, Germany
*Phone:* +49 (0)27 41 9585-0
*Fax:* +49 (0)27 41 6900
*E-mail:* info@messer-cryotherm.de
*Web site:* www.messer-cryotherm.de

**Mettler-Toledo Inc.**
1900 Polaris Parkway, Columbus, OH 43240, USA
*Phone:* 800 METTLER
*Fax:* 614 438 4900
*E-mail:* us@mt-shop.com
*Web site:* www.mt.com

**Millipore Corp.**
80 Ashby Rd., Bedford, MA 01730, USA
*Phone:* 617 275 9200; 800 MILLIPORE
*Fax:* 617 275 5550
*Web site:* www.millipore.com

**Miltex Instruments Company, Inc.**
700 Hicksville Road, Bethpage, NY 11714, USA
*Phone:* 800 645 8000
*Fax:* 516 349 0161
*E-mail:* customerservice@miltex.com
*Web site:* www.miltex.com

**Nalge Nunc International**
2000 N. Aurora Rd., Naperville, IL 60563-1796, USA
*Phone:* 630 983 5700; 800 288 6862; 800 416 6862
*Fax:* 630 416 2556; 630 416 2519
*Web site:* www.nalgenunc.com

**Nalge Nunc**
Postbox 280, DK-4000 Roskilde, Denmark
*Phone:* +45 46 359065
*Fax:* +45 46 350105
*E-mail:* infociety@nunc.dk
*Web site:* www.nalgenunc.com

**NEN Life Science Products**
549 Albany St., Boston, MA 02118, USA
*Phone:* 617 482 9595; 800 551 2121
*Fax:* 617 542 8468
*E-mail:* techsupport@nenlifesci.com
*Web site:* www.nenlifesci.com

**Neomarkers**
47790 Westinhghouse Dr., Fremont, CA 94539, USA
*Phone:* 800 823 1628
*Fax:* 510 991 2826
*E-mail:* LabVision@LabVision.com
*Web site:* www.LabVision.com

**Nikon, Inc.**
1300 Walt Whitman Road, Melville, NY 11747-3064, USA
*Phone:* 516 547 8500
*Fax:* 516 547 0306
*Web site:* www.nikonusa.com

**Novocastra Laboratories Ltd.**
Balliol Business Park West, Benton Lane, Newcastle upon Tyne, NE12 8EW, UK
*Phone:* +44 (0)191 215 0567
*Fax:* +44 (0)191 215 1152
*E-mail:* oem@Novocastra.co.uk
*Web site:* www.novocastra.co.uk

**Novo Nordisk BioLabs Inc.**
ACE BioSciences A/S, Unsbjergvej 2, 5220 Odense SO, Denmark
*Phone:* +45 6565 2121
*Fax:* +45 6565 2122
*E-mail:* mail@acebiosciences.com

**NOYES/ROSS (see Miltex)**

**Nutacon BV**
Tuindeij 25, Post bus 94, 2450 Leimuiden, Netherlands
*Phone:* +31 (0)172 506 214
*Fax:* +31 (0)172 506 514
*E-mail:* info@nutacon.nl
*Web site:* www.nutacon.nl

**NuAire, Inc.**
2100 Fernbrook Ln., Plymouth, MN 55447, USA
*Phone:* 612 553 1270; 800 328 3352
*Fax:* 612 553 0459
*Web site:* www.nuaire.co.uk

**Oncogene Research Products**
84 Rogers Street, Cambridge, MA 02142, USA
*Phone:* 617 577 9333; 800 662 2616
*Fax:* 617 577 8015; 800 828 4871
*E-mail:* customer.service@oncresprod.com
*Web site:* www.apoptosis.com

**PAA Austria Headquarters – PAA Laboratories GmbH**
Wiener Strasse 131, A-4020 Linz, Austria
*Phone:* +43 (0)7229 64 8 65

*Fax:* +43 (0)7229 64 8 66
*E-mail:* info@paa.at
*Web site:* www.paa.at

**PAA UK**
1 Technine Guard Avenue, Houndstone Business Park, Yeovil, Somerset, BA22 8YE, UK
*Phone:* +44 (0)1935 411 418
*Fax:* +44 (0)1935 411 480
*E-mail:* info@paalaboratories.co.uk
*Web site:* www.paa.at

**Pall Gelman Sciences**
Laboratory Products Group, 600 South Wagner Rd., Ann Arbor, MI 48103-9019, USA
*Phone:* 800 521 1520
*Fax:* 313 913 6495
*Web site:* www.gelman.com

**Pel-Freeze Biologicals**
P.O. Box 68, Rogers, AR 72757, USA
*Phone:* 800 643 3426; 479 636 4361
*Fax:* 479 636 3562

**Pfizer**
235 East 42nd Street, New York, NY 10017, USA
*Phone:* 212 733 2323
*Web site:* www.pfizer.com

**Pharmacia (see also Amersham Pharmacia Biotech)**

**Pharmacia Biotech, Inc.**
800 Centennial Ave., Box 1327, Piscataway, NJ 08855-1327, USA
*Phone:* 908 457 8000; 800 526 3593
*Fax:* 908 457 8130; 800 329 3593
*Web site:* www.apbiotech.com

**Pharmacia Animal Health Worldwide Headquarters**
7000 Portage Road, Kalamazoo, MI 49001, USA
*Phone:* 800 793 0596
*Fax:* 800 984 9647
*Web site:* www.pharmacia.com/worldwide

**PharMingen (see also BD Biosciences)**
10975 Torreyana Road, San Diego, CA 92121, USA
*Phone:* 877 232 8995
*Fax:* 800 325 9637
*E-mail:* techserv@pharmingen.com
*Web site:* www.bdbiosciences.com/pharmingen

**Plano W Plannet GmbH**
Ernst Beford Straße 12, D-35523 Wetzlar, Gemany
*Phone:* +49 (0)6441 97650
*Fax:* +49 (0)6441 976565
*E-mail:* plano@t-online.de
*Web site:* www.plano-em.com

**PolySciences**
6600 West Touhy Avenue, P.O. Box 48312, Niles, IL 60714, USA
*Phone:* 847 647 0611; 800 229 7569
*Fax:* 847 647 1155
*E-mail:* polysci@polyscience.com
*Web site:* www.polyscience.com

**Precision Scientific, Inc.**
110-C Industrial Dr., Winchester, VA 22602, USA
*Phone:* 540 869 9892; 800 621 8820

*Fax:* 540 869 0130
*E-mail:* info@precisionsci.com
*Web site:* www.precisionsci.com

**ProSciTek**
P.O. Box 111, Thuringowa, Qld. 4817, Australia
*Fax:* +61 (0)7 4773 2244
*E-mail:* service@proscitech.com
*Web site:* www.proscitech.com.au

**Promega Corp.**
2800 Woods Hollow Rd., Madison, WI 53711, USA
*Phone:* 608 274 4330; 800 356 9526
*Fax:* 608 277 2601; 800 356 1970
*E-mail:* custserv@promega.com
*Web site:* www.promega.com

**R & D Systems**
614 McKinley Place, NE, Minneapolis, MN 55413, USA
*Phone:* 612 379 2956; 800 328 2400
*Fax:* 612 379 6580
*E-mail:* info@rndsystems.com
*Web site:* www.rndsystems.com

**Research Diagnostics, Inc.**
Pleasant Hills Road, Flanders, NJ 07836, USA
*Phone:* 937 584 7093
*Fax:* 973 584 0210
*E-mail:* ResearchD@aol.com
*Web site:* www.researchd.com

**Revco Scientific, Inc.**
275 Aiken Road, Asheville, NC 28804, USA
*Phone:* 704 658 2711; 800 252 7100
*Fax:* 704 645 3368
*E-mail:* sales@revco-sci.com
*Web site:* www.revco-sci.com

**Roche Molecular Biochemicals**
9115 Hague Road, P.O. Box 50414, Indianapolis, IN 46250-0414, USA
*Phone:* 800 262 1640
*Fax:* 317 521 7317
*E-mail:* mannheim.biocheminfo@roche.com
*Web site:* www.biochem.roche.com.

**Roche Deutschland Holding GmbH**
Emil Barell Str. 1, 79639 Grenzach-Wyhlen, Germany
*Phone:* +49 (0)7624 9088 0
*Fax:* +49 (0)7624 9088 3672
*Web site:* www.roche.de

**Roth (see Carl Roth)**

**Santa Cruz Biotechnology, Inc.**
2161 Delaware Avenue, Santa Cruz, CA 95060, USA
*Phone:* 800 457 3801
*Fax:* 831 457 3801
*E-mail:* scbt@scbt.com
*Web site:* www.scbt.com

**Sarstedt AG & Co**
P.O. Box 1220, 51582 Nümbrecht, Germany
*Phone:* +49 (0)2293 30 50; 0800 083 30 50
*Fax:* +49 (0)2293 305 282

*E-mail:* info@sarstedt.com
*Web site:* www.sarstedt.com

**Schärfe System**
Krämerstrasse 22, D-72764 Reutlingen, Germany
*Phone:* +49 (0)7121 38786 0
*Fax:* +49 (0)7121 38786 99
*E-mail:* mail@casy-technology.com
*Web site:* www.casy-technology.com

**Schott Glass Technologies**
400 York Avenue, Duryea, PA 18642, USA
*Phone:* 570 457 7485
*Fax:* 570 457 6960
*E-mail:* www.us.schott.com/sgt
*Web site:* www.us.schott.com/sgt

**Serotec Ltd.**
22 Bankside, Station Approach, Kidlington, Oxford OX5 1JE, England, UK
*Phone:* +44 (0)1865 852 700
*Fax:* +44 (0)1865 373 899
*E-mail:* serotec@serotec.co.uk
*Web site:* www.serotech.co.uk

**Serva**
P.O. Box 105260, D-69ß42 Heidelberg, Germany
*Phone:* +49 (0)0800 73 78 24 62
*Fax:* +49 (0)6221 13840 13
*E-mail:* info@serva.de
*Web site:* www.serva.de

**Shandon (now Thermo Shandon)**
171 Industry Dr., Pittsburg, PA 15275, USA
*Phone:* 800 547 7429
*Fax:* 412 788 1138
*E-mail:* customerserviceusa@thermoshandon.com
*Web site:* www.shandon.com

**Sigma-Aldrich (Austria)**
Hebbelplatz 7, 1100 Vienna, Austria
*Phone:* +43 (0)1 605 81 10
*Fax:* +43 (0)1 605 81 20
*E-mail:* sigma@sigma.co.at
*Web site:* www.sigma-aldrich.com/

**Sigma-Aldrich Company Ltd.**
Fancy Road, Poole, Dorset, BH12 4QH, UK
*Phone:* +44 (0)1202 733114; 0800 717181
*Fax:* +44 (0)1202 715460; 0800 378785
*E-mail:* ukcustsv@eurnotes.sial.com
*Web site:* www.sigma-aldrich.com/

**Sigma-Aldrich Japan (Tokyo)**
*Phone:* +81 (0)3 5821 3594
*Fax:* +81 (0)3 5821 3591
*E-mail:* sialjpts@sial.com
*Web site:* www.sigma-aldrich.com/japan

**Sigma Chemical Co.**
P.O. Box 14508, 3300 South Second St., St Louis, MO 63178, USA
*Phone:* 314 771 5750; 800 325 3010
*Fax:* 314 771 5757; 800 325 5052

*E-mail:* custserv@sial.com; sigma-techserv@sial.com
*Web site:* www.sigma.sial.com

### Spectrum Laboratories, Inc.
18617 Broadwick Street, Rancho Dominguez, CA 90220-6435, USA
*Phone:* 800 445 3300
*Fax:* 800 634 7330
*E-mail:* webmaster@spectrapor.com
*Web site:* www.spectrapor.com

### Stem Cell Technologies, Inc.
777 W. Broadway, Suite 808, Vancouver, BC U5Z 4J7, Canada
*Phone:* 604 877 0713
*Fax:* 604 877 0704
*E-mail:* info@stemcell.com
*Web site:* www.stemcell.com

### Swann Morton
Owerton Green, Sheffield, S6 2BJ, UK
*Phone:* +44 (0)114 234 4231
*Fax:* +44 (0)114 231 4966
*E-mail:* info@swann-morton.com
*Web site:* www.swann-morton.com/contact

### Tekmar-Dohrmann
7143 East Kemper Road, Cincinnati, OH 45249, USA
*Phone:* 513 247 7000; 800 345 2100
*Fax:* 513 247 7050
*E-mail:* td.info@EmersonProcess.com
*Web site:* www.tekmar.com

### Thermo Life Sciences
Unit 5, The Ringway Centre, Edison Road, Basingstoke, Hampshire RG21 6YH, UK
*Phone:* +44 (0)1256 844141
*Fax:* +44 (0)1256 347647
*E-mail:* ukinfo@tmquest.com.
*Web site:* www.tmquest.co.uk.

### Thomas Scientific
99 High Hill Rd., P.O. Box 99, Swedesboro, NJ 08085, USA
*Phone:* 609 467 2000; 800 345 2100
*Fax:* 609 467 3087
*Web site:* www.thomassci.com

### TPP Techno Plastic Products
Zollstrasse 155, CH-8219 Trasadingen, Switzerland
*E-mail:* info@tpp.ch
*Web site:* www.tpp.ch

### Upstate Biotechnology
1100 Winter Street, Suite 2300, Waltham, MA 02451, USA
*Phone:* 781 890 8845; 800 233 3991
*Fax:* 781 890 7738
*E-mail:* info@upstatebiotech.com
*Web site:* www.upstatebiotech.com

### Upstate Biotechnology Inc.
199 Saranac Ave, Lake Placid, NY 12946, USA
*Phone:* 518 523 1518
*Fax:* 518 523 1336
*E-mail:* INFO@upstatebiotech.com
*Web site:* www.upstatebiotech.com

**Vector Laboratories, Inc.**

30 Ingold Road, Burlingame, CA 94010, USA

*Phone:* 650 697 3600

*Fax:* 650 697 0339

*E-mail:* vector@vectorlabs.com

*Web site:* www.vectorlabs.com

**Vector Laboratories Ltd.**

3 Accent Park, Bakewell Road, Orton Southgate, Peterborough, PE2 6XS, UK

*Phone:* +44 (0)1733 237999

*Fax:* +44 (0)1733 237119

*E-mail:* vector@vectorlabs.co.uk

*Web site:* www.vectorlabs.com/

**VWR International**

Frankfurter Str. 259, D- 64293 Darmstadt, Germany

*Phone:* +49 (0)6151 72-0

*Fax:* +49 (0)6151 72-2000

*E-mail:* info@de.vwr.com

*Web site:* www.vwrsp.com

**Wako Pure Chemical Industries, Ltd. (Osaka)**

1-2, Doshomachi 3-Chome, Osaka 540-8605, Japan

*Phone:* +81 (0)120 052 091(free)

*Fax:* +81 (0)120 052 806(free)

*Web site:* www.wako-chem.co.jp

**Worthington Biochemical Corporation**

730 Vassar Avenue, Lakewood, NJ 08701, USA

*Phone:* 732 942 1660

*Fax:* 732 942 9270

*Web site:* www.worthington-biochem.com

**Zeiss, Carl**

P.O. Box 4041, D-37081 Göttingen, Germany

*Phone:* +49 (0)551 5060 660

*Fax:* +49 (0)551 5060 480

*E-mail:* med-mikro-service@zeiss.de

*Web site:* www.teiss.de/micro

**Zeiss Inc.**

Microscope Division, One Zeiss Dr., Thornwood, NY 10594, USA

*Phone:* 914 747 1800; 800 233 2343

*Fax:* 914 681 7446

*E-mail:* micro@zeiss.com; info@zeiss.de

*Web site:* www.zeiss.com/; www.zeiss.de

**Zymed Laboratories, Inc.**

458 Carlton Court, South San Francisco, CA 94080, USA

*Phone:* 650 871 4494; 800 874 4494

*Fax:* 415 871 4499

*E-mail:* info@zymed.com; tech@zymed.com

*Web site:* www.zymed.com

# Index

Acidified ethanol, fixation of malignant brain
    tumor cell lines, 357
Acid-secreting gastric parietal cells, 27
ACL-3 medium, 4–7
Adenocarcinoma:
    bladder cancer cell pathology, 99
    ovarian cancer cell culture, 151–152
    tumor cell culture, 13–14
        development media, ACL-3 medium,
            4–7
        pancreatic metastases, 83
Adenosquamous carcinoma, pancreatic metas-
    tases, 83
Adherent colonies:
    gastric cancer cells:
        cancer-derived cell propagation and
            preservation, 37–38
        rapping technique, 32
        scraping method, 32–33
    neuroendocrine tumor cell cultures, 392
    ovarian cancer cells:
        cell separation, 166–167
        primary culture, 165
Agar cloning:
    cervical carcinoma cell lines, 185
    keratinocytes, 272
    leukemia, 322, 329, 337, 340, 343
    lung, 16
Aggregate transfer:
    cancer-derived cell propagation and preser-
        vation, 35–36
    gastric cancer cell enrichment, 31–32
    mechanical detachment, 32
Aging cells, melanocyte culture, 296
American Type Culture Collection (ATCC),
    neuroendocrine tumor culture, 398
Anchorage-independent growth (see also Agar
    cloning), squamous cell carcinomas of the
    head and neck (SCC-HN), 272–273
Angiogenic inhibitors:
    myoepithelial cell studies, tumor angion-

genesis inhibition, 236–244
    paracrine regulation, 222–223
Anti-angiogenic drugs, pancreatic cell lines,
    orthotopic implantation model, 92
Anticancer drug screening, neuroendocrine
    tumor culture, 396–397
Antigenic phenotype, brain tumor cell culture
    lines, 365–368
    cell surface antigen staining, 368
    cytoplasmic antigen staining, 365–366
    nuclear antigen staining, 367
Anti-involucrin antibody, lung tumor cell dif-
    ferentiation, 16–17
Anti-pancreatic cancer drugs, pancreatic cell
    lines, orthotopic implantation model,
    92
Anti-small proline-rich protein (SPRR1B)
    antibody, lung tumor cell differentiation,
    16–17
Apoptosis, neuroendocrine tumor culture,
    397
Ascitic fluids:
    colon carcinoma cell lines, 76–77
    gastric carcinoma cell lines, 26
    ovarian cancer cell lines:
        collection protocols, 154–157
        monolayer mesothelial cells, epithelial-
            stromal separation, 167–168
        primary culture, 164–165
    pancreatic tumor cell lines, 86–87
    primary cell cultures, 30–31
Astrocytic differentiation, malignant brain
    tumor cell lines, 353–354
Astrocytomas, malignant brain tumor cell
    lines, 351
Attachment medium, myoepithelial cell trans-
    formation, 247
Autocrine/paracrine mechanism, lung tumor
    cell development, 5–7
Autonomous growth, melanoma cell culture,
    311